T0189679

Springer Optimization and Its Applications

Volume 155

Aims and Scope
Optimization has continued to expand in all directions at an astonishing rate. New algorithmic and theoretical techniques are continually developing and the diffusion into other disciplines is proceeding at a rapid pace, with a spot light on machine learning, artificial intelligence, and quantum computing. Our knowledge of all aspects of the field has grown even more profound. At the same time, one of the most striking trends in optimization is the constantly increasing emphasis on the interdisciplinary nature of the field. Optimization has been a basic tool in areas not limited to applied mathematics, engineering, medicine, economics, computer science, operations research, and other sciences.

The series **Springer Optimization and Its Applications (SOIA)** aims to publish state-of-the-art expository works (monographs, contributed volumes, textbooks, handbooks) that focus on theory, methods, and applications of optimization. Topics covered include, but are not limited to, nonlinear optimization, combinatorial optimization, continuous optimization, stochastic optimization, Bayesian optimization, optimal control, discrete optimization, multi-objective optimization, and more. New to the series portfolio include Works at the intersection of optimization and machine learning, artificial intelligence, and quantum computing.

Volumes from this series are indexed by Web of Science, zbMATH, Mathematical Reviews, and SCOPUS.

More information about this series at http://www.springer.com/series/7393

Alexander J. Zaslavski

Convex Optimization with Computational Errors

Alexander J. Zaslavski
Department of Mathematics
Amado Building
Israel Institute of Technology
Haifa, Israel

ISSN 1931-6828 ISSN 1931-6836 (electronic)
Springer Optimization and Its Applications
ISBN 978-3-030-37824-0 ISBN 978-3-030-37822-6 (eBook)
https://doi.org/10.1007/978-3-030-37822-6

Mathematics Subject Classification: 49M37, 65K05, 90C25, 90C26, 90C30

This Springer imprint is published by the registered company Springer Nature Switzerland AG.
The registered company address is: Gewerbestrasse 11, 6330 Cham, Switzerland

Preface

The book is devoted to the study of approximate solutions of optimization problems in the presence of computational errors. It contains a number of results on the convergence behavior of algorithms in a Hilbert space, which are known as important tools for solving optimization problems. The research presented in the book is the continuation and further developments of our book *Numerical Optimization with Computational Errors*, Springer 2016 [92]. In that book as well as in this new one, we study the algorithms taking into account computational errors which always present in practice. In this case the convergence to a solution does not take place. We show that our algorithms generate a good approximate solution, if computational errors are bounded from above by a small positive constant. Clearly, in practice it is sufficient to find a good approximate solution instead of constructing a minimizing sequence. On the other hand in practice computations induce numerical errors, and if one uses methods in order to solve minimization problems these methods usually provide only approximate solutions of the problems. Our main goal is, for a known computational error, to find out what approximate solution can be obtained and how many iterates one needs for this.

The main difference between this new book and the previous one is that here we take into consideration the fact that for every algorithm its iteration consists of several steps and that computational errors for different steps are different, in general. This fact, which was not taken into account in our previous book, is indeed important in practice. For example, the subgradient projection algorithm consists of two steps. The first step is a calculation of a subgradient of the objective function while in the second one we calculate a projection on the feasible set. In each of these two steps there is a computational error, and these two computational errors are different in general. It may happen that the feasible set is simple and the objective function is complicated. As a result, the computational error, made when one calculates the projection, is essentially smaller than the computational error of the calculation of the subgradient. Clearly, an opposite case is possible too.

Another feature of this book is that here we study a number of important algorithms which appeared recently in the literature and which are not discussed in the previous book.

The monograph contains 12 chapters. Chapter 1 is an introduction. In Chap. 2 we study the subgradient projection algorithm for minimization of convex and nonsmooth functions. We begin with minimization problems on bounded sets and generalize Theorem 2.4 of [92] proved in the case when the computational errors for the two steps of an iteration are the same. We also consider minimization problems on unbounded sets and generalize Theorem 2.8 of [92] proved in the case when the computational errors for the two steps of an iteration are the same and prove two results which have no prototype in [92]. Finally, in Chap. 2 we study the subgradient projection algorithm for zero-sum games with two players. For this algorithm each iteration consists of four steps. In each of these steps there is a computational error. We suppose that these computational errors are different and prove two results. In the first theorem, which is a generalization of Theorem 2.11 of [92] obtained in the case when all the computational errors are the same, we study the games on bounded sets. In the second result, which has no prototype in [92], we deal with the games on unbounded sets.

In Chap. 3 we analyze the mirror descent algorithm for minimization of convex and nonsmooth functions, under the presence of computational errors. The problem is described by an objective function and a set of feasible points. For this algorithm each iteration consists of two steps. The first step is a calculation of a subgradient of the objective function while in the second one we solve an auxiliary minimization problem on the set of feasible points. In each of these two steps there is a computational error. In general, these two computational errors are different. We begin with minimization problems on bounded sets and generalize Theorem 3.1 of [92] proved in the case when the computational errors for the two steps of an iteration are the same. We also consider minimization problems on unbounded sets and generalize Theorem 3.3 of [92] proved in the case when the computational errors for the two steps of an iteration are the same and prove two results which have no prototype in [92]. Finally, in Chap. 3 we study the mirror descent algorithm for zero-sum games with two players. For this algorithm each iteration consists of four steps. In each of these steps there is a computational error. We suppose that these computational errors are different and prove two theorems. In the first result, which is a generalization of Theorem 3.4 of [92] obtained in the case when all the computational errors are the same, we study the games on bounded sets. In the second result, which has no prototype in [92], we deal with the games on unbounded sets.

In Chap. 4 we analyze the projected gradient algorithm with a smooth objective function under the presence of computational errors. For this algorithm each iteration also consists of two steps. The first step is a calculation of a gradient of the objective function while in the second one we calculate a projection on the feasible set. In each of these two steps there is a computational error. Our first result in this chapter, for minimization problems on bounded sets, is a generalization of Theorem 4.2 of [92] proved in the case when the computational errors for the two steps of an iteration are the same. We also consider minimization problems on unbounded sets and generalize Theorem 4.5 of [92] proved in the case when the computational errors for the two steps of an iteration are the same and prove two results which have no prototype in [92]. In Chap. 5 we consider an algorithm, which is an extension of the projection gradient algorithm used for solving linear inverse problems arising in signal/image processing. This algorithm is used for minimization of the sum of two given convex functions, and each of its iteration consists of two steps. In each of these two steps there is a computational error. These two computational errors are different. We generalize Theorem 5.1 of [92] obtained in the case when the computational errors for the two steps of an iteration are the same and prove two results which have no prototype in [92].

In Chap. 6 we study continuous subgradient method and continuous subgradient projection algorithm for minimization of convex nonsmooth functions and for computing the saddle points of convex–concave functions, under the presence of computational errors. For these algorithms each iteration consists of a few calculations, and for each of these calculations there is a computational error produced by our computer system. In general, these computational errors are different. All the results of this chapter have no prototype in [92].

In Chaps. 7–12 we analyze several algorithms under the presence of computational errors, which were not considered in [92]. Again, each step of an iteration has a computational error, and we take into account that these errors are, in general, different. An optimization problem with a composite objective function is studied in Chap. 7. A zero-sum game with two players is considered in Chap. 8. A predicted decrease approximation-based method is used in Chap. 9 for constrained convex optimization. Chapter 10 is devoted to minimization of quasiconvex functions. Minimization of sharp weakly convex functions is discussed in Chap. 11. Chapter 12 is devoted to a generalized projected subgradient method for minimization of a convex function over a set which is not necessarily convex.

The author believes that this book will be useful for researches interested in the optimization theory and its applications.

Rishon LeZion, Israel
May 19, 2019

Alexander J. Zaslavski

Contents

Chapter 1
Introduction

In this book we study behavior of algorithms for constrained convex minimization problems in a Hilbert space. Our goal is to obtain a good approximate solution of the problem in the presence of computational errors. It is known that the algorithm generates a good approximate solution, if the sequence of computational errors is bounded from above by a small constant. In our study, presented in this book, we take into consideration the fact that for every algorithm its iteration consists of several steps and that computational errors for different steps are different, in general. In this chapter we discuss several algorithms which are studied in this book.

1.1 Subgradient Projection Method

In Chap. 2 we study the subgradient projection algorithm for minimization of convex and nonsmooth functions and for computing the saddle points of convex–concave functions, under the presence of computational errors. It should be mentioned that the subgradient projection algorithm is one of the most important tools in the optimization theory, nonlinear analysis, and their applications. See, for example, [56, 58, 64, 65, 67–73, 76, 79, 80, 83, 85, 89–92, 94] and the references mentioned therein. The problem is described by an objective function and a set of feasible points. For this algorithm each iteration consists of two steps. The first step is the calculation of a subgradient of the objective function while in the second one we calculate the projection on the feasible set. In each of these two steps there is a computational error. In general, these two computational errors are different. We show that our algorithm generates a good approximate solution, if all the computational errors are bounded from above by a small positive constant. Moreover, if we know the computational errors for the two steps of our algorithm, we find out what approximate solution can be obtained and how many iterates one needs for this.

A. J. Zaslavski, *Convex Optimization with Computational Errors*, Springer Optimization and Its Applications 155, https://doi.org/10.1007/978-3-030-37822-6_1

We use the subgradient projection algorithm for constrained minimization problems in Hilbert spaces equipped with an inner product denoted by $\langle \cdot, \cdot \rangle$ which induces a complete norm $\| \cdot \|$.

In this book we use the following notation. For every $z \in R^1$ denote by $\lfloor z \rfloor$ the largest integer which does not exceed z:

$$\lfloor z \rfloor = \max\{i \in R^1 : i \text{ is an integer and } i \leq z\}.$$

For every nonempty set D, every function $f : D \to R^1$, and every nonempty set $C \subset D$ we set

$$\inf(f, C) = \inf\{f(x) : x \in C\}$$

and

$$\text{argmin}(f, C) = \text{argmin}\{f(x) : x \in C\} = \{x \in C : f(x) = \inf(f, C)\}.$$

Let X be a Hilbert space equipped with an inner product denoted by $\langle \cdot, \cdot \rangle$ which induces a complete norm $\| \cdot \|$. For each $x \in X$ and each $r > 0$ set

$$B_X(x, r) = \{y \in X : \|x - y\| \leq r\}$$

and for each $x \in X$ and each nonempty set $E \subset X$ set

$$d(x, E) = \inf\{\|x - y\| : y \in E\}.$$

For each nonempty open convex set $U \subset X$ and each convex function $f : U \to R^1$, for every $x \in U$ set

$$\partial f(x) = \{l \in X : f(y) - f(x) \geq \langle l, y - x \rangle \text{ for all } y \in U\},$$

which is called the subdifferential of the function f at the point x [61, 62, 81].

Let C be a nonempty closed convex subset of X, U be an open convex subset of X such that $C \subset U$, and $f : U \to R^1$ be a convex function.

Suppose that there exist $L > 0$, $M_0 > 0$ such that

$$C \subset B_X(0, M_0),$$

$$|f(x) - f(y)| \leq L\|x - y\| \text{ for all } x, y \in U.$$

It is not difficult to see that for each $x \in U$,

$$\emptyset \neq \partial f(x) \subset B_X(0, L).$$

For every nonempty closed convex set $D \subset X$ and every $x \in X$ there is a unique point $P_D(x) \in D$ satisfying

$$\|x - P_D(x)\| = \inf\{\|x - y\| : y \in D\}.$$

We consider the minimization problem

$$f(z) \to \min, \ z \in C.$$

Suppose that $\{a_k\}_{k=0}^{\infty} \subset (0, \infty)$. Let us describe our algorithm.

Subgradient projection algorithm
Initialization: select an arbitrary $x_0 \in U$.
Iterative Step: given a current iteration vector $x_t \in U$ calculate

$$\xi_t \in \partial f(x_t)$$

and the next iteration vector $x_{t+1} = P_C(x_t - a_t\xi_t)$.

In [92] we study this algorithm under the presence of computational errors. Namely, in [92] we suppose that $\delta \in (0, 1]$ is a computational error produced by our computer system, and study the following algorithm.

Subgradient projection algorithm with computational errors
Initialization: select an arbitrary $x_0 \in U$.
Iterative Step: given a current iteration vector $x_t \in U$ calculate

$$\xi_t \in \partial f(x_t) + B_X(0, \delta)$$

and the next iteration vector $x_{t+1} \in U$ such that

$$\|x_{t+1} - P_C(x_t - a_t\xi_t)\| \leq \delta.$$

In Chap. 2 of this book we consider more complicated, but more realistic, version of this algorithm. Clearly, for the algorithm each iteration consists of two steps. The first step is a calculation of a subgradient of the objective function f while in the second one we calculate a projection on the set C. In each of these two steps there is a computational error produced by our computer system. In general, these two computational errors are different. This fact is taken into account in the following projection algorithm studied in Chap. 2 of this book.

Suppose that $\{a_k\}_{k=0}^{\infty} \subset (0, \infty)$ and $\delta_f, \delta_C \in (0, 1]$.

Initialization: select an arbitrary $x_0 \in U$.
Iterative Step: given a current iteration vector $x_t \in U$ calculate

$$\xi_t \in \partial f(x_t) + B_X(0, \delta_f)$$

and the next iteration vector $x_{t+1} \in U$ such that

$$\|x_{t+1} - P_C(x_t - a_t\xi_t)\| \le \delta_C.$$

Note that in practice for some problems the set C is simple but the function f is complicated. In this case δ_C is essentially smaller than δ_f. On the other hand, there are cases when f is simple but the set C is complicated and therefore δ_f is much smaller than δ_C.

In Chap. 2 we prove the following result (see Theorem 2.4).

Theorem 1.1 *Let* $\delta_f, \delta_C \in (0, 1]$, $\{a_k\}_{k=0}^{\infty} \subset (0, \infty)$ *and let*

$$x_* \in C$$

satisfy

$$f(x_*) \le f(x) \text{ for all } x \in C.$$

Assume that $\{x_t\}_{t=0}^{\infty} \subset U$, $\{\xi_t\}_{t=0}^{\infty} \subset X$,

$$\|x_0\| \le M_0 + 1$$

and that for each integer $t \ge 0$,

$$\xi_t \in \partial f(x_t) + B_X(0, \delta_f)$$

and

$$\|x_{t+1} - P_C(x_t - a_t\xi_t)\| \le \delta_C.$$

Then for each natural number T,

$$\sum_{t=0}^{T} a_t(f(x_t) - f(x_*))$$

$$\le 2^{-1}\|x_* - x_0\|^2 + \delta_C(T+1)(4M_0+1)$$

$$+\delta_f(2M_0+1)\sum_{t=0}^{T} a_t + 2^{-1}(L+1)^2\sum_{t=0}^{T} a_t^2.$$

Moreover, for each natural number T,

$$f((\sum_{t=0}^{T} a_t)^{-1} \sum_{t=0}^{T} a_t x_t) - f(x_*), \ \min\{f(x_t) : t = 0, \dots, T\} - f(x_*)$$

$$\leq 2^{-1}(\sum_{t=0}^{T} a_t)^{-1} \|x_* - x_0\|^2 + (\sum_{t=0}^{T} a_t)^{-1} \delta_C (T+1)(4M_0+1)$$

$$+\delta_f(2M_0+1) + 2^{-1}(\sum_{t=0}^{T} a_t)^{-1}(L+1)^2 \sum_{t=0}^{T} a_t^2.$$

Theorem 1.1 is proved in Sect. 2.4. It is a generalization of Theorem 2.4 of [92] proved in the case when $\delta_f = \delta_C$.

We are interested in an optimal choice of a_t, $t = 0, 1, \dots$. Let T be a natural number and $A_T = \sum_{t=0}^{T} a_t$ be given. It is shown in Chap. 2 that the best choice is $a_t = (T+1)^{-1} A_T$, $t = 0, \dots, T$.

Let T be a natural number and $a_t = a > 0$, $t = 0, \dots, T$. It is shown in Chap. 2 that the best choice of a is

$$a = (2\delta_C(4M_0+1))^{1/2}(L+1)^{-1}.$$

Now we can think about the best choice of T. It is not difficult to see that it should be at the same order as $\lfloor \delta_C^{-1} \rfloor$.

In Chap. 2 we also study the subgradient algorithm for minimization problems on unbounded sets.

Let D be a nonempty closed convex subset of X, V be an open convex subset of X such that

$$D \subset V,$$

and $f : V \to R^1$ be a convex function which is Lipschitz on all bounded subsets of V. Set

$$D_{min} = \{x \in D : \ f(x) \leq f(y) \text{ for all } y \in D\}.$$

We suppose that

$$D_{min} \neq \emptyset.$$

In Chap. 2 we will prove the following result.

Theorem 1.2 *Let* $\delta_f, \delta_C \in (0, 1]$, $M > 0$ *satisfy*

$$D_{min} \cap B_X(0, M) \neq \emptyset,$$

$$M_0 \geq 4M + 4,$$

$L > 0$ satisfy

$$|f(v_1) - f(v_2)| \le L\|v_1 - v_2\| \text{ for all } v_1, v_2 \in V \cap B_X(0, M_0 + 2),$$

$$0 < \tau_0 \le \tau_1 \le (L + 1)^{-1},$$

$$\epsilon_0 = 2\tau_0^{-1}\delta_C(4M_0 + 1) + 2\delta_f(2M_0 + 1) + 2\tau_1(L + 1)^2$$

and let

$$n_0 = \lfloor \tau_0^{-1}(2M + 2)^2\epsilon_0^{-1} \rfloor.$$

Assume that $\{x_t\}_{t=0}^{\infty} \subset V$, $\{\xi_t\}_{t=0}^{\infty} \subset X$,

$$\{a_t\}_{t=0}^{\infty} \subset [\tau_0, \tau_1],$$

$$\|x_0\| \le M$$

and that for each integer $t \ge 0$,

$$\xi_t \in \partial f(x_t) + B_X(0, \delta_f)$$

and

$$\|x_{t+1} - P_D(x_t - a_t\xi_t)\| \le \delta_C.$$

Then there exists an integer $q \in [1, n_0 + 1]$ such that

$$\|x_i\| \le 3M + 2, \ i = 0, \dots, q$$

and

$$f(x_q) \le f(x) + \epsilon_0 \text{ for all } x \in D.$$

Theorem 1.2 is a generalization of Theorem 2.8 of [92] which was proved in the case when $\delta_C = \delta_f$. It is proved in Sect. 2.6.

We are interested in the best choice of a_t, $t = 0, 1, \dots$. Assume for simplicity that $\tau_1 = \tau_0 = \tau$. In order to meet our goal we need to minimize the function

$$2\tau^{-1}\delta_C(4M_0 + 1) + 2(L + 1)^2\tau, \ \tau \in (0, \infty).$$

This function has a minimizer

$$\tau = (\delta_C(4M_0 + 1))^{1/2}(L + 1)^{-1},$$

the minimal value of ϵ_0 is

$$2\delta_f(2M_0+1)+4(\delta_C(4M_0+1))^{1/2}(L+1),$$

and $n_0 = \lfloor \Delta \rfloor$ where

$$\Delta = (2M+2)^2(4M_0+1)^{-1/2}(L+1)\delta_C^{-1/2}(2\delta_f(2M_0+1)$$
$$+4\delta_C^{1/2}(4M_0+1)^{1/2}(L+1))^{-1}.$$

Note that in the theorem above δ_f, δ_C are the computational errors produced by our computer system. In view of the inequality above, in order to obtain a good approximate solution we need

$$c_1\lfloor \delta_C^{-1/2}\max\{\delta_f, \delta_C^{1/2}\}^{-1}\rfloor + 1$$

iterations, where c_1 is a constant which depends only on M, M_0, L. As a result, we obtain a point $\xi \in V$ such that

$$B_X(\xi, \delta_C) \cap D \neq \emptyset$$

and

$$f(\xi) \le \inf\{f(x) : x \in D\} + 2\delta_f(2M_0+1) + (4M_0+1)^{1/2}4(L+1)\delta_C^{1/2}.$$

The next result, which is proved in Sect. 2.7, does not have a prototype in [92].

Theorem 1.3 *Let $\delta_f, \delta_C \in (0,1)$, $M > 1$, $L > 0$ and let $\{a_t\}_{t=0}^{\infty} \subset (0,\infty)$ be such that*

$$\{x \in V : f(x) \le \inf(f, D) + 3\} \subset B_X(0, M-1),$$

$$|f(v_1) - f(v_2)| \le L\|v_1 - v_2\|$$

$$\textit{for all } v_1, v_2 \in V \cap B_X(0, 4M+8),$$

$$\delta_f \le (6M+5)^{-1} \textit{ and}$$

$$\delta_C(12M+9) \le a_t \le (L+1)^{-2} \textit{ for all integers } t \ge 0.$$

Assume that $\{x_t\}_{t=0}^{\infty} \subset V$, $\{\xi_t\}_{t=0}^{\infty} \subset X$,

$$\|x_0\| \le M,\ B_X(x_0, \delta_C) \cap D \neq \emptyset$$

and that for each integer $t \geq 0$,

$$\xi_t \in \partial f(x_t) + B_X(0, \delta_f)$$

and

$$\|x_{t+1} - P_D(x_t - a_t\xi_t)\| \leq \delta_C.$$

Then

$$\|x_t\| \leq 3M \text{ for all integers } t \geq 0$$

and for each natural number T,

$$\sum_{t=0}^{T} a_t(f(x_t) - \inf(f, D))$$

$$\leq 2M^2 + \delta_C(T+1)(12M+9)$$

$$+2^{-1}(L+1)^2\sum_{t=0}^{T} a_t^2 + \delta_f(6M+5)\sum_{t=0}^{T} a_t.$$

Moreover, for each natural number T,

$$f((\sum_{t=0}^{T} a_t)^{-1}\sum_{t=0}^{T} a_t x_t) - \inf(f, D),$$

$$\min\{f(x_t) : t = 0, \dots, T\} - \inf(f, D)$$

$$\leq 2M^2(\sum_{t=0}^{T} a_t)^{-1} + \delta_C(T+1)(12M+9)(\sum_{t=0}^{T} a_t)^{-1}$$

$$+2^{-1}(L+1)^2\sum_{t=0}^{T} a_t^2(\sum_{t=0}^{T} a_t)^{-1} + \delta_f(6M+5).$$

We are interested in an optimal choice of a_t, $t = 0, 1, \dots$. Let T be a natural number and $A_T = \sum_{t=0}^{T} a_t$ be given. It is shown in Chap. 2 that the best choice is $a_t = (T+1)^{-1}A$. $t = 0, \dots, T$.

Let T be a natural number and $a_t = a > 0$, $t = 0, \dots, T$. It is shown in Chap. 2 that the best choice of a is

$$a = (2\delta_C(12M+9))^{1/2}(L+1)^{-1}.$$

Now we can think about the best choice of T. It is shown that it should be at the same order as $\lfloor \delta_C^{-1} \rfloor$

The next result, which is proved in Sect. 2.8, does not have a prototype in [92].

Theorem 1.4 *Let $\delta_f, \delta_C \in (0, 16^{-1})$, $M > 4$, $L > 0$,*

$$D_{min} \cap B_X(0, M) \neq \emptyset,$$

$$|f(v_1) - f(v_2)| \leq L\|v_1 - v_2\|$$

$$\text{for all } v_1, v_2 \in V \cap B_X(0, 3M+6),$$

$$a = (L+1)^{-1}\delta_C^{1/2}$$

and

$$T = \lfloor 6^{-1} \min\{\delta_C^{-1}, \delta_C^{-1/2}\delta_f^{-1}\}\rfloor - 1.$$

Assume that $\{x_t\}_{t=0}^{\infty} \subset V$, $\{\xi_t\}_{t=0}^{\infty} \subset X$,

$$\|x_0\| \leq M, \; B_X(x_0, \delta_C) \cap D \neq \emptyset$$

and that for each integer $t \geq 0$,

$$\xi_t \in \partial f(x_t) + B_X(0, \delta_f)$$

and

$$\|x_{t+1} - P_D(x_t - a_t\xi_t)\| \leq \delta_C.$$

Then

$$\sum_{t=0}^{T}(T+1)^{-1} f(x_t) - \inf(f, D),$$

$$f\Big(\sum_{t=0}^{T}(T+1)^{-1}x_t\Big) - \inf(f, D), \; \min\{f(x_t) : t = 0, \ldots, T\} - \inf(f, D)$$

$$\leq 12(2M+2)^2(L+1)\max\{\delta_C^{1/2}, \delta_f\}.$$

In Chap. 2 we also study the subgradient projection algorithm for zero-sum games with two players. For this algorithm each iteration consists of four steps. In each of these steps there is a computational error. We suppose that these computational errors are different and prove two results: Theorems 2.15 and 2.16. In Theorem 2.15, which is a generalization of Theorem 2.11 of [92] obtained in the case when all the computational errors are the same, we study the games on bounded sets. In Theorem 2.16, which has no prototype in [92], we deal with the games on unbounded sets.

1.2 The Mirror Descent Method

In Chap. 3 we analyze the mirror descent algorithm for minimization of convex and nonsmooth functions and for computing the saddle points of convex–concave functions, under the presence of computational errors. The problem is described by an objective function and a set of feasible points. For this algorithm each iteration consists of two steps. The first step is a calculation of a subgradient of the objective function while in the second one we solve an auxiliary minimization problem on the set of feasible points. In each of these two steps there is a computational error. In general, these two computational errors are different. We show that our algorithm generates a good approximate solution, if all the computational errors are bounded from above by a small positive constant. Moreover, if we know the computational errors for the two steps of our algorithm, we find out what approximate solution can be obtained and how many iterates one needs for this.

Let X be a Hilbert space equipped with an inner product $\langle \cdot, \cdot \rangle$ which induces a complete norm $\| \cdot \|$. Let C be a nonempty closed convex subset of X, U be an open convex subset of X such that $C \subset U$, and let $f : U \to R^1$ be a convex function. Suppose that there exist $L > 0$, $M_0 > 0$ such that

$$C \subset B_X(0, M_0),$$

$$|f(x) - f(y)| \le L\|x - y\| \text{ for all } x, y \in U.$$

We study the convergence of the mirror descent algorithm under the presence of computational errors. This method was introduced by Nemirovsky and Yudin for solving convex optimization problems [66]. Here we use a derivation of this algorithm proposed by Beck and Teboulle [17].

Let $\delta_f,\ \delta_C \in (0, 1]$ and $\{a_k\}_{k=0}^{\infty} \subset (0, \infty)$.

We describe the inexact version of the mirror descent algorithm.

Mirror descent algorithm
Initialization: select an arbitrary $x_0 \in U$.
Iterative Step: given a current iteration vector $x_t \in U$ calculate

$$\xi_t \in \partial f(x_t) + B_X(0, \delta_f),$$

define

$$g_t(x) = \langle \xi_t, x \rangle + (2a_t)^{-1}\|x - x_t\|^2, \; x \in X$$

and calculate the next iteration vector $x_{t+1} \in U$ such that

$$B_X(x_{t+1}, \delta_C) \cap \text{argmin}\{g_t(y) : \; y \in C\} \neq \emptyset.$$

Here δ_f is a computational error produced by our computer system when we calculate a subgradient of f and δ_C is a computational error produced by our computer system when we calculate a minimizer of the function g_t.

Note that g_t is a convex bounded from below function on X which possesses a minimizer on C.

In Sect. 3.3 we prove the following result.

Theorem 1.5 *Let* $\delta_f, \delta_C \in (0, 1]$, $\{a_k\}_{k=0}^{\infty} \subset (0, \infty)$ *and let*

$$x_* \in C$$

satisfies

$$f(x_*) \leq f(x) \textit{ for all } x \in C.$$

Assume that $\{x_t\}_{t=0}^{\infty} \subset U$, $\{\xi_t\}_{t=0}^{\infty} \subset X$,

$$\|x_0\| \leq M_0 + 1$$

and that for each integer $t \geq 0$,

$$\xi_t \in \partial f(x_t) + B_X(0, \delta_f),$$

$$g_t(z) = \langle \xi_t, z \rangle + (2a_t)^{-1}\|z - x_t\|^2, \; z \in X$$

and

$$B_X(x_{t+1}, \delta_C) \cap \textit{argmin}(g_t, C) \neq \emptyset.$$

Then for each natural number T,

$$\sum_{t=0}^{T} a_t(f(x_t) - f(x_*))$$

$$\leq 2^{-1}(2M_0+1)^2 + (\delta_f(2M_0+1) + \delta_C(L+1)) \sum_{t=0}^{T} a_t$$

$$+4\delta_C(T+1)(2M_0+1) + 2^{-1}(L+1)^2 \sum_{t=0}^{T} a_t^2.$$

Moreover, for each natural number T,

$$f((\sum_{t=0}^{T} a_t)^{-1} \sum_{t=0}^{T} a_t x_t) - f(x_*),$$

$$\min\{f(x_t) : t = 0, \ldots, T\} - f(x_*)$$

$$\leq 2^{-1}(2M_0+1)^2 (\sum_{t=0}^{T} a_t)^{-1} + \delta_f(2M_0+1) + \delta_C(L+1)$$

$$+4\delta_C(T+1)(2M_0+1)(\sum_{t=0}^{T} a_t)^{-1}$$

$$+2^{-1}(L+1)^2 \sum_{t=0}^{T} a_t^2 (\sum_{t=0}^{T} a_t)^{-1}.$$

This theorem is a generalization of Theorem 3.1 of [92] proved in the case when $\delta_f = \delta_C$.

We are interested in an optimal choice of a_t, $t = 0, 1, \ldots$. Let T be a natural number and $A_T = \sum_{t=0}^{T} a_t$ be given. It is shown that the best choice is $a_t = (T+1)^{-1} A_T$, $t = 0, \ldots, T$. Let T be a natural number and $a_t = a$, $t = 0, \ldots, T$. It is shown that the best choice of $a > 0$ is

$$a = (8\delta_C(2M_0+1))^{1/2}(L+1)^{-1}$$

and that the best choice of T should be at the same order as $\lfloor \delta_C^{-1} \rfloor$. As a result, we obtain a point $\xi \in U$ such that

$$B_X(\xi, \delta_C) \cap C \neq \emptyset$$

and

$$f(\xi) \leq f(x_*) + c_1 \delta_C^{1/2} + \delta_f(2M_0+1),$$

where the constant $c_1 > 0$ depends only on L and M_0.

In Chap. 3 we also use the mirror descent algorithm for minimization problems of unbounded sets.

Let D be a nonempty closed convex subset of X, V be an open convex subset of X such that

$$D \subset V,$$

and $f : V \to R^1$ be a convex function which is Lipschitz on all bounded subsets of V. Set

$$D_{min} = \{x \in D : f(x) \leq f(y) \text{ for all } y \in D\}.$$

We suppose that

$$D_{min} \neq \emptyset.$$

In Sect. 3.5 we will prove the following result.

Theorem 1.6 *Let* $\delta_f, \delta_C \in (0, 1]$, $M > 1$ *satisfy*

$$D_{min} \cap B_X(0, M) \neq \emptyset,$$

$$M_0 > 80M + 6,$$

$L > 1$ *satisfy*

$$|f(v_1) - f(v_2)| \leq L\|v_1 - v_2\|$$

$$\textit{for all } v_1, v_2 \in V \cap B_X(0, M_0 + 2),$$

$$0 < \tau_0 \leq \tau_1 \leq (4L + 4)^{-1},$$

$$\epsilon_0 = 8\tau_0^{-1}\delta_C(M_0 + 1) + 2\delta_C(L + 1)$$

$$+2\delta_f(2M_0 + 1) + \tau_1(L + 1)^2$$

and let

$$n_0 = \lfloor \tau_0^{-1}(2M + 2)^2\epsilon_0^{-1} \rfloor + 1.$$

Assume that

$$\{x_t\}_{t=0}^{\infty} \subset V, \ \{\xi_t\}_{t=0}^{\infty} \subset X, \ \{a_t\}_{t=0}^{\infty} \subset [\tau_0, \tau_1],$$

$$\|x_0\| \leq M$$

and that for each integer $t \geq 0$,

$$\xi_t \in \partial f(x_t) + B_X(0, \delta_f),$$

$$g_t(v) = \langle \xi_t, v \rangle + (2a_t)^{-1} \|v - x_t\|^2, \ v \in X$$

and

$$B_X(x_{t+1}, \delta_C) \cap argmin(g_t, D) \neq \emptyset.$$

Then there exists an integer $q \in [1, n_0]$ *such that*

$$f(x_q) \leq \inf(f; D) + \epsilon_0,$$

$$\|x_t\| \leq 15M + 1, \ t = 0, \dots, q.$$

This theorem is a generalization of Theorem 3.3 of [92] proved with $\delta_f = \delta_C$.

We are interested in the best choice of a_t, $t = 0, 1, \dots$. Assume for simplicity that $\tau_1 = \tau_0$. In order to meet our goal we need to minimize ϵ_0 which obtains its minimal value when

$$\tau_0 = (8\delta_C(M_0 + 1))^{1/2}(L + 1)^{-1}$$

and the minimal value of ϵ_0 is

$$2\delta_C(L + 1) + 2\delta_f(2M_0 + 1) + 2(8\delta_C(M_0 + 1))^{1/2}(L + 1).$$

Thus ϵ_0 is at the same order as $\max\{\delta_C^{1/2}, \ \delta_f\}$. It is not difficult to see that n_0 is at the same order as $\delta_C^{-1/2} \max\{\delta_C^{1/2}, \ \delta_f\}^{-1}$.

In Chap. 3 we will also prove the following two theorems which have no prototype in [92].

Theorem 1.7 *Let* $\delta_f, \delta_C \in (0, 1)$, $M > 0$ *satisfy*

$$\delta_f \leq (24M + 48)^{-1}, \ \delta_C \leq 8^{-1}(L + 1)^{-2}(16M + 16)^{-1},$$

$$\{x \in V : \ f(x) \leq \inf(f, D) + 4\} \subset B_X(0, M),$$

$$M_0 = 12M + 12,$$

$L > 1$ *satisfy*

$$|f(v_1) - f(v_2)| \leq L\|v_1 - v_2\|$$

for all $v_1, v_2 \in V \cap B_X(0, M_0 + 2)$

and let $\{a_t\}_{t=0}^{\infty} \subset (0, \infty)$ *satisfy for all integers* $t \geq 0$,

$$4\delta_C(M_0 + 1) \leq a_t \leq (2L + 2)^{-1}(L + 1)^{-1}.$$

Assume that

$$\{x_t\}_{t=0}^{\infty} \subset V, \ \{\xi_t\}_{t=0}^{\infty} \subset X,$$

$$\|x_0\| \leq M$$

and that for each integer $t \geq 0$,

$$\xi_t \in \partial f(x_t) + B_X(0, \delta_f),$$

$$g_t(v) = \langle \xi_t, v\rangle + (2a_t)^{-1}\|v - x_t\|^2, \ v \in X$$

and

$$B_X(x_{t+1}, \delta_C) \cap argmin(g_t, D) \neq \emptyset.$$

Then

$$\|x_t\| \leq 5M + 3 \text{ for all integers } t \geq 0$$

and for each natural number T,

$$\sum_{t=0}^{T} a_t(f(x_t) - \inf(f, D))$$

$$\leq 2M^2 + (\delta_f(2M_0 + 1) + \delta_C(L + 1))\sum_{t=0}^{T} a_t$$

$$+4\delta_C(T + 1)(M_0 + 1) + 2^{-1}(L + 1)^2\sum_{t=0}^{T} a_t^2.$$

Moreover, for each natural number T,

$$f((\sum_{t=0}^{T} a_t)^{-1}\sum_{t=0}^{T} a_t x_t) - \inf(f, D),$$

$$\min\{f(x_t) : t = 0, \ldots, T\} - \inf(f, D)$$

$$\leq 2M^2(\sum_{t=0}^{T} a_t)^{-1} + \delta_f(2M_0 + 1) + \delta_C(L + 1)$$

$$+4\delta_C(T + 1)(M_0 + 1)(\sum_{t=0}^{T} a_t)^{-1} + 2^{-1}(L + 1)^2 \sum_{t=0}^{T} a_t^2 (\sum_{t=0}^{T} a_t)^{-1}.$$

We are interested in the best choice of a_t, $t = 0, 1, \ldots$. Let T be a natural number and $A_T = \sum_{t=0}^{T} a_t$ be given. It is shown that the best choice is $a_t = (T + 1)^{-1} A_T$, $i = 0, \ldots, T$. Let T be a natural number and $a_t = a$, $t = 0, \ldots, T$. The best choice of $a > 0$ is

$$a = (8\delta_C(M_0 + 1))^{1/2}(L + 1)^{-1}$$

and the best choice of T is at the same order as $\lfloor \delta_C^{-1} \rfloor$. As a result, we obtain a point $\xi \in U$ such that

$$B_X(\xi, \delta_C) \cap C \neq \emptyset$$

and

$$f(\xi) \leq \inf(f, D) + c_1 \delta_C^{1/2} + \delta_f(2M_0 + 1),$$

where the constant $c_1 > 0$ depends only on L and M_0.

Theorem 1.8 *Let*

$$\delta_f, \delta_C \in (0, 1), \; M > 8$$

satisfy

$$D_{min} \cap B_X(0, M) \neq \emptyset,$$

$L > 0$ *satisfy*

$$|f(v_1) - f(v_2)| \leq L\|v_1 - v_2\|$$

$$\textit{for all } v_1, v_2 \in V \cap B_X(0, 8M + 8),$$

$$a = (4L + 4)^{-1}\delta_C^{1/2},$$

$$T = \lfloor 8^{-1}\delta_C^{-1/2} \max\{\delta_f, \delta_C^{1/2}\}^{-1} \rfloor.$$

Assume that

$$\{x_t\}_{t=0}^{\infty} \subset V, \ \{\xi_t\}_{t=0}^{\infty} \subset X,$$

$$\|x_0\| \leq M, \ B_X(x_0, \delta_C) \cap D \neq \emptyset$$

and that for each integer $t \geq 0$,

$$\xi_t \in \partial f(x_t) + B_X(0, \delta_f),$$

$$g_t(v) = \langle \xi_t, v \rangle + (2a_t)^{-1} \|v - x_t\|^2, \ v \in X$$

and

$$B_X(x_{t+1}, \delta_C) \cap argmin(g_t, D) \neq \emptyset.$$

Then

$$\|x_t\| \leq 3M + 2 \textit{ for all integers } t \in [0, T]$$

and

$$\sum_{t=0}^{T} (T+1)^{-1} f(x_t) - \inf(f, D),$$

$$f((T+1)^{-1} \sum_{t=0}^{T} x_t) - \inf(f, D), \ \min\{f(x_t) : t = 0, \dots, T\} - \inf(f, D)$$

$$\leq 64M^2(L+1) \max\{\delta_f, \delta_C^{1/2}\}.$$

In Chap. 3 we also study the mirror descent algorithm for zero-sum games with two players. For this algorithm each iteration consists of four steps. In each of these steps there is a computational error. We suppose that these computational errors are different and prove two results: Theorems 3.6 and 3.7. In Theorem 3.6, which is a generalization of Theorem 3.4 of [92] obtained in the case when all the computational errors are the same, we study the games on bounded sets. In Theorem 3.7, which has no prototype in [92], we deal with the games on unbounded sets.

1.3 Gradient Algorithm with a Smooth Objective Function

In Chap. 4 we analyze the convergence of a projected gradient algorithm with a smooth objective function under the presence of computational errors. The problem is described by an objective function and a set of feasible points. For this algorithm each iteration consists of two steps. The first step is a calculation of a gradient of the objective function while in the second one we calculate a projection on the feasible set. In each of these two steps there is a computational error. In general, these two computational errors are different. We show that our algorithm generates a good approximate solution, if all the computational errors are bounded from above by a small positive constant. Moreover, if we know the computational errors for the two steps of our algorithm, we find out what approximate solution can be obtained and how many iterates one needs for this.

Let X be a Hilbert space equipped with an inner product $\langle \cdot, \cdot \rangle$ which induces a complete norm $\| \cdot \|$. Let C be a nonempty closed convex subset of X, U be an open convex subset of X such that $C \subset U$, and let $f : U \to R^1$ be a convex continuous function.

We suppose that the function f is Fréchet differentiable at every point $x \in U$ and for every $x \in U$ we denote by $f'(x) \in X$ the Fréchet derivative of f at x. It is clear that for any $x \in U$ and any $h \in X$

$$\langle f'(x), h \rangle = \lim_{t \to 0} t^{-1}(f(x + th) - f(x)).$$

We suppose that the mapping $f' : U \to X$ is Lipschitz on all bounded subsets of U. It is well known (see Lemma 2.2) that for each nonempty closed convex set $D \subset X$ and each $x \in X$ there exists a unique point $P_D(x) \in D$ such that

$$\|x - P_D(x)\| = \inf\{\|x - y\| : y \in D\}.$$

Suppose that there exist $L > 1$, $M_0 > 0$ such that

$$C \subset B_X(0, M_0),$$

$$|f(v_1) - f(v_2)| \leq L\|v_1 - v_2\| \text{ for all } v_1, v_2 \in U,$$

$$\|f'(v_1) - f'(v_2)\| \leq L\|v_1 - v_2\| \text{ for all } v_1, v_2 \in U.$$

Let $\delta_f, \delta_C \in (0, 1]$. We describe below our algorithm.

Gradient algorithm
Initialization: select an arbitrary $x_0 \in U \cap B_X(0, M_0)$.
Iterative Step: given a current iteration vector $x_t \in U$ calculate

$$\xi_t \in f'(x_t) + B_X(0, \delta_f)$$

and calculate the next iteration vector $x_{t+1} \in U$ such that

$$\|x_{t+1} - P_C(x_t - L^{-1}\xi_t)\| \le \delta_C.$$

In Chap. 4 we prove the following result which is a generalization of Theorem 4.2 of [92] proved in the case when $\delta_f = \delta_C$.

Theorem 1.9 *Let* $\delta_f, \delta_C \in (0, 1]$ *and let*

$$x_0 \in U \cap B_X(0, M_0).$$

Assume that $\{x_t\}_{t=1}^{\infty} \subset U$, $\{\xi_t\}_{t=0}^{\infty} \subset X$ *and that for each integer* $t \ge 0$,

$$\|\xi_t - f'(x_t)\| \le \delta$$

and

$$\|x_{t+1} - P_C(x_t - L^{-1}\xi_t)\| \le \delta_C.$$

Then for each natural number T,

$$\min\{f(x_t) : t = 2, \dots, T+1\} - \inf(f, C),$$

$$f(\sum_{t=2}^{T+1} T^{-1}x_t) - \inf(f, C)$$

$$\le (2T)^{-1}L(2M_0 + 1)^2 + L\delta_C(6M_0 + 7) + \delta_f(4M_0 + 4).$$

We are interested in an optimal choice of T. If we choose T in order to minimize the right-hand side of the estimation in the theorem above, we obtain that T should be at the same order as $\max\{\delta_f, \delta_C\}^{-1}$. In this case the right-hand side of the estimation is at the same order as $\max\{\delta_f, \delta_C\}$.

In Chap. 4 we also study optimization on unbounded sets.

Let D be a nonempty closed convex subset of X, V be an open convex subset of X such that

$$D \subset V,$$

and $f : V \to R^1$ be a convex Fréchet differentiable function which is Lipschitz on all bounded subsets of V. Set

$$D_{min} = \{x \in D : f(x) \le f(y) \text{ for all } y \in D\}.$$

We suppose that

$$D_{min} \neq \emptyset.$$

We will prove the following result which is a generalization of Theorem 4.5 of [92] proved in the case when $\delta_f = \delta_C$.

Theorem 1.10 *Let* $\delta_f, \delta_C \in (0, 1]$, $M > 0$,

$$D_{min} \cap B_X(0, M) \neq \emptyset,$$

$$M_0 = 4M + 8,$$

$L \geq 1$ *satisfy*

$$|f(v_1) - f(v_2)| \leq L\|v_1 - v_2\|$$

$$\textit{for all } v_1, v_2 \in V \cap B_X(0, M_0 + 2),$$

$$\|f'(v_1) - f'(v_2)\| \leq L\|v_1 - v_2\|$$

$$\textit{for all } v_1, v_2 \in V \cap B_X(0, M_0 + 2),$$

$$\epsilon_0 = \delta_f(4M_0 + 6) + \delta_C L(6M_0 + 8)$$

and let

$$n_0 = \lfloor 2^{-1}L(2M + 1)^2(\delta_f + L\delta_C)^{-1} \rfloor + 1.$$

Assume that $\{x_t\}_{t=0}^{\infty} \subset V$, $\{\xi_t\}_{t=0}^{\infty} \subset X$,

$$\|x_0\| \leq M$$

and that for each integer $t \geq 0$,

$$\|\xi_t - f'(x_t)\| \leq \delta_f$$

and

$$\|x_{t+1} - P_D(x_t - L^{-1}\xi_t)\| \leq \delta_C.$$

Then there exists an integer $q \in [1, n_0 + 1]$ *such that*

$$f(x_q) \leq \inf(f, D) + \epsilon_0,$$

$$\|x_i\| \le 3M + 3,\ i = 0, \dots, q.$$

The following two results have no prototype in [92].

Theorem 1.11 *Let* $\delta_f, \delta_C \in (0, 1]$, $M > 2$ *satisfy*

$$\{x \in V:\ f(x) \le \inf(f, D) + 3\} \subset B_X(0, M - 2),$$

$L \ge 1$ *satisfy*

$$|f(v_1) - f(v_2)| \le L\|v_1 - v_2\| \text{ for all } v_1, v_2 \in V \cap B_X(0, 4M + 8),$$

$$\|f'(v_1) - f'(v_2)\| \le L\|v_1 - v_2\| \text{ for all } v_1, v_2 \in V \cap B_X(0, 4M + 8),$$

$$\delta_C \le (L(18M + 19))^{-1},\ \delta_f \le (12M + 3)^{-1}.$$

Assume that $\{x_t\}_{t=0}^{\infty} \subset V$, $\{\xi_t\}_{t=0}^{\infty} \subset X$,

$$\|x_0\| \le M$$

and that for each integer $t \ge 0$,

$$\|\xi_t - f'(x_t)\| \le \delta_f$$

and

$$\|x_{t+1} - P_D(x_t - L^{-1}\xi_t)\| \le \delta_C.$$

Then

$$\|x_t\| \le 3M \text{ for all integers } t \ge 0$$

and for each natural number T,

$$\min\{f(x_t):\ t = 1, \dots, T + 1\} - \inf(f, D),\ f(\sum_{t=1}^{T+1}(T + 1)^{-1}x_t) - \inf(f, D)$$

$$\le L\delta_C(18M + 19) + \delta_f(12M + 13) + 2(T + 1)^{-1}M^2L.$$

It is clear that a best choice of T should be at the same order as $\max\{\delta_C, \delta_f\}^{-1}$.

Theorem 1.12 *Let* $\delta_f, \delta_C \in (0, 1)$, $M > 1$,

$$\delta_C \le (160M)^{-1},\ \delta_f \le 120^{-1},$$

$$D_{min} \cap B_X(0, M) \neq \emptyset,$$

$L \geq 1$ satisfy

$$|f(v_1) - f(v_2)| \leq L\|v_1 - v_2\| \text{ for all } v_1, v_2 \in V \cap B_X(0, 4M + 10),$$

$$\|f'(v_1) - f'(v_2)\| \leq L\|v_1 - v_2\| \text{ for all } v_1, v_2 \in V \cap B_X(0, 4M + 10),$$

$$T = \lfloor 36^{-1} \min\{\delta_C^{-1},\ L\delta_f^{-1}\}\rfloor.$$

Assume that $\{x_t\}_{t=0}^{\infty} \subset V$, $\{\xi_t\}_{t=0}^{\infty} \subset X$,

$$\|x_0\| \leq M$$

and that for each integer $t \geq 0$,

$$\|\xi_t - f'(x_t)\| \leq \delta_f$$

and

$$\|x_{t+1} - P_D(x_t - L^{-1}\xi_t)\| \leq \delta_C.$$

Then

$$\|x_t\| \leq 3M \text{ for all integers } t = 0, \dots, T + 1$$

and

$$\sum_{t=0}^{T}(T + 1)^{-1}(f(x_{t+1}) - \inf(f, D))$$

$$\min\{f(x_t) :\ t = 1, \dots, T + 1\} - \inf(f, D),\ f(\sum_{t=1}^{T+1}(T + 1)^{-1}x_t) - \inf(f, D)$$

$$\leq 72(M + 1)^2 \max\{\delta_C L, \delta_f\}.$$

1.4 Examples

Example 1.13 Let $X = R^n$, $C = \{x = (x_1, \dots, x_n) \in R^n :\ x_i \in [0, 1],\ i = 1, \dots, n\}$,

$$U = \{x \in R^n :\ \|x\| < 2n^{1/2}\}$$

and let for all $x \in R^n$,

$$f(x) = \langle Ax, x\rangle + \langle b, x\rangle,$$

where $A = (a_{i,j})_{i,j=1}^n$ is a symmetric positive semidefinite matrix and $b \in R^n$. Clearly, the function f is convex and $\nabla f(x) = b + 2Ax$ for all $x \in R^n$. We consider the minimization problem

$$f(x) \to \min, \; x \in C$$

and use the subgradient projection method taking into account computational errors (see Sect. 1.2). It is clear that for all $x = (x_1, \ldots, x_n) \in R^n$,

$$P_C(x) = (\min\{\max\{x_1, 0\}, 1\}, \ldots,$$

$$\min\{\max\{x_i, 0\}, 1\}, \ldots, \min\{\max\{x_n, 0\}, 1\}).$$

Assume that our computer system has an accuracy δ_*. Then it is not difficult to see that in our case $\delta_C = \delta_* n^{1/2}$, $M_0 = 2n^{1/2}$,

$$\delta_f = 2\delta_* n \max\{|a_{i,j}| : \; i, j = 1, \ldots, n\},$$

$$L = \|b\| + 4(\sum_{i,j=1}^n a_{i,j}^2)^{1//2} n^{1/2}.$$

We apply Theorem 1.1 and obtain that for all integers $t \geq 0$,

$$a_t = a = (2\delta_C(4M_0 + 1))^{1/2}(L+1)^{-1}$$

$$= (2\delta_* n^{1/2}(8n^{1/2} + 1))^{1/2}(\|b\| + 1 + 4(\sum_{i,j=1}^n a_{i,j}^2)^{1/2} n^{1/2})^{-1}$$

and

$$T = \lfloor \delta_C^{-1} \rfloor = \lfloor \delta_*^{-1} n^{-1/2} \rfloor.$$

Example 1.14 Consider the previous example and apply the gradient projection algorithm with the smooth objective function (see Theorem 1.9) and with the same C, Uf, A, b, δ_*. It is not difficult to see that

$$M_0 = 2n^{1/2}, \; L = \|b\| + 4(\sum_{i,j=1}^n a_{i,j}^2)^{1//2} n^{1/2},$$

$$\delta_C = \delta_* n^{1/2},$$

$$\delta_f = 2\delta_* n \max\{|a_{i,j}| : i, j = 1, \dots, n\}.$$

According to Theorem 1.9, $\lfloor \max\{\delta_C, \delta_f\}^{-1} \rfloor$ iterates should be done. As a result we obtain a point $z \in R^n$ such that $d(z, C) \le \delta_C$ and $f(z) \le \inf(f, C) + c_0 \max\{\delta_f, \delta_C\}$, where c_0 is a positive constant. Its value can be easily calculated using Theorem 1.9.

Example 1.15 Consider the minimization problem from the previous two examples applying Theorem 1.9. Assume that we do 10^3 iterates. By Theorem 1.9, we obtain $z \in R^n$ such that $d(z, C) \le \delta_C$ and

$$f(z) \le \inf(f, C) + 2^{-1}10^{-3}L(2M_0 + 1)^2 + L\delta_C(6M_0 + 7) + \delta_f(4M_0 + 4).$$

Chapter 2
Subgradient Projection Algorithm

In this chapter we study the subgradient projection algorithm for minimization of convex and nonsmooth functions and for computing the saddle points of convex–concave functions, under the presence of computational errors. The problem is described by an objective function and a set of feasible points. For this algorithm each iteration consists of two steps. The first step is a calculation of a subgradient of the objective function while in the second one we calculate a projection on the feasible set. In each of these two steps there is a computational error. In general, these two computational errors are different. We show that our algorithm generates a good approximate solution, if all the computational errors are bounded from above by a small positive constant. Moreover, if we know the computational errors for the two steps of our algorithm, we find out what approximate solution can be obtained and how many iterates one needs for this.

2.1 Preliminaries

The subgradient projection algorithm is one of the most important tools in the optimization theory, nonlinear analysis, and their applications. See, for example, [1–3, 11, 15, 17, 22, 27, 29, 30, 32, 35, 36, 38, 39, 43, 46, 53–55] and the references mentioned therein. In this chapter we use this method for constrained minimization problems in Hilbert spaces.

Let X be a Hilbert space equipped with an inner product $\langle \cdot, \cdot \rangle$ which induces a complete norm $\| \cdot \|$. Let C be a nonempty closed convex subset of X, U be an open convex subset of X such that $C \subset U$, and $f : U \to R^1$ be a convex function. Suppose that there exist $L > 0$, $M_0 > 0$ such that

A. J. Zaslavski, *Convex Optimization with Computational Errors*, Springer Optimization and Its Applications 155, https://doi.org/10.1007/978-3-030-37822-6_2

$$C \subset B_X(0, M_0), \tag{2.1}$$

$$|f(x) - f(y)| \le L\|x - y\| \text{ for all } x, y \in U. \tag{2.2}$$

In view of (2.2), for each $x \in U$,

$$\emptyset \ne \partial f(x) \subset B_X(0, L). \tag{2.3}$$

It is easy to see that the following result is true.

Lemma 2.1 *Let $z, y_0, y_1 \in X$. Then*

$$\|z - y_0\|^2 - \|z - y_1\|^2 - \|y_0 - y_1\|^2 = 2\langle z - y_1, y_1 - y_0\rangle.$$

The next result is well known in the literature [12, 13, 92, 93].

Lemma 2.2 *Let D be a nonempty closed convex subset of X. Then for each $x \in X$ there is a unique point $P_D(x) \in D$ satisfying*

$$\|x - P_D(x)\| = \inf\{\|x - y\| : y \in D\}.$$

Moreover,

$$\|P_D(x) - P_D(y)\| \le \|x - y\| \textit{ for all } x, y \in X$$

and for each $x \in X$ and each $z \in D$,

$$\langle z - P_D(x), x - P_D(x)\rangle \le 0,$$

$$\|z - P_D(x)\|^2 + \|x - P_D(x)\|^2 \le \|z - x\|^2.$$

For the proof of the next simple result see Lemma 2.3 of [92].

Lemma 2.3 *Let $A > 0$ and $n \ge 2$ be an integer. Then the minimization problem*

$$\sum_{i=1}^{n} a_i^2 \to \min$$

$$a = (a_1, \dots, a_n) \in R^n \textit{ and } \sum_{i=1}^{n} a_i = A$$

has a unique solution $a^* = (a_1^*, \dots, a_n^*)$ where $a_i^* = n^{-1}A$, $i = 1, \dots, n$.

2.2 A Convex Minimization Problem

Suppose that $\delta_f, \delta_C \in (0, 1]$ and $\{a_k\}_{k=0}^{\infty} \subset (0, \infty)$.

Let us describe our algorithm.

Subgradient projection algorithm
Initialization: select an arbitrary $x_0 \in U$.
Iterative Step: given a current iteration vector $x_t \in U$ calculate

$$\xi_t \in \partial f(x_t) + B_X(0, \delta_f)$$

and the next iteration vector $x_{t+1} \in U$ such that

$$\|x_{t+1} - P_C(x_t - a_t\xi_t)\| \le \delta_C.$$

Here δ_f is a computational error produced by our computer system when we calculate a subgradient of f and δ_C is a computational error produced by our computer system when we calculate a projection on the set C.

In this chapter we prove the following result.

Theorem 2.4 *Let $\delta_f, \delta_C \in (0, 1]$, $\{a_k\}_{k=0}^{\infty} \subset (0, \infty)$ and let*

$$x_* \in C \tag{2.4}$$

satisfy

$$f(x_*) \le f(x) \text{ for all } x \in C. \tag{2.5}$$

Assume that $\{x_t\}_{t=0}^{\infty} \subset U$, $\{\xi_t\}_{t=0}^{\infty} \subset X$,

$$\|x_0\| \le M_0 + 1 \tag{2.6}$$

and that for each integer $t \ge 0$,

$$\xi_t \in \partial f(x_t) + B_X(0, \delta_f) \tag{2.7}$$

and

$$\|x_{t+1} - P_C(x_t - a_t\xi_t)\| \le \delta_C. \tag{2.8}$$

Then for each natural number T,

$$\sum_{t=0}^{T} a_t(f(x_t) - f(x_*))$$

$$\leq 2^{-1}\|x_* - x_0\|^2 + \delta_C(T+1)(4M_0+1)$$

$$+ \delta_f(2M_0+1)\sum_{t=0}^{T} a_t + 2^{-1}(L+1)^2\sum_{t=0}^{T} a_t^2. \tag{2.9}$$

Moreover, for each natural number T,

$$f((\sum_{t=0}^{T} a_t)^{-1}\sum_{t=0}^{T} a_t x_t) - f(x_*),\ \min\{f(x_t) : t = 0, \dots, T\} - f(x_*)$$

$$\leq 2^{-1}(\sum_{t=0}^{T} a_t)^{-1}\|x_* - x_0\|^2 + (\sum_{t=0}^{T} a_t)^{-1}\delta_C(T+1)(4M_0+1)$$

$$+ \delta_f(2M_0+1) + 2^{-1}(\sum_{t=0}^{T} a_t)^{-1}(L+1)^2\sum_{t=0}^{T} a_t^2. \tag{2.10}$$

Theorem 2.4 is proved in Sect. 2.4. It is a generalization of Theorem 2.4 of [92] proved in the case when $\delta_f = \delta_C$.

We are interested in an optimal choice of a_t, $t = 0, 1, \dots$. Let T be a natural number and $A_T = \sum_{t=0}^{T} a_t$ be given. By Theorem 2.4, in order to make the best choice of a_t, $t = 0, \dots, T$, we need to minimize the function

$$\phi(a_0, \dots, a_T) := 2^{-1}A_T^{-1}\|x_* - x_0\|^2 + A_T^{-1}\delta_C(T+1)(4M_0+1)$$

$$+\delta_f(2M_0+1) + 2^{-1}A_T^{-1}(L+1)^2\sum_{t=0}^{T} a_t^2$$

on the set

$$\{a = (a_0, \dots, a_T) \in R^{T+1} : a_i \geq 0,\ i = 0, \dots, T,\ \sum_{i=0}^{T} a_i = A_T\}.$$

By Lemma 2.3, this function has a unique minimizer $a^* = (a_0^*, \dots, a_T^*)$ where $a_i^* = (T+1)^{-1}A_T$, $i = 0, \dots, T$. This is the best choice of a_t, $t = 0, 1, \dots, T$.

Theorem 2.4 implies the following result.

Theorem 2.5 *Let $\delta_f, \delta_C \in (0, 1]$, $a > 0$ and let $x_* \in C$ satisfy*

$$f(x_*) \le f(x) \text{ for all } x \in C.$$

Assume that $\{x_t\}_{t=0}^{\infty} \subset U$, $\{\xi_t\}_{t=0}^{\infty} \subset X$,

$$\|x_0\| \le M_0 + 1$$

and that for each integer $t \ge 0$,

$$\xi_t \in \partial f(x_t) + B_X(0, \delta_f)$$

and

$$\|x_{t+1} - P_C(x_t - a\xi_t)\| \le \delta_C.$$

Then for each natural number T,

$$f((T+1)^{-1}\sum_{t=0}^{T} x_t) - f(x_*), \ \min\{f(x_t) : t = 0, \dots, T\} - f(x_*)$$

$$\le 2^{-1}(T+1)^{-1}a^{-1}(2M_0+1)^2 + a^{-1}\delta_C(4M_0+1)$$

$$+\delta_f(2M_0+1) + 2^{-1}(L+1)^2 a.$$

Now we will find the best $a > 0$. Since T can be arbitrarily large, we need to find a minimizer of the function

$$\phi(a) := a^{-1}\delta_C(4M_0+1) + 2^{-1}(L+1)^2 a, \ a \in (0, \infty).$$

Clearly, the minimizer a satisfies

$$a^{-1}\delta_C(4M_0+1) = 2^{-1}(L+1)^2 a,$$

$$a = (2\delta_C(4M_0+1))^{1/2}(L+1)^{-1}$$

and the minimal value of ϕ is

$$(2\delta_C(4M_0+1))^{1/2}(L+1).$$

Theorem 2.5 implies the following result.

Theorem 2.6 *Let* $\delta_f, \delta_C \in (0, 1]$,

$$a = (2\delta_C(4M_0 + 1))^{1/2}(L + 1)^{-1},$$

$x_* \in C$ *satisfy*

$$f(x_*) \leq f(x) \text{ for all } x \in C.$$

Assume that $\{x_t\}_{t=0}^{\infty} \subset U$, $\{\xi_t\}_{t=0}^{\infty} \subset X$,

$$\|x_0\| \leq M_0 + 1$$

and that for each integer $t \geq 0$,

$$\xi_t \in \partial f(x_t) + B_X(0, \delta_f)$$

and

$$\|x_{t+1} - P_C(x_t - a\xi_t)\| \leq \delta_C.$$

Then for each natural number T,

$$f((T + 1)^{-1}\sum_{t=0}^{T} x_t) - f(x_*),\ \min\{f(x_t) : t = 0, \ldots, T\} - f(x_*)$$

$$\leq 2^{-1}(T + 1)^{-1}(2M_0 + 1)^2(L + 1)(2\delta_C(4M_0 + 1))^{-1/2} + \delta_f(2M_0 + 1)$$

$$+2^{-1}(2\delta_C(4M_0 + 1))^{1/2}(L + 1) + \delta_C^{1/2}(4M_0 + 1)(L + 1)(2(4M_0 + 1))^{-1/2}.$$

Now we can think about the best choice of T. It is clear that it should be at the same order as $\lfloor \delta_C^{-1} \rfloor$. Putting $T = \lfloor \delta^{-1} \rfloor$, we obtain that

$$f((T + 1)^{-1}\sum_{t=0}^{T} x_t) - f(x_*),\ \min\{f(x_t) : t = 0, \ldots, T\} - f(x_*)$$

$$\leq 2^{-1}(2M_0 + 1)^2(L + 1)(8M_0 + 2)^{-1/2}\delta_C^{1/2} + \delta_f(2M_0 + 1)$$

$$+(L + 1)\delta_C^{1/2}(8M_0 + 2)^{1/2}.$$

Note that in the theorems above δ_f is the computational error, produced by our computer system when we calculate a subgradient of f and δ_C is the computational error, produced by our computer system when we calculate a projection on C subgradient.

In view of the inequality above, which has the right-hand side bounded by $c_1\delta_C^{1/2} + \delta_f(2M_0 + 1)$ with a constant $c_1 > 0$, we conclude that after $T = \lfloor \delta_C^{-1} \rfloor$ iterations we obtain a point $\xi \in U$ such that

$$B_X(\xi, \delta_C) \cap C \neq \emptyset$$

and

$$f(\xi) \le f(x_*) + c_1\delta_C^{1/2} + \delta_f(2M_0 + 1),$$

where the constant $c_1 > 0$ depends only on L and M_0. It is interesting that the number of iterations depends only on δ_C.

2.3 The Main Lemma

Lemma 2.7 *Let* $\delta_f, \delta_C \in (0, 1]$, $a > 0$ *and let*

$$z \in C. \tag{2.11}$$

Assume that

$$x \in U \cap B_X(0, M_0 + 1), \tag{2.12}$$

$$\xi \in \partial f(x) + B_X(0, \delta_f) \tag{2.13}$$

and that

$$u \in U \tag{2.14}$$

satisfies

$$\|u - P_C(x - a\xi)\| \le \delta_C. \tag{2.15}$$

Then

$$a(f(x) - f(z)) \le 2^{-1}\|z - x\|^2 - 2^{-1}\|z - u\|^2$$

$$+\delta_C(4M_0 + 1) + a\delta_f(2M_0 + 1) + 2^{-1}a^2(L + 1)^2.$$

Proof In view of (2.13), there exists

$$l \in \partial f(x) \tag{2.16}$$

such that

$$\|l - \xi\| \le \delta_f. \tag{2.17}$$

By Lemmas 2.1 and 2.2 and (2.11),

$$\begin{aligned} 0 &\le \langle z - P_C(x - a\xi), P_C(x - a\xi) - (x - a\xi)\rangle \\ &= \langle z - P_C(x - a\xi), P_C(x - a\xi) - x\rangle \\ &\quad + \langle a\xi, z - P_C(x - a\xi)\rangle \\ &= 2^{-1}[\|z - x\|^2 - \|z - P_C(x - a\xi)\|^2 - \|x - P_C(x - a\xi)\|^2] \\ &\quad + \langle a\xi, z - x\rangle + \langle a\xi, x - P_C(x - a\xi)\rangle. \end{aligned} \tag{2.18}$$

Clearly,

$$|\langle a\xi, x - P_C(x - a\xi)\rangle| \le 2^{-1}(\|a\xi\|^2 + \|x - P_C(x - a\xi)\|^2). \tag{2.19}$$

It follows from (2.18) and (2.19) that

$$\begin{aligned} 0 &\le 2^{-1}[\|z - x\|^2 - \|z - P_C(x - a\xi)\|^2 - \|x - P_C(x - a\xi)\|^2] \\ &\quad + \langle a\xi, z - x\rangle + 2^{-1}a^2\|\xi\|^2 + 2^{-1}\|x - P_C(x - a\xi)\|^2 \\ &\le 2^{-1}\|z - x\|^2 - 2^{-1}\|z - P_C(x - a\xi)\|^2 + 2^{-1}a^2\|\xi\|^2 + \langle a\xi, z - x\rangle. \end{aligned} \tag{2.20}$$

Relations (2.1), (2.11) and (2.15) imply that

$$\begin{aligned} &|\|z - P_C(x - a\xi)\|^2 - \|z - u\|^2| \\ &= |\|z - P_C(x - a\xi)\| - \|z - u\||(\|z - P_C(x - a\xi)\| + \|z - u\|) \\ &\le \|u - P_C(x - a\xi)\|(4M_0 + 1) \le (4M_0 + 1)\delta_C. \end{aligned} \tag{2.21}$$

By (2.1), (2.11), (2.12), and (2.17),

$$\begin{aligned} \langle a\xi, z - x\rangle &= \langle al, z - x\rangle + \langle a(\xi - l), z - x\rangle \\ &\le \langle al, z - x\rangle + a\|\xi - l\|\|z - x\| \end{aligned}$$

$$\leq \langle al, z - x\rangle + a\delta_f(2M_0 + 1). \tag{2.22}$$

It follows from (2.3), (2.16), (2.17), (2.20), (2.21), and (2.22) that

$$0 \leq 2^{-1}\|z - x\|^2 - 2^{-1}\|z - P_C(x - a\xi)\|^2 + 2^{-1}a^2\|\xi\|^2 + \langle a\xi, z - x\rangle$$

$$\leq 2^{-1}\|z - x\|^2 - 2^{-1}\|z - u\|^2 + \delta_C(4M_0 + 1) + 2^{-1}a^2(L + 1)^2$$

$$+ \langle al, z - x\rangle + a\delta_f(2M_0 + 1). \tag{2.23}$$

By (2.16) and (2.23),

$$a(f(z) - f(x)) \geq \langle al, z - x\rangle$$

and

$$a(f(x) - f(z)) \leq \langle al, x - z\rangle$$

$$\leq 2^{-1}\|z - x\|^2 - 2^{-1}\|z - u\|^2 + \delta_C(4M_0 + 1) + 2^{-1}a^2(L + 1)^2$$

$$+ a\delta_f(2M_0 + 1).$$

This completes the proof of Lemma 2.7. □

2.4 Proof of Theorem 2.4

It is clear that

$$\|x_t\| \leq M_0 + 1,\ t = 0, 1, \ldots.$$

Let $t \geq 0$ be an integer. Applying Lemma 2.7 with

$$z = x_*,\ a = a_t,\ x = x_t,\ \xi = \xi_t,\ u = x_{t+1}$$

we obtain that

$$a_t(f(x_t) - f(x_*)) \leq 2^{-1}\|x_* - x_t\|^2 - 2^{-1}\|x_* - x_{t+1}\|^2$$

$$+ \delta_C(4M_0 + 1) + a_t\delta_f(2M_0 + 1) + 2^{-1}a_t^2(L + 1)^2. \tag{2.24}$$

By (2.24), for each natural number T,

$$\sum_{t=0}^{T} a_t(f(x_t) - f(x_*))$$

$$\leq \sum_{t=0}^{T}(2^{-1}\|x_* - x_t\|^2 - 2^{-1}\|x_* - x_{t+1}\|^2)$$

$$+\delta_C(4M_0 + 1) + a_t\delta_f(2M_0 + 1) + 2^{-1}a_t^2(L + 1)^2$$

$$\leq 2^{-1}\|x_* - x_0\|^2 + \delta_C(T + 1)(4M_0 + 1)$$

$$+\delta_f(2M_0 + 1)\sum_{t=0}^{T} a_t + 2^{-1}(L + 1)^2\sum_{t=0}^{T} a_t^2.$$

Thus (2.9) is true. Evidently, (2.10) implies (2.9). Theorem 2.4 is proved. □

2.5 Subgradient Algorithm on Unbounded Sets

Let X be a Hilbert space with an inner product $\langle \cdot, \cdot \rangle$ which induces a complete norm $\| \cdot \|$, D be a nonempty closed convex subset of X, V be an open convex subset of X such that

$$D \subset V, \tag{2.25}$$

and $f : V \to R^1$ be a convex function which is Lipschitz on all bounded subsets of V. Set

$$D_{min} = \{x \in D : \ f(x) \leq f(y) \text{ for all } y \in D\}. \tag{2.26}$$

We suppose that

$$D_{min} \neq \emptyset. \tag{2.27}$$

In this chapter we will prove the following result.

Theorem 2.8 *Let* $\delta_f, \delta_C \in (0, 1]$, $M > 0$ *satisfy*

$$D_{min} \cap B_X(0, M) \neq \emptyset, \tag{2.28}$$

$$M_0 \geq 4M + 4, \tag{2.29}$$

$L > 0$ *satisfy*

$$|f(v_1) - f(v_2)| \leq L\|v_1 - v_2\| \text{ for all } v_1, v_2 \in V \cap B_X(0, M_0 + 2), \tag{2.30}$$

$$0 < \tau_0 \leq \tau_1 \leq (L+1)^{-1}, \tag{2.31}$$

$$\epsilon_0 = 2\tau_0^{-1}\delta_C(4M_0 + 1) + 2\delta_f(2M_0 + 1) + 2\tau_1(L+1)^2 \tag{2.32}$$

and let

$$n_0 = \lfloor \tau_0^{-1}(2M+2)^2\epsilon_0^{-1} \rfloor. \tag{2.33}$$

Assume that $\{x_t\}_{t=0}^{\infty} \subset V$, $\{\xi_t\}_{t=0}^{\infty} \subset X$,

$$\{a_t\}_{t=0}^{\infty} \subset [\tau_0, \tau_1], \tag{2.34}$$

$$\|x_0\| \leq M \tag{2.35}$$

and that for each integer $t \geq 0$,

$$\xi_t \in \partial f(x_t) + B_X(0, \delta_f) \tag{2.36}$$

and

$$\|x_{t+1} - P_D(x_t - a_t\xi_t)\| \leq \delta_C. \tag{2.37}$$

Then there exists an integer $q \in [1, n_0 + 1]$ *such that*

$$\|x_i\| \leq 3M + 2, \ i = 0, \ldots, q$$

and

$$f(x_q) \leq f(x) + \epsilon_0 \text{ for all } x \in D.$$

Theorem 2.8 is a generalization of Theorem 2.8 of [92] which was proved in the case when $\delta_C = \delta_f$. It is proved in Sect. 2.6.

We are interested in the best choice of a_t, $t = 0, 1, \ldots$. Assume for simplicity that $\tau_1 = \tau_0 = \tau$. In order to meet our goal we need to minimize the function

$$2\tau^{-1}\delta_C(4M_0 + 1) + 2(L+1)^2\tau, \ \tau \in (0, \infty).$$

This function has a minimizer

$$\tau = (\delta_C(4M_0 + 1))^{1/2}(L+1)^{-1},$$

the minimal value of ϵ_0 is

$$2\delta_f(2M_0+1)+4(\delta_C(4M_0+1))^{1/2}(L+1)$$

and $n_0 = \lfloor \Delta \rfloor$ where

$$\Delta = (2M+2)^2(4M_0+1)^{-1/2}(L+1)\delta_C^{-1/2}(2\delta_f(2M_0+1)$$

$$+4\delta_C^{1/2}(4M_0+1)^{1/2}(L+1))^{-1}.$$

Note that in the theorem above δ_f, δ_C are the computational errors produced by our computer system. In view of the inequality above, in order to obtain a good approximate solution we need

$$c_1\lfloor \delta_C^{-1/2}\max\{\delta_f, \delta_C^{1/2}\}^{-1}\rfloor + 1$$

iterations, where c_1 is a constant which depends only on M, M_0, L. As a result, we obtain a point $\xi \in V$ such that

$$B_X(\xi, \delta_C) \cap D \neq \emptyset$$

and

$$f(\xi) \leq \inf\{f(x) : x \in D\} + 2\delta_f(2M_0+1) + (4M_0+1)^{1/2}4(L+1)\delta_C^{1/2}.$$

The next result, which is proved in Sect. 2.7, does not have a prototype in [92].

Theorem 2.9 *Let $\delta_f, \delta_C \in (0,1)$, $M > 1$, $L > 0$ and let $\{a_t\}_{t=0}^{\infty} \subset (0,\infty)$ be such that*

$$\{x \in V : f(x) \leq \inf(f, D) + 3\} \subset B_X(0, M-1), \tag{2.38}$$

$$|f(v_1) - f(v_2)| \leq L\|v_1 - v_2\|$$

$$\textit{for all } v_1, v_2 \in V \cap B_X(0, 4M+8), \tag{2.39}$$

$$\delta_f \leq (6M+5)^{-1} \textit{ and}$$

$$\delta_C(12M+9) \leq a_t \leq (L+1)^{-2} \textit{ for all integers } t \geq 0. \tag{2.40}$$

Assume that $\{x_t\}_{t=0}^{\infty} \subset V$, $\{\xi_t\}_{t=0}^{\infty} \subset X$,

$$\|x_0\| \leq M,\ B_X(x_0, \delta_C) \cap D \neq \emptyset \tag{2.41}$$

and that for each integer $t \geq 0$,

$$\xi_t \in \partial f(x_t) + B_X(0, \delta_f) \tag{2.42}$$

and

$$\|x_{t+1} - P_D(x_t - a_t\xi_t)\| \le \delta_C. \tag{2.43}$$

Then

$$\|x_t\| \le 3M \text{ for all integers } t \ge 0$$

and for each natural number T,

$$\sum_{t=0}^{T} a_t(f(x_t) - \inf(f, D))$$

$$\le 2M^2 + \delta_C(T+1)(12M+9)$$

$$+2^{-1}(L+1)^2\sum_{t=0}^{T} a_t^2 + \delta_f(6M+5)\sum_{t=0}^{T} a_t.$$

Moreover, for each natural number T,

$$f((\sum_{t=0}^{T} a_t)^{-1}\sum_{t=0}^{T} a_t x_t) - \inf(f, D),$$

$$\min\{f(x_t) : t = 0, \dots, T\} - \inf(f, D)$$

$$\le 2M^2(\sum_{t=0}^{T} a_t)^{-1} + \delta_C(T+1)(12M+9)(\sum_{t=0}^{T} a_t)^{-1}$$

$$+2^{-1}(L+1)^2\sum_{t=0}^{T} a_t^2(\sum_{t=0}^{T} a_t)^{-1} + \delta_f(6M+5).$$

We are interested in an optimal choice of a_t, $t = 0, 1, \dots$. Let T be a natural number and $A_T = \sum_{t=0}^{T} a_t$ be given. By Theorem 2.9, in order to make the best choice of a_t, $t = 0, \dots, T$, we need to minimize the function

$$\phi(a_0, \dots, a_T) := 2^{-1}A_T^{-1}M^2 + A_T^{-1}\delta_C(T+1)(12M+9)$$

$$+\delta_f(6M+5)+2^{-1}A_T^{-1}(L+1)^2\sum_{t=0}^{T}a_t^2$$

on the set

$$\{a=(a_0,\ldots,a_T)\in R^{T+1}:\ a_i\geq 0,\ i=0,\ldots,T,\ \sum_{i=0}^{T}a_i=A_T\}.$$

Theorem 2.9 and Lemma 2.3 imply the following result.

Theorem 2.10 *Let $\delta_f, \delta_C \in (0,1)$, $a>0$, $M>1$ and $L>0$ be such that*

$$\{x\in V:\ f(x)\leq \inf(f,D)+3\}\subset B_X(0,M-1),$$

$$|f(v_1)-f(v_2)|\leq L\|v_1-v_2\|$$

$$\textit{for all } v_1,v_2\in V\cap B_X(0,4M+8),$$

$$\delta_f\leq (6M+5)^{-1} \textit{ and } \delta_C(12M+9)\leq a\leq (L+1)^{-2}.$$

Assume that $\{x_t\}_{t=0}^{\infty}\subset V$, $\{\xi_t\}_{t=0}^{\infty}\subset X$,

$$\|x_0\|\leq M,\ B_X(x_0,\delta_C)\cap D\neq\emptyset$$

and that for each integer $t\geq 0$,

$$\xi_t\in\partial f(x_t)+B_X(0,\delta_f)$$

and

$$\|x_{t+1}-P_D(x_t-a\xi_t)\|\leq\delta_C.$$

Then

$$\|x_t\|\leq 3M \textit{ for all integers } t\geq 0$$

and for each natural number T,

$$f((T+1)^{-1}\sum_{t=0}^{T}x_t)-\inf(f,D),$$

$$\min\{f(x_t):\ t=0,\ldots,T\}-\inf(f,D)$$

$$\leq 2M^2(T+1)^{-1}a^{-1} + \delta_C(12M+9)a^{-1}$$

$$+2^{-1}(L+1)^2 a + \delta_f(6M+5).$$

Now we will find the best $a > 0$. Since T can be arbitrarily large, we need to find a minimizer of the function

$$\phi(a) := a^{-1}\delta_C(12M+9) + 2^{-1}(L+1)^2 a,\ a \in (0,\infty).$$

Clearly, the minimizer a satisfies

$$a^{-1}\delta_C(12M+9) = 2^{-1}(L+1)^2 a,$$

$$a = (2\delta_C(12M+9))^{1/2}(L+1)^{-1}$$

and the minimal value of ϕ is

$$(2\delta_C(12M+9))^{1/2}(L+1).$$

Clearly, we need to check if the minimizer a defined above satisfies (2.40). It is not difficult to see that this takes place if

$$\delta_C \leq 2^{-1}(12M+9)^{-1}(L+1)^{-2}.$$

Theorem 2.10 implies the following result.

Theorem 2.11 *Let $\delta_f, \delta_C \in (0,1)$, $M > 1$ and $L > 0$ be such that*

$$\{x \in V : f(x) \leq \inf(f, D) + 3\} \subset B_X(0, M-1),$$

$$|f(v_1) - f(v_2)| \leq L\|v_1 - v_2\|$$

for all $v_1, v_2 \in V \cap B_X(0, 4M+8)$,

$$\delta_C \leq 2^{-1}(12M+9)^{-1}(L+1)^{-2}$$

$$\delta_f \leq (6M+5)^{-1}$$

and

$$a = (2\delta_C(12M+9))^{1/2}(L+1)^{-1}.$$

Assume that $\{x_t\}_{t=0}^{\infty} \subset V$, $\{\xi_t\}_{t=0}^{\infty} \subset X$,

$$\|x_0\| \leq M,\ B_X(x_0, \delta_C) \cap D \neq \emptyset$$

and that for each integer $t \geq 0$,

$$\xi_t \in \partial f(x_t) + B_X(0, \delta_f)$$

and

$$\|x_{t+1} - P_D(x_t - a_t \xi_t)\| \leq \delta_C.$$

Then

$$\|x_t\| \leq 3M \textit{ for all integers } t \geq 0$$

and for each natural number T,

$$f((T+1)^{-1}\sum_{t=0}^{T} x_t) - \inf(f, D),$$

$$\min\{f(x_t) : t = 0, \dots, T\} - \inf(f, D)$$

$$\leq 2M^2(T+1)^{-1}(L+1)(2\delta_C(12M+9))^{-1/2}$$

$$+(L+1)(2\delta_C(12M+9))^{1/2} + \delta_f(6M+5).$$

Now we can think about the best choice of T. It is clear that it should be at the same order as $\lfloor \delta_C^{-1} \rfloor$. Putting $T = \lfloor \delta_C^{-1} \rfloor$, we obtain that

$$f((T+1)^{-1}\sum_{t=0}^{T} x_t) - \inf(f; D), \ \min\{f(x_t) : t = 0, \dots, T\} - \inf(f; D)$$

$$\leq 2M^2(L+1)\delta_C^{1/2}(24M+18)^{-1/2}$$

$$+\delta_C^{1/2}(12M+9)(L+1)(24M+18)^{-1/2}$$

$$+2^{-1}(L+1)\delta_C^{1/2}(24M+18)^{1/2} + \delta_f(6M+5).$$

In view of the inequality above, which has the right-hand side bounded by $c_1\delta_C^{1/2} + \delta_f(6M+5)$ with a constant $c_1 > 0$, we conclude that after $T = \lfloor \delta_C^{-1} \rfloor$ iterations we obtain a point $\xi \in U$ such that

$$B_X(\xi, \delta_C) \cap D \neq \emptyset$$

and

$$f(\xi) \le \inf(f, D) + c_1\delta_C^{1/2} + \delta_f(6M + 5),$$

where the constant $c_1 > 0$ depends only on L and M.

The next result, which is proved in Sect. 2.8, does not have a prototype in [92].

Theorem 2.12 *Let* $\delta_f, \delta_C \in (0, 16^{-1})$, $M > 4$, $L > 0$,

$$D_{min} \cap B_X(0, M) \neq \emptyset,$$

$$|f(v_1) - f(v_2)| \le L\|v_1 - v_2\|$$

$$\textit{for all } v_1, v_2 \in V \cap B_X(0, 3M + 6), \tag{2.44}$$

$$a = (L + 1)^{-1}\delta_C^{1/2}$$

and

$$T = \lfloor 6^{-1} \min\{\delta_C^{-1}, \delta_C^{-1/2}\delta_f^{-1}\}\rfloor - 1.$$

Assume that $\{x_t\}_{t=0}^{\infty} \subset V$, $\{\xi_t\}_{t=0}^{\infty} \subset X$,

$$\|x_0\| \le M, \ B_X(x_0, \delta_C) \cap D \neq \emptyset \tag{2.45}$$

and that for each integer $t \ge 0$,

$$\xi_t \in \partial f(x_t) + B_X(0, \delta_f) \tag{2.46}$$

and

$$\|x_{t+1} - P_D(x_t - a_t\xi_t)\| \le \delta_C. \tag{2.47}$$

Then

$$\sum_{t=0}^{T}(T + 1)^{-1} f(x_t) - \inf(f, D),$$

$$f(\sum_{t=0}^{T}(T + 1)^{-1}x_t) - \inf(f, D), \ \min\{f(x_t) : \ t = 0, \dots, T\} - \inf(f, D)$$

$$\le 12(2M + 2)^2(L + 1)\max\{\delta_C^{1/2}, \ \delta_f\}.$$

2.6 Proof of Theorem 2.8

By (2.28) there exists

$$z \in D_{min} \cap B_X(0, M). \tag{2.48}$$

Lemma 2.2, (2.18), (2.34), (2.35), and (2.37) imply that

$$\begin{aligned} &\|x_1 - z\| \\ &\leq \|x_1 - P_D(x_0 - a_0\xi_0)\| + \|P_D(x_0 - a_0\xi_0) - z\| \\ &\leq \delta_C + \|x_0 - z\| + a_0\|\xi_0\| \\ &\leq 1 + 2M + \tau_1\|\xi_0\|. \end{aligned} \tag{2.49}$$

In view of (2.30), (2.35), and (2.36),

$$\xi_0 \in \partial f(x_0) + B_X(0, 1) \subset B_X(0, L) + 1,$$

$$\|\xi_0\| \leq L + 1. \tag{2.50}$$

It follows from (2.31) and (2.48)–(2.50) that

$$\|x_1 - z\| \leq 2M + 2, \ \|x_1\| \leq 3M + 2. \tag{2.51}$$

Assume that T is a natural number and that

$$f(x_t) - f(z) > \epsilon_0, \ t = 1, \dots, T. \tag{2.52}$$

Set

$$U = V \cap \{v \in X : \ \|v\| < M_0 + 2\} \tag{2.53}$$

and

$$C = D \cap B_X(0, M_0). \tag{2.54}$$

By induction we show that for every integer $t \in [1, T]$,

$$\|x_t - z\| \leq 2M + 2, \tag{2.55}$$

$$f(x_t) - f(z)$$

$$\leq (2\tau_0)^{-1}(\|z - x_t\|^2 - \|z - x_{t+1}\|^2)$$

$$+ \tau_0^{-1}\delta_C(4M_0 + 1) + \delta_f(2M_0 + 1) + 2^{-1}\tau_1(L + 1)^2. \tag{2.56}$$

In view of (2.51), (2.55) holds for $t = 1$.

Assume that an integer $t \in [1, T]$ and that (2.55) holds. It follows from (2.29), (2.48), and (2.53)–(2.55) that

$$z \in C \subset B_X(0, M_0), \tag{2.57}$$

$$x_t \in U \cap B_X(0, M_0 + 1). \tag{2.58}$$

Relation (2.37) implies that $x_{t+1} \in V$ satisfies

$$\|x_{t+1} - P_D(x_t - a_t\xi_t)\| \leq 1. \tag{2.59}$$

By (2.30), (2.36) and (2.58),

$$\xi_t \in \partial f(x_t) + B_X(0, 1) \subset B_X(0, L + 1). \tag{2.60}$$

It follows from (2.31), (2.34), (2.48), (2.55), (2.60), and Lemma 2.2 that

$$\|z - P_D(x_t - a_t\xi_t)\|$$

$$\leq \|z - x_t + a_t\xi_t\|$$

$$\leq \|z - x_t\| + \|\xi_t\|a_t \leq 2M + 3,$$

$$\|P_D(x_t - a_t\xi_t)\| \leq 3M + 3. \tag{2.61}$$

In view of (2.29), (2.54), and (2.61),

$$P_D(x_t - a_t\xi_t) \in C, \tag{2.62}$$

and

$$P_D(x_t - a_t\xi_t) = P_C(x_t - a_t\xi_t). \tag{2.63}$$

Relations (2.29), (2.53), (2.59), and (2.61) imply that

$$\|x_{t+1}\| \leq 3M + 4, \ x_{t+1} \in U. \tag{2.64}$$

By (2.25), (2.30), (2.36), (2.37), (2.53), (2.54), (2.57), (2.58), (2.63), (2.64), and Lemma 2.7 which holds with

$$x = x_t,\ a = a_t,\ \xi = \xi_t,\ u = x_{t+1},$$

we have

$$a_t(f(x_t) - f(z))$$

$$\leq 2^{-1}\|z - x_t\|^2 - 2^{-1}\|z - x_{t+1}\|^2$$

$$+\delta_C(4M_0 + 1) + a_t\delta_f(2M_0 + 1) + 2^{-1}a_t^2(L + 1)^2.$$

It follows from the relation above, (2.31) and (2.34) that

$$f(x_t) - f(z)$$

$$\leq (2\tau_0)^{-1}\|z - x_t\|^2 - (2\tau_0)^{-1}\|z - x_{t+1}\|^2$$

$$+ \tau_0^{-1}\delta_C(4M_0 + 1) + (2M_0 + 1)\delta_f + 2^{-1}\tau_1(L + 1)^2. \tag{2.65}$$

By (2.30), (2.37), (2.48), (2.62), and (2.63),

$$f(x_t) \geq f(P_D(x_t - a_t\xi_t)) - \delta_C L \geq f(z) - \delta_C L.$$

In view of the relation above, (2.32), (2.52), (2.55), (2.65), and the inclusion $t \in [1, T]$,

$$\|z - x_t\|^2 - \|z - x_{t+1}\|^2 \geq 0,$$

$$\|z - x_{t+1}\| \leq \|z - x_t\| \leq 2M + 2. \tag{2.66}$$

Therefore we assumed that (2.55) is true and showed that (2.65) and (2.66) hold. Hence by induction we showed that (2.65) holds for all $t = 1, \dots, T$ and (2.55) holds for all $t = 1, \dots, T + 1$.

It follows from (2.65) which holds for all $t = 1, \dots, T$, (2.51) and (2.52) that

$$T\epsilon_0 < T(\min\{f(x_t) : t = 1, \dots, T\} - f(z))$$

$$\leq \sum_{t=1}^{T}(f(x_t) - f(z))$$

$$\leq (2\tau_0)^{-1}\sum_{t=1}^{T}(\|z - x_t\|^2 - \|z - x_{t+1}\|^2)$$

$$+T\tau_0^{-1}\delta_C(4M_0 + 1) + T(2M_0 + 1)\delta_f + 2^{-1}T\tau_1(L + 1)^2$$

$$\leq (2\tau_0)^{-1}(2M+2)^2 + T\tau_0^{-1}\delta_C(4M_0+1)$$

$$+T(2M_0+1)\delta_f + 2^{-1}T\tau_1(L+1)^2.$$

Together with (2.32) and (2.33) this implies that

$$\epsilon_0 < (2\tau_0 T)^{-1}(2M+2)^2 + \tau_0^{-1}\delta_C(4M_0+1)$$

$$+(2M_0+1)\delta_f + 2^{-1}\tau_1(L+1)^2,$$

$$2^{-1}\epsilon_0 < (2\tau_0 T)^{-1}(2M+2)^2,$$

$$T < \tau_0^{-1}(2M+2)^2\epsilon_0^{-1} \leq n_0 + 1.$$

Thus we have shown that if an integer $T \geq 1$ satisfies (2.52), then $T \leq n_0$ and

$$\|z - x_t\| \leq 2M+2,\ t = 1, \dots, T+1,$$

$$\|x_t\| \leq 3M+2,\ t = 0, \dots, T+1.$$

This implies that there exists an integer $q \in [1, n_0+1]$ such that

$$\|x_t\| \leq 3M+2,\ t = 0, \dots, q$$

and

$$f(x_q) - f(z) \leq \epsilon_0.$$

Theorem 2.8 is proved. □

2.7 Proof of Theorem 2.9

Fix

$$z \in D_{min}. \tag{2.67}$$

Set

$$U = V \cap \{x \in X : \|x\| < 3M+2\} \tag{2.68}$$

and

$$C = D \cap B_X(0, 3M + 1). \tag{2.69}$$

In view of (2.38) and (2.67),

$$\|z\| \le M - 1. \tag{2.70}$$

By (2.41) and (2.70),

$$\|x_0 - z\| \le 2M. \tag{2.71}$$

Assume that $t \ge 0$ is an integer and that

$$\|x_t - z\| \le 2M. \tag{2.72}$$

It follows from (2.38), (2.67)–(2.69), and (2.72) that

$$z \in C \subset B_X(0, 3M + 1), \tag{2.73}$$

$$x_t \in U \cap B_X(0, 3M). \tag{2.74}$$

Relation (2.43) implies that $x_{t+1} \in V$ satisfies

$$\|x_{t+1} - P_D(x_t - a_t\xi_t)\| \le 1. \tag{2.75}$$

By (2.39), (2.42), and (2.74),

$$\xi_t \in \partial f(x_t) + B_X(0, 1) \subset B_X(0, L + 1). \tag{2.76}$$

It follows from (2.38), (2.40), (2.67), (2.70), (2.72), (2.76), and Lemma 2.2 that

$$\|z - P_D(x_t - a_t\xi_t)\|$$

$$\le \|z - x_t + a_t\xi_t\|$$

$$\le \|z - x_t\| + \|\xi_t\|a_t$$

$$\le 2M + (L + 1)a_t \le 2M + 1,$$

$$\|P_D(x_t - a_t\xi_t)\| \le 3M + 1. \tag{2.77}$$

In view of (2.69) and (2.77),

$$P_D(x_t - a_t\xi_t) \in C, \tag{2.78}$$

and

$$P_D(x_t - a_t\xi_t) = P_C(x_t - a_t\xi_t). \tag{2.79}$$

Relations (2.43), (2.68), and (2.77) imply that

$$\|x_{t+1}\| \le \|P_D(x_t - a_t\xi_t)\| + \delta_C < 3M + 2, \tag{2.80}$$

$$x_{t+1} \in U. \tag{2.81}$$

By (2.43), (2.73), (2.74), (2.79), (2.81), and Lemma 2.7 which holds with

$$x = x_t,\ a = a_t,\ \xi = \xi_t,\ u = x_{t+1},\ M_0 = 3M + 2,$$

we have

$$a_t(f(x_t) - f(z))$$

$$\le 2^{-1}\|z - x_t\|^2 - 2^{-1}\|z - x_{t+1}\|^2$$

$$+ \delta_C(12M + 9) + a_t\delta_f(6M + 5) + 2^{-1}a_t^2(L + 1)^2. \tag{2.82}$$

There are two cases:

$$\|z - x_{t+1}\| \le \|z - x_t\|; \tag{2.83}$$

$$\|z - x_{t+1}\| > \|z - x_t\|. \tag{2.84}$$

If (2.83) holds, then in view of (2.72),

$$\|z - x_{t+1}\| \le 2M.$$

Assume that (2.84) holds. It follows from (2.40), (2.82), and (2.84) that

$$f(x_t) - f(z)$$

$$< a_t^{-1}\delta_C(12M + 9) + \delta_f(6M + 5) + 2^{-1}a_t(L + 1)^2 \le 3.$$

By the inequality above, (2.38) and (2.67),

$$\|x_t\| \le M - 1. \tag{2.85}$$

Relations (2.70) and (2.85) imply that

$$\|x_t - z\| \le 2M - 2. \tag{2.86}$$

Lemma 2.2, (2.40), (2.43), (2.67), and (2.76) imply that

$$\|x_{t+1} - z\|$$
$$\leq \|x_{t+1} - P_D(x_t - a_t\xi_t)\| + \|P_D(x_t - a_t\xi_t) - z\|$$
$$\leq \delta_C + \|x_t - a_t\xi_t - z\|$$
$$\leq \delta_C + \|x_t - z\| + a_t\|\xi_t\|$$
$$\leq 1 + \|x_t - z\| + (L+1)a_t$$
$$\leq \|x_t - z\| + 2 \leq 2M.$$

Thus in both cases

$$\|x_{t+1} - z\| \leq 2M$$

and (2.82) holds. Therefore by induction we showed that for all integers $t \geq 0$,

$$\|x_t\| \leq 3M$$

and that (2.82) holds.

Let T be a natural number. By (2.67), (2.72), and (2.82),

$$\sum_{t=0}^{T} a_t(f(x_t) - \inf(f, D))$$
$$\leq \sum_{t=0}^{T} (2^{-1}\|z - x_t\|^2 - 2^{-1}\|z - x_{t+1}\|^2)$$
$$+\delta_C(12M + 9)(T + 1)$$
$$+\delta_f(6M + 5)\sum_{t=0}^{T} a_t + 2^{-1}(L+1)^2 \sum_{t=0}^{T} a_t^2$$
$$\leq 2M^2 + \delta_C(12M + 9)(T + 1)$$
$$+\delta_f(6M + 5)\sum_{t=0}^{T} a_t + 2^{-1}(L+1)^2 \sum_{t=0}^{T} a_t^2.$$

This completes the proof of Theorem 2.9. □

2.8 Proof of Theorem 2.12

Clearly, there exists

$$z \in D_{min} \cap B_X(0, M). \tag{2.87}$$

In view of the theorem assumptions and (2.87),

$$\|x_0 - z\| \le 2M. \tag{2.88}$$

Set

$$U = V \cap \{v \in X : \ \|v\| < 3M + 6\} \tag{2.89}$$

and

$$C = D \cap B_X(0, 3M + 4). \tag{2.90}$$

Assume that $t \ge 0$ is an integer and that

$$\|x_t - z\| \le 2M + 2. \tag{2.91}$$

It follows from (2.85) and (2.91) that

$$\|x_t\| \le 3M + 2. \tag{2.92}$$

In view of (2.89) and (2.92),

$$x_t \in U \cap B_X(0, 3M + 2). \tag{2.93}$$

By (2.44), (2.46), and (2.92),

$$\|\xi_t\| \le L + 1. \tag{2.94}$$

Lemma 2.2, (2.45), (2.47), (2.85), (2.89), (2.91), and (2.94) imply that

$$\|x_{t+1} - z\|$$

$$\le \|x_{t+1} - P_D(x_t - a\xi_t)\| + \|P_D(x_t - a\xi_t) - z\|$$

$$\le \delta_C + \|x_t - a\xi_t - z\|$$

$$\le \delta_C + \|x_t - z\| + a\|\xi_t\|$$

$$\leq \delta_C + 2M + 2 + a(L+1) \leq 2M + 4,$$

$$\|x_{t+1}\| \leq 3M + 4 \tag{2.95}$$

and

$$x_{t+1} \in U. \tag{2.96}$$

It follows from (2.45), (2.85), (2.90), (2.91), (2.94), and Lemma 2.2 that

$$\|P_D(x_t - a\xi_t)\| \leq \|z\| + \|P_D(x_t - a\xi_t) - z\|$$

$$\leq M + \|z - x_t\| + a\|\xi_t\|$$

$$\leq M + 2M + 2 + 1,$$

$$P_D(x_t - a\xi_t) \in C, \tag{2.97}$$

and

$$P_D(x_t - a\xi_t) = P_C(x_t - a\xi_t). \tag{2.98}$$

By (2.46), (2.47), (2.85), (2.89), (2.90), (2.93), (2.96), (2.98), and Lemma 2.7 which holds with $M_0 = 3M + 4$,

$$x = x_t, \ \xi = \xi_t, \ u = x_{t+1},$$

we have

$$a(f(x_t) - f(z))$$

$$\leq 2^{-1}\|z - x_t\|^2 - 2^{-1}\|z - x_{t+1}\|^2$$

$$+ \delta_C(12M + 17) + a\delta_f(6M + 9) + 2^{-1}a^2(L+1)^2. \tag{2.99}$$

In view of (2.44), (2.47), (2.85), (2.92), and the relation

$$B_X(x_0, \delta_C) \cap D \neq \emptyset,$$

we have

$$f(x_t) \geq f(z) - \delta_C L.$$

It follows from (2.85), (2.99), and the inequality above that

$$\|z - x_{t+1}\|^2 - \|z - x_t\|^2$$

$$\leq 2\delta_C(12M + 18) + a^2(L + 1)^2 + 2a\delta_f(6M + 9). \tag{2.100}$$

Thus we have shown that if for an integer $t \geq 0$ relation (2.91) holds, then (2.99) and (2.100) are true.

Let us show that for all integers $t = 0, \ldots, T$,

$$\|z - x_t\|^2 \leq \|z - x_0\|^2$$

$$+t[2\delta_C(12M + 18) + a^2(L + 1)^2 + 2a\delta_f(6M + 9)]$$

$$\leq \|z - x_0\|^2$$

$$+T[2\delta_C(12M + 18) + a^2(L + 1)^2 + 2a\delta_f(6M + 9)]$$

$$\leq 4M^2 + 8M + 4. \tag{2.101}$$

First, note that by (2.45),

$$T[2\delta_C(12M + 18) + a^2(L + 1)^2 + 2a\delta_f(6M + 9)]$$

$$= 2T\delta_C(12M + 18) + 2T\delta_C + 2Ta\delta_f(6M + 9)$$

$$= 2T\delta_C(12M + 19) + 2T\delta_f(6M + 9)\delta_C^{1/2}(L + 1)^{-1}$$

$$\leq 6^{-1}\min\{\delta_C^{-1}, \delta_C^{-1/2}\delta_f^{-1}\}(2\delta_C(12M + 19)$$

$$+2\delta_f(6M + 9)\delta_C^{1/2}(L + 1)^{-1})$$

$$\leq 6^{-1}(24M + 38 + 12M + 18) \leq 8M + 4. \tag{2.102}$$

It follows from (2.86) and (2.102) that

$$\|z - x_0\|^2 + T[2\delta_C(12M + 18)$$

$$+a^2(L + 1)^2 + 2a\delta_f(6M + 9)]$$

$$\leq 4M^2 + 8M + 4 = (2M + 2)^2. \tag{2.103}$$

Assume that $t \in \{0, \ldots, T\} \setminus \{T\}$ and that (2.101) holds. Then

$$\|x_t - z\| \le 2M + 2$$

and (2.99) and (2.100) hold. By (2.100), (2.101), and (2.103),

$$\|z - x_{t+1}\|^2$$

$$\le \|z - x_t\|^2 + 2\delta_C(12M + 18) + a^2(L + 1)^2 + 2a\delta_f(6M + 9)$$

$$\le \|z - x_0\|^2$$

$$+(t + 1)[2\delta_C(12M + 18) + a^2(L + 1)^2 + 2a\delta_f(6M + 9)]$$

$$\le \|z - x_0\|^2$$

$$+T[2\delta_C(12M + 18) + a^2(L + 1)^2 + 2a\delta_f(6M + 9)]$$

$$\le 4M^2 + 8M + 4$$

and

$$\|z - x_{t+1}\| \le 2M + 2.$$

Thus we have shown by induction that (2.99) and (2.101) hold for all $t = 0, \dots, T$. It follows from (2.45), (2.85), (2.86), and (2.99) that

$$\sum_{t=0}^{T} a(f(x_t) - \inf(f, D))$$

$$\le 2^{-1}\|z - x_0\|^2 + (T + 1)(\delta_C(12M + 17)$$

$$+a\delta_f(6M + 9) + 2^{-1}a^2(L + 1)^2)$$

$$\le 2M^2 + \delta_C(12M + 18)(6\delta_C)^{-1}$$

$$+\delta_C^{1/2}\delta_f(6M + 9)(6\delta_f\delta_C^{1/2})^{-1}$$

$$\le 2M^2 + 2M + 3 + M + 2$$

$$= 2M^2 + 3M + 5.$$

Together with (2.45) this implies that

$$\sum_{t=0}^{T}(T+1)^{-1}f(x_t) - \inf(f, D),$$

$$f(\sum_{t=0}^{T}(T+1)^{-1}x_t) - \inf(f, D),$$

$$\min\{f(x_t) : t = 0, \ldots, T\} - \inf(f, D)$$

$$\leq (2M+2)^2 a^{-1}(T+1)^{-1}$$

$$\leq 12(2M+2)^2(L+1)\delta_C^{-1/2} \max\{\delta_C, \delta_C^{1/2}\delta_f\}$$

$$\leq 12(2M+2)^2(L+1) \max\{\delta_C^{1/2}, \delta_f\}.$$

Theorem 2.12 is proved. □

2.9 Zero-Sum Games with Two Players

Let $(X, \langle\cdot,\cdot\rangle)$, $(Y, \langle\cdot,\cdot\rangle)$ be Hilbert spaces equipped with the complete norms $\|\cdot\|$ which are induced by their inner products. Let C be a nonempty closed convex subset of X, D be a nonempty closed convex subset of Y, U be an open convex subset of X, and V be an open convex subset of Y such that

$$C \subset U, \ D \subset V \tag{2.104}$$

and let a function $f : U \times V \to R^1$ possess the following properties:

(i) for each $v \in V$, the function $f(\cdot, v) : U \to R^1$ is convex;
(ii) for each $u \in U$, the function $f(u, \cdot) : V \to R^1$ is concave.

Assume that a function $\phi : R^1 \to [0, \infty)$ is bounded on all bounded sets and that positive numbers M_1, M_2, L_1, L_2 satisfy

$$C \subset B_X(0, M_1),$$

$$D \subset B_Y(0, M_2), \tag{2.105}$$

$$|f(u_1, v) - f(u_2, v)| \leq L_1\|u_1 - u_2\|$$

$$\text{for all } v \in V \text{ and all } u_1, u_2 \in U, \tag{2.106}$$

$$|f(u, v_1) - f(u, v_2)| \leq L_2\|v_1 - v_2\|$$

$$\text{for all } u \in U \text{ and all } v_1, v_2 \in V. \tag{2.107}$$

Let

$$x_* \in C \text{ and } y_* \in D \tag{2.108}$$

satisfy

$$f(x_*, y) \le f(x_*, y_*) \le f(x, y_*) \tag{2.109}$$

for each $x \in C$ and each $y \in D$.

The following result is proved in Sect. 2.10.

Proposition 2.13 *Let T be a natural number, δ_C, $\delta_D \in (0, 1]$, $\{a_t\}_{t=0}^T \subset (0, \infty)$ and let $\{b_{t,1}\}_{t=0}^T$, $\{b_{t,2}\}_{t=0}^T \subset (0, \infty)$. Assume that $\{x_t\}_{t=0}^{T+1} \subset U$, $\{y_t\}_{t=0}^{T+1} \subset V$, for each $t \in \{0, \dots, T+1\}$,*

$$B_X(x_t, \delta_C) \cap C \neq \emptyset,$$

$$B_Y(y_t, \delta_D) \cap D \neq \emptyset, \tag{2.110}$$

for each $z \in C$ and each $t \in \{0, \dots, T\}$,

$$a_t(f(x_t, y_t) - f(z, y_t))$$

$$\le \phi(\|z - x_t\|) - \phi(\|z - x_{t+1}\|) + b_{t,1} \tag{2.111}$$

and that for each $v \in D$ and each $t \in \{0, \dots, T\}$,

$$a_t(f(x_t, v) - f(x_t, y_t))$$

$$\le \phi(\|v - y_t\|) - \phi(\|v - y_{t+1}\|) + b_{t,2}. \tag{2.112}$$

Let

$$\widehat{x}_T = (\sum_{i=0}^T a_i)^{-1} \sum_{t=0}^T a_t x_t,$$

$$\widehat{y}_T = (\sum_{i=0}^T a_i)^{-1} \sum_{t=0}^T a_t y_t. \tag{2.113}$$

Then

$$B_X(\widehat{x}_T, \delta_C) \cap C \neq \emptyset, \ B_Y(\widehat{y}_T, \delta_D) \cap D \neq \emptyset, \tag{2.114}$$

$$|(\sum_{t=0}^{T} a_t)^{-1} \sum_{t=0}^{T} a_t f(x_t, y_t) - f(x_*, y_*)|$$

$$\leq (\sum_{t=0}^{T} a_t)^{-1} \max\{\sum_{t=0}^{T} b_{t,1}, \sum_{t=0}^{T} b_{t,2}\}$$

$$+ \max\{L_1\delta_C, L_2\delta_D\}$$

$$+ (\sum_{t=0}^{T} a_t)^{-1} \sup\{\phi(s) : s \in [0, \max\{2M_1, 2M_2\} + 1]\}, \tag{2.115}$$

$$|f(\widehat{x}_T, \widehat{y}_T) - (\sum_{t=0}^{T} a_t)^{-1} \sum_{t=0}^{T} a_t f(x_t, y_t)|$$

$$\leq (\sum_{t=0}^{T} a_t)^{-1} \sup\{\phi(s) : s \in [0, \max\{2M_1, 2M_2\} + 1]\}$$

$$+(\sum_{t=0}^{T} a_t)^{-1} \max\{\sum_{t=0}^{T} b_{t,1}, \sum_{t=0}^{T} b_{t,2}\}$$

$$+ \max\{L_1\delta_C, L_2\delta_D\} \tag{2.116}$$

and for each $z \in C$ and each $v \in D$,

$$f(z, \widehat{y}_T) \geq f(\widehat{x}_T, \widehat{y}_T)$$

$$-2(\sum_{t=0}^{T} a_t)^{-1} \sup\{\phi(s) : s \in [0, \max\{2M_1, 2M_2\} + 1]\}$$

$$-2(\sum_{t=0}^{T} a_t)^{-1} \max\{\sum_{t=0}^{T} b_{t,1}, \sum_{t=0}^{T} b_{t,2}\}$$

$$- \max\{L_1\delta_C, L_2\delta_D\}, \tag{2.117}$$

$$f(\widehat{x}_T, v) \leq f(\widehat{x}_T, \widehat{y}_T)$$

$$+2(\sum_{t=0}^{T} a_t)^{-1} \sup\{\phi(s) : s \in [0, \max\{2M_1, 2M_2\} + 1]\}$$

$$+2(\sum_{t=0}^{T} a_t)^{-1} \max\{\sum_{t=0}^{T} b_{t,1}, \sum_{t=0}^{T} b_{t,2}\}$$

$$+\max\{L_1\delta_C, L_2\delta_D\}. \tag{2.118}$$

Corollary 2.14 *Suppose that all the assumptions of Proposition 2.13 hold and that*

$$\tilde{x} \in C, \tilde{y} \in D$$

satisfy

$$\|\widehat{x}_T - \tilde{x}\| \leq \delta_C, \|\widehat{y}_T - \tilde{y}\| \leq \delta_D. \tag{2.119}$$

Then

$$|f(\tilde{x}, \tilde{y}) - f(\widehat{x}_T, \widehat{y}_T)| \leq L_1\delta_C + L_2\delta_D \tag{2.120}$$

and for each $z \in C$ and each $v \in D$,

$$f(z, \tilde{y}) \geq f(\tilde{x}, \tilde{y})$$

$$-2(\sum_{t=0}^{T} a_t)^{-1} \sup\{\phi(s) : s \in [0, \max\{2M_1, 2M_2\} + 1]\}$$

$$-2(\sum_{t=0}^{T} a_t)^{-1} \max\{\sum_{t=0}^{T} b_{t,1}, \sum_{t=0}^{T} b_{t,2}\} - 4\max\{L_1\delta_C, L_2\delta_D\}$$

and

$$f(\tilde{x}, v) \leq f(\tilde{x}, \tilde{y})$$

$$+2(\sum_{t=0}^{T} a_t)^{-1} \sup\{\phi(s) : s \in [0, \max\{2M_1, 2M_2\} + 1]\}$$

$$+2(\sum_{t=0}^{T} a_t)^{-1} \max\{\sum_{t=0}^{T} b_{t,1}, \sum_{t=0}^{T} b_{t,2}\} + 4\max\{L_1\delta_C, L_2\delta_D\}.$$

Proof In view of (2.106), (2.107), and (2.119),

$$|f(\tilde{x}, \tilde{y}) - f(\widehat{x}_T, \widehat{y}_T)|$$

$$\leq |f(\tilde{x}, \tilde{y}) - f(\tilde{x}, \widehat{y}_T)| + |f(\tilde{x}, \widehat{y}_T) - f(\widehat{x}_T, \widehat{y}_T)|$$

$$\leq L_2\|\tilde{y} - \widehat{y}_T\| + L_1\|\tilde{x} - \widehat{x}_T\| \leq L_1\delta_C + L_2\delta_D$$

and (2.120) holds.

Let $z \in C$ and $v \in D$. Relations (2.106), (2.107), and (2.119) imply that

$$|f(z, \tilde{y}) - f(z, \widehat{y}_T)| \leq L_2\delta_D,$$

$$|f(\tilde{x}, v) - f(\widehat{x}_T, v)| \leq L_1\delta_C.$$

By the relation above and (2.117)–(2.120),

$$f(z, \tilde{y}) \geq f(z, \widehat{y}_T) - L_2\delta_D$$

$$\geq f(\widehat{x}_T, \widehat{y}_T) - 2(\sum_{t=0}^{T} a_t)^{-1} \sup\{\phi(s) : s \in [0, \max\{2M_1, 2M_2\} + 1]\}$$

$$-2(\sum_{t=0}^{T} a_t)^{-1} \max\{\sum_{t=0}^{T} b_{t,1}, \sum_{t=0}^{T} b_{t,2}\}$$

$$- \max\{L_1\delta_C, L_2\delta_D\} - L_2\delta_D$$

$$\geq f(\tilde{x}, \tilde{y}) - 2(\sum_{t=0}^{T} a_t)^{-1} \sup\{\phi(s) : s \in [0, \max\{2M_1, 2M_2\} + 1]\}$$

$$-2(\sum_{t=0}^{T} a_t)^{-1} \max\{\sum_{t=0}^{T} b_{t,1}, \sum_{t=0}^{T} b_{t,2}\}$$

$$-4 \max\{L_1\delta_C, L_2\delta_D\}$$

and

$$f(\tilde{x}, v) \leq f(\widehat{x}_T, v) + L_1\delta_C$$

$$\leq f(\widehat{x}_T, \widehat{y}_T) + 2(\sum_{t=0}^{T} a_t)^{-1} \sup\{\phi(s) : s \in [0, \max\{2M_1, 2M_2\} + 1]\}$$

$$+2(\sum_{t=0}^{T} a_t)^{-1} \max\{\sum_{t=0}^{T} b_{t,1}, \sum_{t=0}^{T} b_{t,2}\}$$

$$+\max\{L_1\delta_C,\ L_2\delta_D\}+L_1\delta_C$$

$$\leq f(\tilde{x},\tilde{y})+2(\sum_{t=0}^{T}a_t)^{-1}\sup\{\phi(s):\ s\in[0,\max\{2M_1,\ 2M_2\}+1]\}$$

$$+2(\sum_{t=0}^{T}a_t)^{-1}\max\{\sum_{t=0}^{T}b_{t,1},\ \sum_{t=0}^{T}b_{t,2}\}$$

$$+4\max\{L_1\delta_C,\ L_2\delta_D\}.$$

This completes the proof of Corollary 2.14. □

2.10 Proof of Proposition 2.13

It is clear that (2.114) is true. In view of (2.110), for each $t\in\{0,\dots,T+1\}$, there exist

$$\tilde{x}_t\in C,\ \tilde{y}_t\in D \tag{2.121}$$

such that

$$\|x_t-\tilde{x}_t\|\leq\delta_C,\ \|y_t-\tilde{y}_t\|\leq\delta_D. \tag{2.122}$$

Let $t\in\{0,\dots,T\}$. By (2.106)–(2.109), (2.111), (2.112), and (2.122),

$$a_t(f(x_t,y_t)-f(x_*,y_*))$$

$$\leq a_t(f(x_t,y_t)-f(x_*,\tilde{y}_t))$$

$$\leq a_t(f(x_t,y_t)-f(x_*,y_t))+a_t(f(x_*,y_t)-f(x_*,\tilde{y}_t))$$

$$\leq\phi(\|x_*-x_t\|)-\phi(\|x_*-x_{t+1}\|)$$

$$+b_{t,1}+a_tL_2\|y_t-\tilde{y}_t\|$$

$$\leq\phi(\|x_*-x_t\|)-\phi(\|x_*-x_{t+1}\|)$$

$$+b_{t,1}+a_tL_2\delta_D \tag{2.123}$$

and

$$
\begin{aligned}
&a_t(f(x_*, y_*) - f(x_t, y_t))\\
&\le a_t(f(\tilde{x}_t, y_*) - f(x_t, y_t))\\
&= a_t(f(\tilde{x}_t, y_*) - f(x_t, y_*)) + a_t(f(x_t, y_*) - f(x_t, y_t))\\
&\le a_t L_1 \|\tilde{x}_t - x_t\|\\
&+\phi(\|y_* - y_t\|) - \phi(\|y_* - y_{t+1}\|) + b_{t,2}\\
&\le a_t L_1 \delta_C + \phi(\|y_* - y_t\|) - \phi(\|y_* - y_{t+1}\|) + b_{t,2}.
\end{aligned}
\tag{2.124}
$$

In view of (2.123) and (2.124),

$$
\begin{aligned}
&\sum_{t=0}^{T} a_t f(x_t, y_t) - \sum_{t=0}^{T} a_t f(x_*, y_*)\\
&\le \sum_{t=0}^{T} (\phi(\|x_* - x_t\|) - \phi(\|x_* - x_{t+1}\|))\\
&+\sum_{t=0}^{T} b_{t,1} + \delta_D L_2 \sum_{t=0}^{T} a_t\\
&\le \phi(\|x_* - x_0\|) + \sum_{t=0}^{T} b_{t,1} + \delta_D L_2 \sum_{t=0}^{T} a_t
\end{aligned}
\tag{2.125}
$$

and

$$
\begin{aligned}
&\sum_{t=0}^{T} a_t f(x_*, y_*) - \sum_{t=0}^{T} a_t f(x_t, y_t)\\
&\le \sum_{t=0}^{T} (\phi(\|y_* - y_t\|) - \phi(\|y_* - y_{t+1}\|))\\
&+\sum_{t=0}^{T} b_{t,2} + L_1 \delta_C \sum_{t=0}^{T} a_t\\
&\le \phi(\|y_* - y_0\|) + \sum_{t=0}^{T} b_{t,2} + L_1 \delta_C \sum_{t=0}^{T} a_t.
\end{aligned}
\tag{2.126}
$$

Relations (2.105), (2.108), (2.110), (2.125), and (2.126) imply that

$$|(\sum_{t=0}^{T} a_t)^{-1} \sum_{t=0}^{T} a_t f(x_t, y_t) - f(x_*, y_*)|$$

$$\leq (\sum_{t=0}^{T} a_t)^{-1} \sup\{\phi(s) : \ s \in [0, \max\{2M_1, \ 2M_2\} + 1]\}$$

$$+(\sum_{t=0}^{T} a_t)^{-1} \max\{\sum_{t=0}^{T} b_{t,1}, \ \sum_{t=0}^{T} b_{t,2}\}$$

$$+ \max\{L_1\delta_C, \ L_2\delta_D\}. \tag{2.127}$$

By (2.214), there exists

$$z_T \in C \tag{2.128}$$

such that

$$\|z_T - \widehat{x}_T\| \leq \delta_C. \tag{2.129}$$

In view of (2.128), we apply (2.111) with $z = z_T$ and obtain that for all $t = 0, \ldots, T$,

$$a_t(f(x_t, y_t) - f(z_T, y_t))$$

$$\leq \phi(\|z_T - x_t\|) - \phi(\|z_T - x_{t+1}\|) + b_{t,1}. \tag{2.130}$$

It follows from (2.106) and (2.129) that for all $t = 0, \ldots, T$,

$$|f(z_T, y_t) - f(\widehat{x}_T, y_t)| \leq L_1\|z_T - \widehat{x}_T\| \leq L_1\delta_C. \tag{2.131}$$

By (2.130) and (2.131), for all $t = 0, \ldots, T$,

$$a_t(f(x_t, y_t) - f(\widehat{x}_T, y_t))$$

$$\leq a_t(f(x_t, y_t) - f(z_T, y_t)) + a_t L_1\delta_C$$

$$\leq \phi(\|z_T - x_t\|) - \phi(\|z_T - x_{t+1}\|) + b_{t,1} + a_t L_1\delta_C. \tag{2.132}$$

Combined with (2.105), (2.110), and (2.128) this implies that

$$\sum_{t=0}^{T} a_t f(x_t, y_t) - \sum_{t=0}^{T} a_t f(\widehat{x}_T, y_t)$$

$$\leq \sum_{t=0}^{T} (\phi(\|z_T - x_t\|) - \phi(\|z_T - x_{t+1}\|)$$

$$+ \sum_{t=0}^{T} b_{t,1} + \sum_{t=0}^{T} a_t L_1 \delta_C$$

$$\leq \phi(\|z_T - x_0\|) + \sum_{t=0}^{T} b_{t,1} + \sum_{t=0}^{T} a_t L_1 \delta_C$$

$$\leq \sup\{\phi(s) : s \in [0, 2M_1 + 1]\} + \sum_{t=0}^{T} b_{t,1} + \sum_{t=0}^{T} a_t L_1 \delta_C. \tag{2.133}$$

Property (ii) and (2.113) imply that

$$\sum_{t=0}^{T} a_t f(\widehat{x}_T, y_t) = (\sum_{i=0}^{T} a_i) \sum_{t=0}^{T} (a_t (\sum_{i=0}^{T} a_i)^{-1} f(\widehat{x}_T, y_t))$$

$$\leq (\sum_{t=0}^{T} a_t) f(\widehat{x}_T, \widehat{y}_T). \tag{2.134}$$

By (2.133) and (2.134),

$$\sum_{t=0}^{T} a_t f(x_t, y_t) - (\sum_{t=0}^{T} a_t) f(\widehat{x}_T, \widehat{y}_T)$$

$$\leq \sum_{t=0}^{T} a_t f(x_t, y_t) - \sum_{t=0}^{T} a_t f(\widehat{x}_T, y_t)$$

$$\leq \sup\{\phi(s) : s \in [0, 2M_1 + 1]\}$$

$$+ \sum_{t=0}^{T} b_{t,1} + \sum_{t=0}^{T} a_t L_1 \delta_C. \tag{2.135}$$

By (2.114), there exists

$$h_T \in D \tag{2.136}$$

such that

$$\|h_T - \widehat{y}_T\| \le \delta_D. \tag{2.137}$$

In view of (2.136), we apply (2.112) with $v = h_T$ and obtain that for all $t = 0, \dots, T$,

$$a_t(f(x_t, h_T) - f(x_t, y_t))$$

$$\le \phi(\|h_T - y_t\|) - \phi(\|h_T - y_{t+1}\|) + b_{t,2}. \tag{2.138}$$

It follows from (2.107) and (2.137) that for all $t = 0, \dots, T$,

$$|f(x_t, h_T) - f(x_t, \widehat{y}_T)| \le L_2\|h_T - \widehat{y}_T\| \le L_2\delta_D. \tag{2.139}$$

By (2.138) and (2.139), for all $t = 0, \dots, T$,

$$a_t(f(x_t, \widehat{y}_T) - f(x_t, y_t))$$

$$\le a_t(f(x_t, \widehat{y}_T) - f(x_t, h_T)) + a_t(f(x_t, h_T) - f(x_t, y_t))$$

$$\le \phi(\|h_T - y_t\|) - \phi(\|h_T - y_{t+1}\|) + b_{t,2} + a_t L_2\delta_D. \tag{2.140}$$

In view of (2.140),

$$\sum_{t=0}^{T} a_t f(x_t, \widehat{y}_T) - \sum_{t=0}^{T} a_t f(x_t, y_t)$$

$$\le \sum_{t=0}^{T}(\phi(\|h_T - y_t\|) - \phi(\|h_T - y_{t+1}\|)) + \sum_{t=0}^{T} b_{t,2} + \sum_{t=0}^{T} a_t L_2\delta_D. \tag{2.141}$$

Property (i) and (2.113) imply that

$$\sum_{t=0}^{T} a_t f(x_t, \widehat{y}_T) = (\sum_{i=0}^{T} a_i)\sum_{t=0}^{T}(a_t(\sum_{i=0}^{T} a_i)^{-1} f(x_t, \widehat{y}_T))$$

$$\ge \sum_{t=0}^{T} a_t f(\widehat{x}_T, \widehat{y}_T). \tag{2.142}$$

By (2.90), (2.110), (2.136), (2.141), and (2.142),

$$\sum_{t=0}^{T} a_t f(\widehat{x}_T, \widehat{y}_T) - \sum_{t=0}^{T} a_t f(x_t, y_t)$$

$$\leq \sum_{t=0}^{T} a_t f(x_t, \widehat{y}_T) - \sum_{t=0}^{T} a_t f(x_t, y_t)$$

$$\leq \sum_{t=0}^{T} (\phi(\|h_T - y_t\|) - \phi(\|h_T - y_{t+1}\|)) + \sum_{t=0}^{T} b_{t,2} + \sum_{t=0}^{T} a_t L_2 \delta_D$$

$$\leq \sup\{\phi(s) : s \in [0, 2M_2 + 1]\} + \sum_{t=0}^{T} b_{t,2} + \sum_{t=0}^{T} a_t L_2 \delta_D. \tag{2.143}$$

It follows from (2.135) and (2.143) that

$$|\sum_{t=0}^{T} a_t f(\widehat{x}_T, \widehat{y}_T) - \sum_{t=0}^{T} a_t f(x_t, y_t)|$$

$$\leq \sup\{\phi(s) : s \in [0, \max\{2M_1, 2M_2\} + 1]\}$$

$$+ \max\{\sum_{t=0}^{T} b_{t,1}, \sum_{t=0}^{T} b_{t,2}\} + \sum_{t=0}^{T} a_t \max\{L_1 \delta_C, L_2 \delta_D\}.$$

This relation implies (2.116).

Let $z \in C$. By (2.111),

$$\sum_{t=0}^{T} a_t (f(x_t, y_t) - f(z, y_t))$$

$$\leq \sum_{t=0}^{T} [\phi(\|z - x_t\|) - \phi(\|z - x_{t+1}\|)] + \sum_{t=0}^{T} b_{t,1}. \tag{2.144}$$

By property (ii) and (2.113),

$$\sum_{t=0}^{T} a_t f(z, y_t) = (\sum_{i=0}^{T} a_i) \sum_{t=0}^{T} (a_t (\sum_{i=0}^{T} a_i)^{-1} f(z, y_t))$$

$$\leq (\sum_{t=0}^{T} a_t) f(z, \widehat{y}_T). \tag{2.145}$$

In view of (2.144) and (2.145),

$$\sum_{t=0}^{T} a_t f(x_t, y_t) - \sum_{t=0}^{T} a_t f(z, \widehat{y}_T)$$

$$\leq \sum_{t=0}^{T} a_t (f(x_t, y_t) - f(z, y_t))$$

$$\leq \sum_{t=0}^{T} [\phi(\|z - x_t\|) - \phi(\|z - x_{t+1}\|)] + \sum_{t=0}^{T} b_{t,1}$$

$$\leq \phi(\|z - x_0\|) + \sum_{t=0}^{T} b_{t,1}. \tag{2.146}$$

It follows from (2.105) and (2.110) that

$$f(z, \widehat{y}_T) \geq (\sum_{i=0}^{T} a_i)^{-1} \sum_{t=0}^{T} a_t f(x_t, y_t)$$

$$-(\sum_{i=0}^{T} a_i)^{-1} \sup\{\phi(s) : s \in [0, 2M_1 + 1]\} - (\sum_{i=0}^{T} a_i)^{-1} \sum_{t=0}^{T} b_{t,1}$$

$$\geq f(\widehat{x}_T, \widehat{y}_T) - 2(\sum_{t=0}^{T} a_t)^{-1} \sup\{\phi(s) : s \in [0, \max\{2M_1, 2M_2\} + 1]\}$$

$$-2(\sum_{t=0}^{T} a_t)^{-1} \max\{\sum_{t=0}^{T} b_{t,1}, \sum_{t=0}^{T} b_{t,2}\} - \max\{L_1\delta_C, L_2\delta_D\}$$

and (2.117) holds.

Let $v \in D$. By (2.112),

$$\sum_{t=0}^{T} a_t (f(x_t, v) - f(x_t, y_t))$$

$$\leq \sum_{t=0}^{T}[\phi(\|v - y_t\|) - \phi(\|v - y_{t+1}\|)] + \sum_{t=0}^{T} b_{t,2}. \tag{2.147}$$

By property (i) and (2.113),

$$\sum_{t=0}^{T} a_t f(x_t, v) = (\sum_{i=0}^{T} a_i)\sum_{t=0}^{T}(a_t(\sum_{i=0}^{T} a_i)^{-1} f(x_t, v))$$

$$\geq (\sum_{t=0}^{T} a_t) f(\widehat{x}_T, v). \tag{2.148}$$

In view of (2.147) and (2.148),

$$\sum_{t=0}^{T} a_t f(\widehat{x}_T, v) - \sum_{t=0}^{T} a_t f(x_t, y_t)$$

$$\leq \sum_{t=0}^{T} a_t f(x_t, v) - \sum_{t=0}^{T} a_t f(x_t, y_t)$$

$$\leq \phi(\|v - y_0\|) + \sum_{t=0}^{T} b_{t,2}.$$

Together with (2.105), (2.110), and (2.116) this implies that

$$f(\widehat{x}_T, v) \leq (\sum_{t=0}^{T} a_t)^{-1} \sum_{t=0}^{T} a_t f(x_t, y_t)$$

$$+(\sum_{t=0}^{T} a_t)^{-1} \sup\{\phi(s) :\ s \in [0, 2M_2 + 1]\} + (\sum_{t=0}^{T} a_t)^{-1} \sum_{t=0}^{T} b_{t,2}$$

$$\leq f(\widehat{x}_T, \widehat{y}_T) + 2(\sum_{t=0}^{T} a_t)^{-1} \sup\{\phi(s) :\ s \in [0, 2\max\{M_1, M_2\} + 1]\}$$

$$+2(\sum_{t=0}^{T} a_t)^{-1} \max\{\sum_{t=0}^{T} b_{t,1},\ \sum_{t=0}^{T} b_{t,2}\} + \max\{L_1\delta_C,\ L_2\delta_D\}.$$

Therefore (2.118) holds. This completes the proof of Proposition 2.13. □

2.11 Zero-Sum Games on Bounded Sets

Let $(X, \langle \cdot, \cdot \rangle)$, $(Y, \langle \cdot, \cdot \rangle)$ be Hilbert spaces equipped with the complete norms $\| \cdot \|$ which are induced by their inner products. Let C be a nonempty closed convex subset of X, D be a nonempty closed convex subset of Y, U be an open convex subset of X, and V be an open convex subset of Y such that

$$C \subset U, \ D \subset V. \tag{2.149}$$

For each concave function $g : V \to R^1$ and each $x \in V$ set

$$\partial g(x) = \{l \in Y : \langle l, y - x \rangle \geq g(y) - g(x) \text{ for all } y \in V\}. \tag{2.150}$$

Clearly, for each $x \in V$,

$$\partial g(x) = -(\partial(-g)(x)). \tag{2.151}$$

Suppose that there exist $L_1, L_2, M_1, M_2 > 0$ such that

$$C \subset B_X(0, M_1), \ D \subset B_Y(0, M_2), \tag{2.152}$$

a function $f : U \times V \to R^1$ possesses the following properties:

(i) for each $v \in V$, the function $f(\cdot, v) : U \to R^1$ is convex;
(ii) for each $u \in U$, the function $f(u, \cdot) : V \to R^1$ is concave,

for each $v \in V$,

$$|f(u_1, v) - f(u_2, v)| \leq L_1 \|u_1 - u_2\|$$
$$\text{for all } u_1, u_2 \in U \tag{2.153}$$

and that for each $u \in U$,

$$|f(u, v_1) - f(u, v_2)| \leq L_2 \|v_1 - v_2\|$$
$$\text{for all } v_1, v_2 \in V. \tag{2.154}$$

For each $(\xi, \eta) \in U \times V$, set

$$\partial_x f(\xi, \eta) = \{l \in X :$$
$$f(y, \eta) - f(\xi, \eta) \geq \langle l, y - \xi \rangle \text{ for all } y \in U\}, \tag{2.155}$$
$$\partial_y f(\xi, \eta) = \{l \in Y :$$

$$\langle l, y - \eta \rangle \geq f(\xi, y) - f(\xi, \eta) \text{ for all } y \in V\}. \tag{2.156}$$

In view of properties (i) and (ii), (2.153), and (2.154), for each $\xi \in U$ and each $\eta \in V$,

$$\emptyset \neq \partial_x f(\xi, \eta) \subset B_X(0, L_1), \tag{2.157}$$

$$\emptyset \neq \partial_y f(\xi, \eta) \subset B_Y(0, L_2). \tag{2.158}$$

Let

$$x_* \in C \text{ and } y_* \in D \tag{2.159}$$

satisfy

$$f(x_*, y) \leq f(x_*, y_*) \leq f(x, y_*) \tag{2.160}$$

for each $x \in C$ and each $y \in D$.

Let $\delta_{f,1}, \delta_{f,2}, \delta_C, \delta_D \in (0, 1]$ and $\{a_k\}_{k=0}^{\infty} \subset (0, \infty)$.

Let us describe our algorithm.

Subgradient projection algorithm for zero-sum games
Initialization: select arbitrary $x_0 \in U$ and $y_0 \in V$.
Iterative Step: given current iteration vectors $x_t \in U$ and $y_t \in V$ calculate

$$\xi_t \in \partial_x f(x_t, y_t) + B_X(0, \delta_{f,1}),$$

$$\eta_t \in \partial_y f(x_t, y_t) + B_Y(0, \delta_{f,2})$$

and the next pair of iteration vectors $x_{t+1} \in U$, $y_{t+1} \in V$ such that

$$\|x_{t+1} - P_C(x_t - a_t \xi_t)\| \leq \delta_C,$$

$$\|y_{t+1} - P_D(y_t + a_t \eta_t)\| \leq \delta_D.$$

In this chapter we prove the following result.

Theorem 2.15 *Let $\delta_{f,1}, \delta_{f,2}, \delta_C, \delta_D \in (0, 1]$ and $\{a_k\}_{k=0}^{\infty} \subset (0, \infty)$. Assume that $\{x_t\}_{t=0}^{\infty} \subset U$, $\{y_t\}_{t=0}^{\infty} \subset V$, $\{\xi_t\}_{t=0}^{\infty} \subset X$, $\{\eta_t\}_{t=0}^{\infty} \subset Y$,*

$$B_X(x_0, \delta_C) \cap C \neq \emptyset, \; B_Y(y_0, \delta_D) \cap D \neq \emptyset \tag{2.161}$$

and that for each integer $t \geq 0$,

$$\xi_t \in \partial_x f(x_t, y_t) + B_X(0, \delta_{f,1}), \tag{2.162}$$

$$\eta_t \in \partial_y f(x_t, y_t) + B_Y(0, \delta_{f,2}), \tag{2.163}$$

$$\|x_{t+1} - P_C(x_t - a_t \xi_t)\| \le \delta_C \tag{2.164}$$

and

$$\|y_{t+1} - P_D(y_t + a_t \eta_t)\| \le \delta_D. \tag{2.165}$$

Let for each natural number T,

$$\widehat{x}_T = (\sum_{i=0}^{T} a_t)^{-1} \sum_{t=0}^{T} a_t x_t, \ \widehat{y}_T = (\sum_{i=0}^{T} a_t)^{-1} \sum_{t=0}^{T} a_t y_t. \tag{2.166}$$

Then for each natural number T,

$$B_X(\widehat{x}_T, \delta_C) \cap C \ne \emptyset, \ B_Y(\widehat{y}_T, \delta_D) \cap D \ne \emptyset,$$

$$|(\sum_{t=0}^{T} a_t)^{-1} \sum_{t=0}^{T} a_t f(x_t, y_t) - f(x_*, y_*)|$$

$$\le (2 \sum_{t=0}^{T} a_t)^{-1} (\max\{2M_1, 2M_2\} + 1)^2 + \max\{L_1 \delta_C, \ L_2 \delta_D\}$$

$$+ (\sum_{t=0}^{T} a_t)^{-1} (T+1) \max\{(4M_1 + 1)\delta_C, \ (4M_2 + 1)\delta_D\}$$

$$+ 2^{-1} (\max\{L_1, L_2\} + 1)^2 (\sum_{t=0}^{T} a_t)^{-1} \sum_{t=0}^{T} a_t^2$$

$$+ \max\{\delta_{f,1}, \delta_{f,2}\}(2 \max\{M_1, M_2\} + 1), \tag{2.167}$$

$$|f(\widehat{x}_T, \widehat{y}_T) - (\sum_{t=0}^{T} a_t)^{-1} \sum_{t=0}^{T} a_t f(x_t, y_t)|$$

$$\le (2 \sum_{t=0}^{T} a_t)^{-1} (\max\{2M_1, 2M_2\} + 1)^2 + \max\{L_1 \delta_C, \ L_2 \delta_D\}$$

$$+ (\sum_{t=0}^{T} a_t)^{-1} (T+1) \max\{(4M_1 + 1)\delta_C, \ (4M_2 + 1)\delta_D\}$$

$$+2^{-1}(\max\{L_1, L_2\}+1)^2(\sum_{t=0}^{T} a_t)^{-1}\sum_{t=0}^{T} a_t^2$$

$$+\max\{\delta_{f,1}, \delta_{f,2}\}(2\max\{M_1, M_2\}+1) \tag{2.168}$$

and for each natural number T, each $z \in C$, and each $u \in D$,

$$f(z, \widehat{y}_T) \geq f(\widehat{x}_T, \widehat{y}_T)$$

$$-(\sum_{t=0}^{T} a_t)^{-1}(\max\{2M_1, 2M_2\}+1)^2 - \max\{L_1\delta_C,\ L_2\delta_D\}$$

$$-2(\sum_{t=0}^{T} a_t)^{-1}(T+1)\max\{(4M_1+1)\delta_C,\ (4M_2+1)\delta_D\}$$

$$-(\max\{L_1, L_2\}+1)^2(\sum_{t=0}^{T} a_t)^{-1}\sum_{t=0}^{T} a_t^2$$

$$-2\max\{\delta_{f,1}, \delta_{f,2}\}(2\max\{M_1, M_2\}+1),$$

$$f(\widehat{x}_T, u) \leq f(\widehat{x}_T, \widehat{y}_T)$$

$$+(\sum_{t=0}^{T} a_t)^{-1}(\max\{2M_1, 2M_2\}+1)^2 + \max\{L_1\delta_C,\ L_2\delta_D\}$$

$$+2(\sum_{t=0}^{T} a_t)^{-1}(T+1)\max\{(4M_1+1)\delta_C,\ (4M_2+1)\delta_D\}$$

$$+(\max\{L_1, L_2\}+1)^2(\sum_{t=0}^{T} a_t)^{-1}\sum_{t=0}^{T} a_t^2$$

$$+2\max\{\delta_{f,1}, \delta_{f,2}\}(2\max\{M_1, M_2\}+1).$$

Proof By (2.152), (2.164), and (2.165), for all integers $t \geq 0$,

$$\|x_t\| \leq M_1+1,\ \ \|y_t\| \leq M_2+1. \tag{2.169}$$

Let $t \geq 0$ be an integer. Applying Lemma 2.7 with

$$a = a_t,\ x = x_t,\ f = f(\cdot, y_t),\ \xi = \xi_t,\ u = x_{t+1}$$

we obtain that for each $z \in C$,

$$a_t(f(x_t, y_t) - f(z, y_t)) \le 2^{-1}\|z - x_t\|^2 - 2^{-1}\|z - x_{t+1}\|^2$$

$$+ \delta_C(4M_1 + 1) + a_t\delta_{f,1}(2M_1 + 1) + 2^{-1}a_t^2(L_1 + 1)^2. \quad (2.170)$$

Applying Lemma 2.7 with

$$a = a_t, \ x = y_t, \ f = -f(x_t, \cdot), \ \xi = -\eta_t, \ u = y_{t+1}$$

we obtain that for each $v \in D$,

$$a_t(f(x_t, v) - f(x_t, y_t)) \le 2^{-1}\|v - y_t\|^2 - 2^{-1}\|v - y_{t+1}\|^2$$

$$+ \delta_D(4M_2 + 1) + a_t\delta_{f,2}(2M_2 + 1) + 2^{-1}a_t^2(L_2 + 1)^2. \quad (2.171)$$

For all integers $t \ge 0$ set

$$b_{t,1} = \delta_C(4M_1 + 1) + a_t\delta_{f,1}(2M_1 + 1) + 2^{-1}a_t^2(L_1 + 1)^2,$$

$$b_{t,2} = \delta_D(4M_2 + 1) + a_t\delta_{f,2}(2M_2 + 1) + 2^{-1}a_t^2(L_2 + 1)^2,$$

and define

$$\phi(s) = 2^{-1}s^2, \ s \in R^1.$$

It is easy to see that all the assumptions of Proposition 2.13 hold and it implies Theorem 2.15. □

Theorem 2.15 is a generalization of Theorem 2.11 of [92] obtained in the case when

$$\delta_{f,1} = \delta_{f,2} = \delta_C = \delta_D.$$

We are interested in the optimal choice of a_t, $t = 0, 1, \dots, T$. For simplicity assume that $L_1 = L_2$. Let T be a natural number and $A_T = \sum_{t=0}^{T} a_t$ be given. By Theorem 2.15, in order to make the best choice of a_t, $t = 0, \dots, T$, we need to minimize the function $\sum_{t=0}^{T} a_t^2$ on the set

$$\{a = (a_0, \dots, a_T) \in R^{T+1} : \ a_i \ge 0, \ i = 0, \dots, T, \ \sum_{i=0}^{T} a_i = A_T\}.$$

By Lemma 2.3, this function has a unique minimizer $a^* = (a_0^*, \dots, a_T^*)$ where $a_i^* = (T + 1)^{-1}A_T$, $i = 0, \dots, T$ which is the best choice of a_t, $t = 0, 1, \dots, T$.

Let T be a natural number and $a_t = a$ for all $t = 0, \ldots, T$. Now we will find the best $a > 0$. In order to meet this goal we need to choose a which is a minimizer of the function

$$\Psi_T(a) = ((T+1)a)^{-1}(2(\max\{2M_1, 2M_2\} + 1)^2)$$

$$+2\max\{(4M_1+1)\delta_C,\ (4M_2+1)\delta_D\} + (\max\{L_1, L_2\}+1)^2 a.$$

Since T can be arbitrarily large, we need to find a minimizer of the function

$$\phi(a) := 2a^{-1}\max\{(4M_1+1)\delta_C,\ (4M_2+1)\delta_D\}+(\max\{L_1, L_2\}+1)^2 a,\ a \in (0, \infty).$$

It is not difficult to see that the minimizer is

$$a = (2\max\{(4M_1+1)\delta_C,\ (4M_2+1)\delta_D\})^{1/2}(\max\{L_1, L_2\}+1)^{-1}$$

and the minimal value of ϕ is

$$(8\max\{(4M_1+1)\delta_C,\ (4M_2+1)\delta_D\})^{1/2}(\max\{L_1, L_2\}+1).$$

Now our goal is to find the best integer $T > 0$ which gives us an appropriate value of $\Psi_T(a)$. Since in view of the inequalities above, this value is bounded from below by $c_0 \max\{\delta_C,\ \delta_D\}^{1/2}$, with the constant c_0 depending on L_1, L_2, M_1, M_2, it is clear that in order to make the best choice of T, it should be at the same order as $\lfloor \max\{\delta_C,\ \delta_D\}^{-1} \rfloor$.

Let $T = \lfloor \max\{\delta_C,\ \delta_D\}^{-1} \rfloor$. In this case we obtain a pair of points $\widehat{x} \in U, \widehat{y} \in V$ such that

$$B_X(\widehat{x}, \delta_C) \cap C \neq \emptyset,\ B_Y(\widehat{y}, \delta_D) \cap D \neq \emptyset$$

and for each $z \in C$ and each $v \in D$,

$$f(z, \widehat{y}) \geq f(\widehat{x}, \widehat{y}) - c\max\{\delta_C,\ \delta_D\}^{1/2} - 2\max\{\delta_{f,1}, \delta_{f,2}\}(2\max\{M_1, M_2\}+1),$$

$$f(\widehat{x}, v) \leq f(\widehat{x}, \widehat{y}) + c\max\{\delta_C,\ \delta_D\}^{1/2} + 2\max\{\delta_{f,1}, \delta_{f,2}\}(2\max\{M_1, M_2\}+1),$$

where the constant $c > 0$ depends only on L_1, L_2, M_1, M_2.

2.12 Zero-Sum Games on Unbounded Sets

Let $(X, \langle\cdot,\cdot\rangle)$, $(Y, \langle\cdot,\cdot\rangle)$ be Hilbert spaces equipped with the complete norms $\|\cdot\|$ which are induced by their inner products. Let A be a nonempty closed convex subset of X, D be a nonempty closed convex subset of Y, W be an open convex subset of X, and V be an open convex subset of Y such that

$$A \subset W, \ D \subset V \tag{2.172}$$

and let a function $f : W \times V \to R^1$ possess the following properties:

(i) for each $v \in V$, the function $f(\cdot, v) : W \to R^1$ is convex and continuous;
(ii) for each $w \in W$, the function $f(w, \cdot) : V \to R^1$ is concave and continuous.

Suppose that

$$L_1 > 1, L_2 > 0, M_0, M_1, M_2 > 0, \tag{2.173}$$

$$\theta_0 \in A \cap B_X(0, M_0), \tag{2.174}$$

$$D \subset B_Y(0, M_0), \tag{2.175}$$

$$\sup\{f(\theta_0, y) : \ y \in V \cap B_Y(0, M_0 + 1)\} < M_1, \tag{2.176}$$

for each $y \in V \cap B_Y(0, M_0 + 1)$,

$$\{x \in W : \ f(x, y) \le M_1 + 4\} \subset B_X(0, M_2), \tag{2.177}$$

for each $v \in V \cap B_Y(0, M_0 + 1)$,

$$|f(u_1, v) - f(u_2, v)| \le L_1\|u_1 - u_2\|$$

$$\text{for all } u_1, u_2 \in W \cap B_X(0, M_0 + 3M_2 + 4) \tag{2.178}$$

and that for each $u \in W \cap B_X(0, M_0 + 3M_2 + 4)$,

$$|f(u, v_1) - f(u, v_2)| \le L_2\|v_1 - v_2\|$$

$$\text{for all } v_1, v_2 \in V \cap B_Y(0, M_0 + 2). \tag{2.179}$$

Let

$$x_* \in A \text{ and } y_* \in D \tag{2.180}$$

satisfy

$$f(x_*, y) \le f(x_*, y_*) \le f(x, y_*) \tag{2.181}$$

for each $x \in A$ and each $y \in D$. In view of (2.172), (2.174)–(2.176), (2.180), and (2.181),

$$f(x_*, y_*) \le f(\theta_0, y_*) < M_1, \tag{2.182}$$

$$\|x_*\| \le M_2, \ \|\theta_0\| \le M_2. \tag{2.183}$$

In this chapter we prove the following result which does not have an analog in [92]. (Note that in [92] we consider only games on bounded sets.)

Theorem 2.16 *Let* $\delta_{f,1}, \delta_{f,2}, \delta_A, \delta_D \in (0, 1/2)$ *and* $\{a_k\}_{k=0}^{\infty} \subset (0, \infty)$ *satisfy*

$$\delta_A \le (12M_2 + 9)^{-1}(L_1 + 1)^{-2}, \ \delta_{f,1} \le (6M_2 + 5)^{-1}, \tag{2.184}$$

$$\delta_A(12M_2 + 9) \le a_t \le (L_1 + 1)^{-2} \text{ for all integers } t \ge 0. \tag{2.185}$$

Assume that $\{x_t\}_{t=0}^{\infty} \subset W$, $\{y_t\}_{t=0}^{\infty} \subset V$, $\{\xi_t\}_{t=0}^{\infty} \subset X$, $\{\eta_t\}_{t=0}^{\infty} \subset Y$,

$$B_X(x_0, \delta_A) \cap A \ne \emptyset, \ B_Y(y_0, \delta_D) \cap D \ne \emptyset, \tag{2.186}$$

$$\|x_0\| \le M_2 + 1 \tag{2.187}$$

and that for each integer $t \ge 0$,

$$\xi_t \in \partial_x f(x_t, y_t) + B_X(0, \delta_{f,1}), \tag{2.188}$$

$$\eta_t \in \partial_y f(x_t, y_t) + B_Y(0, \delta_{f,2}), \tag{2.189}$$

$$\|x_{t+1} - P_A(x_t - a_t\xi_t)\| \le \delta_A \tag{2.190}$$

and

$$\|y_{t+1} - P_D(y_t + a_t\eta_t)\| \le \delta_D. \tag{2.191}$$

Let for each natural number T,

$$\widehat{x}_T = (\sum_{i=0}^{T} a_t)^{-1} \sum_{t=0}^{T} a_t x_t, \ \widehat{y}_T = (\sum_{i=0}^{T} a_t)^{-1} \sum_{t=0}^{T} a_t y_t. \tag{2.192}$$

Then for all integers $t \ge 0$,

$$\|x_t\| \le 3M_2 + 1 \tag{2.193}$$

and for each natural number T,

$$B_X(\widehat{x}_T, \delta_A) \cap A \ne \emptyset, \ B_Y(\widehat{y}_T, \delta_D) \cap D \ne \emptyset, \tag{2.194}$$

$$|(\sum_{t=0}^{T} a_t)^{-1} \sum_{t=0}^{T} a_t f(x_t, y_t) - f(x_*, y_*)|$$

$$\leq \max\{L_1\delta_A,\ L_2\delta_D\}$$

$$+(2\sum_{t=0}^{T} a_t)^{-1}(\max\{2M_0, 6M_2+4\}+1)^2$$

$$+\max\{\delta_{f,1}, \delta_{f,2}\}(2\max\{2M_0+1, 6M_2+5\})$$

$$+2^{-1}(\max\{L_1, L_2\}+1)^2(\sum_{t=0}^{T} a_t)^{-1}\sum_{t=0}^{T} a_t^2$$

$$+(\sum_{t=0}^{T} a_t)^{-1}(T+1)\max\{(12M_2+9)\delta_A,\ (4M_0+1)\delta_D\}, \quad (2.195)$$

$$|f(\widehat{x}_T, \widehat{y}_T) - (\sum_{t=0}^{T} a_t)^{-1} \sum_{t=0}^{T} a_t f(x_t, y_t)|$$

$$\leq \max\{L_1\delta_A,\ L_2\delta_D\}$$

$$+(2\sum_{t=0}^{T} a_t)^{-1}(\max\{2M_0, 6M_2+4\}+1)^2$$

$$+\max\{\delta_{f,1}, \delta_{f,2}\}(\max\{2M_0+1, 6M_2+5\})$$

$$+2^{-1}(\max\{L_1, L_2\}+1)^2(\sum_{t=0}^{T} a_t)^{-1}\sum_{t=0}^{T} a_t^2$$

$$+(\sum_{t=0}^{T} a_t)^{-1}(T+1)\max\{(12M_2+9)\delta_A,\ (4M_0+1)\delta_D\} \quad (2.196)$$

and for each natural number T, each $z \in A$, and each $u \in D$,

$$f(z, \widehat{y}_T) \geq f(\widehat{x}_T, \widehat{y}_T)$$

$$-\max\{L_1\delta_A,\ L_2\delta_D\}$$

$$-(\sum_{t=0}^{T} a_t)^{-1}(\max\{2M_0, 6M_2+4\}+1)^2$$

$$-2\max\{\delta_{f,1}, \delta_{f,2}\}(\max\{2M_0+1, 6M_2+5\})$$

$$-(\max\{L_1, L_2\}+1)^2(\sum_{t=0}^{T} a_t)^{-1}\sum_{t=0}^{T} a_t^2$$

$$-2(\sum_{t=0}^{T} a_t)^{-1}(T+1)\max\{(12M_2+9)\delta_A,\ (4M_0+1)\delta_D\}, \tag{2.197}$$

$$f(\widehat{x}_T, u) \le f(\widehat{x}_T, \widehat{y}_T)$$

$$+\max\{L_1\delta_A,\ L_2\delta_D\}$$

$$+(\sum_{t=0}^{T} a_t)^{-1}(\max\{2M_0, 6M_2+4\}+1)^2$$

$$+\max\{\delta_{f,1}, \delta_{f,2}\}(2\max\{2M_0+1, 6M_2+5\})$$

$$+(\max\{L_1, L_2\}+1)^2(\sum_{t=0}^{T} a_t)^{-1}\sum_{t=0}^{T} a_t^2$$

$$+2(\sum_{t=0}^{T} a_t)^{-1}(T+1)\max\{(12M_2+9)\delta_A,\ (4M_0+1)\delta_D\}. \tag{2.198}$$

We are interested in the optimal choice of a_t, $t = 0, 1, \ldots, T$. Let T be a natural number and $A_T = \sum_{t=0}^{T} a_t$ be given. By Theorem 2.16, in order to make the best choice of a_t, $t = 0, \ldots, T$, we need to minimize the function $\sum_{t=0}^{T} a_t^2$ on the set

$$\{a = (a_0, \ldots, a_T) \in R^{T+1} : \ a_i \ge 0,\ i = 0, \ldots, T,\ \sum_{i=0}^{T} a_i = A_T\}.$$

By Lemma 2.3, this function has a unique minimizer $a^* = (a_0^*, \ldots, a_T^*)$ where $a_i^* = (T+1)^{-1}A_T$, $i = 0, \ldots, T$ which is the best choice of a_t, $t = 0, 1, \ldots, T$.

Let T be a natural number and $a_t = a$ for all $t = 0, \ldots, T$. Now we will find the best $a > 0$. In order to meet this goal we need to choose a which is a minimizer of the function

$$\Psi_T(a) = ((T+1)a)^{-1}(\max\{2M_0, 6M_2+2\}+1)^2$$

$$+a^{-1}\max\{(12M_2+9)\delta_A,\ (4M_0+1)\delta_D\} + (\max\{L_1, L_2\}+1)^2 a,\ a > 0.$$

Since T can be arbitrarily large, we need to find a minimizer of the function

$$\phi(a) := a^{-1}\max\{(12M_2+9)\delta_A,\ (4M_0+1)\delta_D\}$$

$$+(\max\{L_1, L_2\}+1)^2 a,\ a \in (0,\infty).$$

It is not difficult to see that the minimizer is

$$a = (\max\{(12M_2+9)\delta_A,\ (4M_0+1)\delta_D\})^{1/2}(\max\{L_1, L_2\}+1)^{-1}$$

and the minimal value of ϕ is

$$(2\max\{(12M_2+9)\delta_A,\ (4M_0+1)\delta_D\})^{1/2}(\max\{L_1, L_2\}+1).$$

Now our goal is to find the best integer $T > 0$ which gives us an appropriate value of $\Psi_T(a)$. In view of the inequalities above, this value is bounded from below by $c_0\max\{\delta_C,\ \delta_D\}^{1/2}$, with the constant c_0 depending on L_1, L_2, M_0, M_1, M_2, and it is clear that in order to make the best choice of T, it should be at the same order as $\lfloor \max\{\delta_C,\ \delta_D\}^{-1}\rfloor$. In this case we obtain a pair of points $\widehat{x} \in W \cap B_X(0, 3M_2+4)$, $\widehat{y} \in V$ such that

$$B_X(\widehat{x}, \delta_A) \cap A \cap B_X(0, 3M_2+4) \neq \emptyset,$$

$$B_Y(\widehat{y}, \delta_D) \cap D \neq \emptyset$$

and for each $z \in A$ and each $v \in D$,

$$f(z, \widehat{y}) \geq f(\widehat{x}, \widehat{y})$$

$$-c\max\{\delta_A,\ \delta_D\}^{1/2} - 2\max\{\delta_{f,1}, \delta_{f,2}\}\max\{6M_2+5, 2M_0+1\},$$

$$f(\widehat{x}, v) \leq f(\widehat{x}, \widehat{y})$$

$$+c\max\{\delta_A,\ \delta_D\}^{1/2} + 2\max\{\delta_{f,1}, \delta_{f,2}\}\max\{6M_2+5, 2M_0+1\},$$

where the constant $c > 0$ depends only on L_1, L_2, M_0, M_1, M_2.

2.13 Proof of Theorem 2.16

By (2.175), (2.186), and (2.191), for all integers $t \geq 0$,

$$\|y_t\| \leq M_0 + 1. \tag{2.199}$$

Set

$$C = A \cap B_X(0, 3M_2 + 2) \tag{2.200}$$

and

$$U = W \cap \{x \in X : \|x\| < 3M_2 + 4\}. \tag{2.201}$$

We show that

$$\|x_t - \theta_0\| \leq 2M_2 + 1$$

for all integers $t \geq 0$. In view of (2.183) and (2.187),

$$\|x_0 - \theta_0\| \leq \|x_0\| + \|\theta_0\| \leq M_2 + 1 + M_2 = 2M_2 + 1. \tag{2.202}$$

Assume that $t \geq 0$ is an integer and that

$$\|x_t - \theta_0\| \leq 2M_2 + 1. \tag{2.203}$$

It follows from (2.183), (2.201), and (2.203) that

$$x_t \in U. \tag{2.204}$$

Relation (2.190) implies

$$\|x_{t+1} - P_A(x_t - a_t\xi_t)\| \leq 1. \tag{2.205}$$

By (2.178), (2.188), (2.199), (2.201), and (2.204),

$$\xi_t \in \partial_x f(x_t, y_t) + B_X(0, 1) \subset B_X(0, L_1 + 1). \tag{2.206}$$

It follows from (2.174), (2.185), (2.203), (2.206), and Lemma 2.2 that

$$\begin{aligned}\|\theta_0 - P_A(x_t - a_t\xi_t)\| &\leq \|\theta_0 - x_t + a_t\xi_t\| \\ &\leq \|\theta_0 - x_t\| + \|\xi_t\|a_t \leq 2M_2 + 2.\end{aligned} \tag{2.207}$$

In view of (2.183) and (2.207),

$$\|P_A(x_t - a_t\xi_t)\| \le \|\theta_0\| + 2M_2 + 2 \le 3M_2 + 2. \tag{2.208}$$

By (2.205) and (2.208),

$$\|x_{t+1}\| \le \|P_A(x_t - a_t\xi_t)\| + 1 \le 3M_2 + 3. \tag{2.209}$$

It follows from (2.201) and (2.209) that

$$x_{t+1} \in U. \tag{2.210}$$

Relations (2.200) and (2.208) imply that

$$P_A(x_t - a_t\xi_t) \in C, \tag{2.211}$$

and

$$P_A(x_t - a_t\xi_t) = P_C(x_t - a_t\xi_t). \tag{2.212}$$

By (2.178), (2.183), (2.188), (2.190), (2.200), (2.201), (2.203), (2.204), and (2.210)–(2.212), we apply Lemma 2.7 with

$$M_0 = 3M_2 + 2,\ a = a_t,\ x = x_t,\ f = f(\cdot, y_t),\ \xi = \xi_t,\ u = x_{t+1}$$

and obtain that for each $z \in C$,

$$\begin{aligned} &a_t(f(x_t, y_t) - f(z, y_t)) \\ &\le 2^{-1}\|x_t - z\|^2 - 2^{-1}\|x_{t+1} - z\|^2 \\ &+ \delta_A(12M_2 + 9) + a_t\delta_{f,1}(6M_2 + 5) + 2^{-1}a_t^2(L_1 + 1)^2. \end{aligned} \tag{2.213}$$

In view of (2.213) with $z = \theta_0$,

$$\begin{aligned} &a_t(f(x_t, y_t) - f(\theta_0, y_t)) \\ &\le 2^{-1}\|x_t - \theta_0\|^2 - 2^{-1}\|x_{t+1} - \theta_0\|^2 \\ &+ \delta_A(12M_2 + 9) + a_t\delta_{f,1}(6M_2 + 5) + 2^{-1}a_t^2(L_1 + 1)^2. \end{aligned} \tag{2.214}$$

There are two cases:

$$\|\theta_0 - x_{t+1}\| \le \|\theta_0 - x_t\|; \tag{2.215}$$

$$\|\theta_0 - x_{t+1}\| > \|\theta_0 - x_t\|. \tag{2.216}$$

Assume that (2.215) holds. By (2.203) and (2.215),

$$\|\theta_0 - x_{t+1}\| \le 2M_2 + 1. \tag{2.217}$$

Assume that (2.216) is true. It follows from (2.184), (2.185), (2.214), and (2.216) that

$$f(x_t, y_t) - f(\theta_0, y_t)$$

$$\le a_t^{-1}\delta_A(12M_2 + 9) + \delta_{f,1}(6M_2 + 5) + 2^{-1}a_t(L_1 + 1)^2 \le 3. \tag{2.218}$$

By (2.176), (2.199), and (2.218),

$$f(x_t, y_t) \le M_1 + 3. \tag{2.219}$$

In view of (2.177), (2.199), and (2.219),

$$\|x_t\| \le M_2.$$

Together with (2.183) this implies that

$$\|x_t - \theta_0\| \le 2M_2.$$

It follows from the inequality above, (2.174), (2.185), (2.190), (2.206), and Lemma 2.2 that

$$\|\theta_0 - x_{t+1}\|$$

$$\le \|x_{t+1} - P_A(x_t - a_t\xi_t)\| + \|\theta_0 - P_A(x_t - a_t\xi_t)\|$$

$$\le \delta_A + \|\theta_0 - P_A(x_t - a_t\xi_t)\|$$

$$\delta + \|\theta_0 - x_t + a_t\xi_t\|$$

$$\le \delta_A + \|\theta_0 - x_t\| + \|\xi_t\|a_t$$

$$\le \delta_A + 2M_2 + (L_1 + 1)a_t \le 2M_2 + 1$$

and (2.217) holds in both cases. Thus by induction we have showed that for all integers $t \ge 0$,

$$\|x_t - \theta_0\| \le 2M_2 + 1$$

and that (2.210)–(2.212) hold. Now we can apply Theorem 2.15 with C, D, U and

$$V \cap \{y \in Y : \|y\| < M_0 + 2\}$$

and obtain that (2.195), (2.196) are true and that for each natural number T, each $z \in C$, and each $u \in D$, (2.197) and (2.198) hold. In order to complete the proof it is sufficient to show that (2.197) holds for all $z \in A \setminus C$.

By (2.176), (2.177), (2.194), (2.197) (with $z = \theta_0$), and (2.200),

$$f(z, \widehat{y}_T) > M_1 + 4 > 4 + f(\theta_0, \widehat{y}_T)$$

and (2.197) holds for all $z \in A$. Theorem 2.16 is proved. □

2.14 An Example for Theorem 2.16

Let X, Y be Hilbert spaces, A be a nonempty closed convex subset of X, D be a nonempty closed convex subset of Y, W be an open convex subset of X, and V be an open convex subset of Y such that

$$A \subset W, \; D \subset V,$$

let a function $g : W \to R^1$ be convex and continuous, a function $h : V \to R^1$ be concave and continuous, and

$$f(w, v) = g(w)h(v), \; w \in W, \; v \in V.$$

Suppose that $c_0, M_0, M_1, M_2, L_0, L_1 > 0$,

$$g(w) \geq c_0, \; w \in W,$$

$$h(v) \geq c_0, \; v \in V,$$

$$D \subset B_Y(0, M_0),$$

$$\theta_0 \in A \cap B_X(0, M_0),$$

$$|h(v_1) - h(v_2)| \leq L_0\|v_1 - v_2\|$$

$$\text{for all } v_1, v_2 \in V \cap B_Y(0, M_0 + 2),$$

$$\sup\{h(y) : \; y \in V \cap B_Y(0, M_0 + 1)\} < M_1,$$

$$\{x \in W : \; g(x) \leq c_0^{-1}(M_1(g(\theta_0) + 1) + 4)\} \subset B_X(0, M_2),$$

$$|g(u_1) - g(u_2)| \leq L_1 \|u_1 - u_2\|$$

$$\text{for all } u_1, u_2 \in W \cap B_X(0, 3M_2 + M_0 + 4)$$

and that

$$x_* \in A \text{ and } y_* \in D$$

satisfy

$$f(x_*, y) \leq f(x_*, y_*) \leq f(x, y_*)$$

for all $x \in A$ and all $y \in D$. It is not difficult to see that all the assumptions made in Sect. 2.12 hold.

Chapter 3
The Mirror Descent Algorithm

In this chapter we analyze the mirror descent algorithm for minimization of convex and nonsmooth functions and for computing the saddle points of convex–concave functions, under the presence of computational errors. The problem is described by an objective function and a set of feasible points. For this algorithm each iteration consists of two steps. The first step is a calculation of a subgradient of the objective function while in the second one we solve an auxiliary minimization problem on the set of feasible points. In each of these two steps there is a computational error. In general, these two computational errors are different. We show that our algorithm generates a good approximate solution, if all the computational errors are bounded from above by a small positive constant. Moreover, if we know the computational errors for the two steps of our algorithm, we find out what approximate solution can be obtained and how many iterates one needs for this.

3.1 Optimization on Bounded Sets

Let X be a Hilbert space equipped with an inner product $\langle \cdot, \cdot \rangle$ which induces a complete norm $\| \cdot \|$. Let C be a nonempty closed convex subset of X, U be an open convex subset of X such that $C \subset U$ and let $f : U \to R^1$ be a convex function. Suppose that there exist $L > 0$, $M_0 > 0$ such that

$$C \subset B_X(0, M_0), \tag{3.1}$$

$$|f(x) - f(y)| \leq L\|x - y\| \text{ for all } x, y \in U. \tag{3.2}$$

In view of (3.2), for each $x \in U$,

$$\emptyset \neq \partial f(x) \subset B_X(0, L). \tag{3.3}$$

A. J. Zaslavski, *Convex Optimization with Computational Errors*, Springer Optimization and Its Applications 155, https://doi.org/10.1007/978-3-030-37822-6_3

For each nonempty set $D \subset X$ and each function $h : D \to R^1$ put

$$\inf(h, D) = \inf\{h(y) : y \in D\}$$

and

$$\operatorname{argmin}(h, D) = \operatorname{argmin}\{h(y) : y \in D\} = \{y \in D : h(y) = \inf(h, D)\}.$$

We study the convergence of the mirror descent algorithm under the presence of computational errors. This method was introduced by Nemirovsky and Yudin for solving convex optimization problems [66]. Here we use a derivation of this algorithm proposed by Beck and Teboulle [17].

Let $\delta_f, \delta_C \in (0, 1]$ and $\{a_k\}_{k=0}^{\infty} \subset (0, \infty)$.

We describe the inexact version of the mirror descent algorithm.

Mirror descent algorithm
Initialization: select an arbitrary $x_0 \in U$.
Iterative Step: given a current iteration vector $x_t \in U$ calculate

$$\xi_t \in \partial f(x_t) + B_X(0, \delta_f),$$

define

$$g_t(x) = \langle \xi_t, x \rangle + (2a_t)^{-1}\|x - x_t\|^2, \ x \in X$$

and calculate the next iteration vector $x_{t+1} \in U$ such that

$$B_X(x_{t+1}, \delta_C) \cap \operatorname{argmin}\{g_t(y) : y \in C\} \neq \emptyset.$$

Here δ_f is a computational error produced by our computer system when we calculate a subgradient of f and δ_C is a computational error produced by our computer system when we calculate a minimizer of the function g_t.

Note that g_t is a convex bounded from below function on X which possesses a minimizer on C.

In this chapter we prove the following result.

Theorem 3.1 *Let $\delta_f, \delta_C \in (0, 1]$, $\{a_k\}_{k=0}^{\infty} \subset (0, \infty)$ and let*

$$x_* \in C \tag{3.4}$$

satisfies

$$f(x_*) \le f(x) \text{ for all } x \in C. \tag{3.5}$$

Assume that $\{x_t\}_{t=0}^{\infty} \subset U$, $\{\xi_t\}_{t=0}^{\infty} \subset X$,

$$\|x_0\| \le M_0 + 1 \tag{3.6}$$

and that for each integer $t \ge 0$,

$$\xi_t \in \partial f(x_t) + B_X(0, \delta_f), \tag{3.7}$$

$$g_t(z) = \langle \xi_t, z \rangle + (2a_t)^{-1}\|z - x_t\|^2, \ z \in X \tag{3.8}$$

and

$$B_X(x_{t+1}, \delta_C) \cap argmin(g_t, C) \neq \emptyset. \tag{3.9}$$

Then for each natural number T,

$$\sum_{t=0}^{T} a_t(f(x_t) - f(x_*))$$

$$\le 2^{-1}(2M_0 + 1)^2 + (\delta_f(2M_0 + 1) + \delta_C(L + 1)) \sum_{t=0}^{T} a_t$$

$$+ 4\delta_C(T + 1)(2M_0 + 1) + 2^{-1}(L + 1)^2 \sum_{t=0}^{T} a_t^2. \tag{3.10}$$

Moreover, for each natural number T,

$$f((\sum_{t=0}^{T} a_t)^{-1} \sum_{t=0}^{T} a_t x_t) - f(x_*),$$

$$\min\{f(x_t) : \ t = 0, \dots, T\} - f(x_*)$$

$$\le 2^{-1}(2M_0 + 1)^2(\sum_{t=0}^{T} a_t)^{-1} + \delta_f(2M_0 + 1) + \delta_C(L + 1)$$

$$+4\delta_C(T + 1)(2M_0 + 1)(\sum_{t=0}^{T} a_t)^{-1}$$

$$+ 2^{-1}(L + 1)^2 \sum_{t=0}^{T} a_t^2(\sum_{t=0}^{T} a_t)^{-1}. \tag{3.11}$$

Theorem 3.1 is proved in Sect. 3.3. It is a generalization of Theorem 3.1 of [92] proved in the case when $\delta_f = \delta_C$.

We are interested in an optimal choice of a_t, $t = 0, 1, \ldots$. Let T be a natural number and $A_T = \sum_{t=0}^{T} a_t$ be given. By Theorem 3.1, in order to make the best choice of a_t, $t = 0, \ldots, T$, we need to minimize the function $\sum_{t=0}^{T} a_t^2$ on the set

$$\{a = (a_0, \ldots, a_T) \in R^{T+1} : a_i \geq 0,\ i = 0, \ldots, T,\ \sum_{i=0}^{T} a_i = A_T\}.$$

By Lemma 2.3, this function has a unique minimizer $a^* = (a_0^*, \ldots, a_T^*)$ where $a_i^* = (T+1)^{-1}A_T$, $i = 0, \ldots, T$. This is the best choice of a_t, $t = 0, 1, \ldots, T$.

Let T be a natural number and $a_t = a$, $t = 0, \ldots, T$. Now we will find the best $a > 0$. By Theorem 3.1, we need to choose a which is a minimizer of the function

$$2^{-1}((T+1)a)^{-1}(2M_0+1)^2 + \delta_f(2M_0+1) + \delta_C(L+1)$$

$$+4a^{-1}\delta_C(2M_0+1) + 2^{-1}(L+1)^2 a.$$

Since T can be arbitrarily large, we need to find a minimizer of the function

$$\phi(a) := 4a^{-1}\delta_C(2M_0+1) + 2^{-1}(L+1)^2 a,\ a \in (0, \infty).$$

Clearly, the minimizer is

$$a = (8\delta_C(2M_0+1))^{1/2}(L+1)^{-1}$$

and the minimal value of ϕ is

$$(8\delta_C(2M_0+1))^{1/2}(L+1).$$

Now we can think about the best choice of T. It is clear that it should be at the same order as $\lfloor \delta_C^{-1} \rfloor$. As a result, we obtain a point $\xi \in U$ such that

$$B_X(\xi, \delta_C) \cap C \neq \emptyset$$

and

$$f(\xi) \leq f(x_*) + c_1\delta_C^{1/2} + \delta_f(2M_0+1),$$

where the constant $c_1 > 0$ depends only on L and M_0.

3.2 The Main Lemma

Lemma 3.2 *Let* $\delta_f, \delta_C \in (0, 1]$, $a > 0$ *and let*

$$z \in C. \tag{3.12}$$

Assume that

$$x \in U \cap B_X(0, M_0 + 1), \tag{3.13}$$

$$\xi \in \partial f(x) + B_X(0, \delta_f), \tag{3.14}$$

$$g(v) = \langle \xi, v \rangle + (2a)^{-1} \|v - x\|^2, \ v \in X \tag{3.15}$$

and that

$$u \in U \tag{3.16}$$

satisfies

$$B_X(u, \delta_C) \cap \{v \in C : \ g(v) = \inf(g, C)\} \neq \emptyset. \tag{3.17}$$

Then

$$a(f(x) - f(z)) \le \delta_f a(2M_0 + 1) + \delta_C a(L + 1) + 4\delta_C(M_0 + 1)$$
$$+2^{-1}a^2(L + 1)^2 + 2^{-1}\|z - x\|^2 - 2^{-1}\|z - u\|^2.$$

Proof In view of (3.14), there exists

$$l \in \partial f(x) \tag{3.18}$$

such that

$$\|l - \xi\| \le \delta. \tag{3.19}$$

Clearly, the function g is Fréchet differentiable on X. We denote by $g'(v)$ its Fréchet derivative at $v \in X$. It is easy to see that

$$g'(v) = \xi + a^{-1}(v - x), \ v \in X. \tag{3.20}$$

By (3.17), there exists

$$\widehat{u} \in B_X(u, \delta_C) \cap C \tag{3.21}$$

such that

$$g(\widehat{u}) = \inf(g, C). \tag{3.22}$$

It follows from (3.20)–(3.22) that for all $v \in C$,

$$0 \le \langle g'(\widehat{u}), v - \widehat{u} \rangle = \langle \xi + a^{-1}(\widehat{u} - x), v - \widehat{u} \rangle. \tag{3.23}$$

By (3.1), (3.12), (3.13), (3.18), and (3.19),

$$\begin{aligned} a(f(x) - f(z)) &\le a\langle x - z, l \rangle \\ &= a\langle x - z, \xi \rangle + a\langle x - z, l - \xi \rangle \\ &\le a\langle x - z, \xi \rangle + a\|x - z\|\|l - \xi\| \\ &\le \delta_f a(2M_0 + 1) + a\langle x - z, \xi \rangle \\ &= \delta_f a(2M_0 + 1) + a\langle \xi, x - \widehat{u} \rangle + a\langle \xi, \widehat{u} - z \rangle \\ &\le \delta_f a(2M_0 + 1) + \langle x - \widehat{u}, a\xi \rangle \\ &+ \langle z - \widehat{u}, x - \widehat{u} - a\xi \rangle + \langle z - \widehat{u}, \widehat{u} - x \rangle. \end{aligned} \tag{3.24}$$

Relations (3.12) and (3.23) imply that

$$\langle z - \widehat{u}, x - \widehat{u} - a\xi \rangle \le 0. \tag{3.25}$$

In view of Lemma 2.1,

$$\langle z - \widehat{u}, \widehat{u} - x \rangle = 2^{-1}[\|z - x\|^2 - \|z - \widehat{u}\|^2 - \|\widehat{u} - x\|^2]. \tag{3.26}$$

It follows from (3.3), (3.18), (3.19), and (3.21) that

$$\begin{aligned} \langle x - \widehat{u}, a\xi \rangle &= \langle x - u, a\xi \rangle + \langle u - \widehat{u}, a\xi \rangle \\ &\le a\delta_C\|\xi\| + \langle x - u, a\xi \rangle \\ &\le a\delta_C(L + 1) + \langle x - u, a\xi \rangle \\ &\le a\delta_C(L + 1) + 2^{-1}\|x - u\|^2 + 2^{-1}a^2\|\xi\|^2. \end{aligned} \tag{3.27}$$

By (3.3), (3.18), (3.19), and (3.24)–(3.27),

$$
\begin{aligned}
& a(f(x) - f(z)) \\
& \le \delta_f a(2M_0 + 1) + \langle x - \widehat{u}, a\xi \rangle \\
& + \langle z - \widehat{u}, x - \widehat{u} - a\xi \rangle + \langle z - \widehat{u}, \widehat{u} - x \rangle \\
& \le \delta_f a(2M_0 + 1) + a\delta_C(L + 1) \\
& + 2^{-1}\|x - u\|^2 + 2^{-1}a^2\|\xi\|^2 \\
& + 2^{-1}\|z - x\|^2 - 2^{-1}\|z - \widehat{u}\|^2 - 2^{-1}\|\widehat{u} - x\|^2 \\
& \le \delta_f a(2M_0 + 1) + \delta_C a(L + 1) + 2^{-1}a^2(L + 1)^2 \\
& + 2^{-1}\|z - x\|^2 - 2^{-1}\|z - \widehat{u}\|^2 \\
& + 2^{-1}\|x - u\|^2 - 2^{-1}\|\widehat{u} - x\|^2.
\end{aligned} \tag{3.28}
$$

In view of (3.1), (3.13), (3.17), and (3.21),

$$
\begin{aligned}
& |\|x - u\|^2 - \|\widehat{u} - x\|^2| \\
& \le |\|x - u\| - \|\widehat{u} - x\||(\|x - u\| + \|\widehat{u} - x\|) \\
& \le 4\|u - \widehat{u}\|(M_0 + 1) \le 4(M_0 + 1)\delta_C.
\end{aligned} \tag{3.29}
$$

Relations (3.1), (3.12), (3.13), and (3.21) imply that

$$
\begin{aligned}
& |\|z - \widehat{u}\|^2 - \|z - u\|^2| \\
& \le |\|z - \widehat{u}\| - \|z - u\||(\|z - \widehat{u}\| + \|z - u\|) \\
& \le \|u - \widehat{u}\|(4M_0 + 1) \le (4M_0 + 1)\delta_C.
\end{aligned} \tag{3.30}
$$

By (3.28), (3.29), and (3.30),

$$
\begin{aligned}
& a(f(x) - f(z)) \\
& \le \delta_f a(2M_0 + 1) + \delta_C(L + 1) + +2^{-1}a^2(L + 1)^2 \\
& + 2^{-1}\|z - x\|^2 - 2^{-1}\|z - u\|^2 + 4(M_0 + 1)\delta_C.
\end{aligned}
$$

This completes the proof of Lemma 3.2. □

3.3 Proof of Theorem 3.1

In view of (3.1), (3.6), and (3.9),

$$\|x_t\| \le M_0 + 1,\ t = 0, 1, \ldots. \tag{3.31}$$

Let $t \ge 0$ be an integer. Applying Lemma 3.2 with

$$g = g_t,\ z = x_*,\ a = a_t,\ x = x_t,\ \xi = \xi_t,\ u = x_{t+1}$$

we obtain that

$$\begin{aligned} &a_t(f(x_t) - f(x_*)) \\ &\le 2^{-1}\|x_* - x_t\|^2 - 2^{-1}\|x_* - x_{t+1}\|^2 \\ &+4\delta_C(M_0+1) + a_t\delta_f(2M_0+1) \\ &+ a_t\delta_C(L+1) + 2^{-1}a_t^2(L+1)^2. \end{aligned} \tag{3.32}$$

By (3.1), (3.4), (3.31), and (3.32), for each natural number T,

$$\begin{aligned} &\sum_{t=0}^{T} a_t(f(x_t) - f(x_*)) \\ &\le \sum_{t=0}^{T}(2^{-1}\|x_* - x_t\|^2 - 2^{-1}\|x_* - x_{t+1}\|^2) \\ &+\delta_f(2M_0+1)\sum_{t=0}^{T} a_t + \delta_C(L+1)\sum_{t=0}^{T} a_t \\ &+4(T+1)\delta_C(M_0+1) + 2^{-1}(L+1)^2\sum_{t=0}^{T} a_t^2 \\ &\le 2^{-1}(2M_0+1)^2 + (\delta_f(2M_0+1) + \delta_C(L+1))\sum_{t=0}^{T} a_t \\ &+4(T+1)\delta_C(M_0+1) + 2^{-1}(L+1)^2\sum_{t=0}^{T} a_t^2. \end{aligned}$$

Thus (3.10) is true. Evidently, (3.10) implies (3.11). This completes the proof of Theorem 3.1. □

3.4 Optimization on Unbounded Sets

Let X be a Hilbert space equipped with an inner product $\langle \cdot, \cdot \rangle$ which induces a complete norm $\| \cdot \|$. Let D be a nonempty closed convex subset of X, V be an open convex subset of X such that

$$D \subset V, \tag{3.33}$$

and $f : V \to R^1$ be a convex function which is Lipschitz on all bounded subsets of V. Set

$$D_{min} = \{x \in D : \ f(x) \le f(y) \text{ for all } y \in D\}. \tag{3.34}$$

We suppose that

$$D_{min} \neq \emptyset.$$

In Sect. 3.5 we will prove the following result.

Theorem 3.3 *Let* $\delta_f, \delta_C \in (0, 1]$, $M > 1$ *satisfy*

$$D_{min} \cap B_X(0, M) \neq \emptyset, \tag{3.35}$$

$$M_0 > 80M + 6, \tag{3.36}$$

$L > 1$ *satisfy*

$$|f(v_1) - f(v_2)| \le L\|v_1 - v_2\|$$

$$\textit{for all } v_1, v_2 \in V \cap B_X(0, M_0 + 2), \tag{3.37}$$

$$0 < \tau_0 \le \tau_1 \le (4L + 4)^{-1}, \tag{3.38}$$

$$\epsilon_0 = 8\tau_0^{-1}\delta_C(M_0 + 1) + 2\delta_C(L + 1)$$

$$+ 2\delta_f(2M_0 + 1) + \tau_1(L + 1)^2 \tag{3.39}$$

and let

$$n_0 = \lfloor \tau_0^{-1}(2M + 2)^2\epsilon_0^{-1} \rfloor + 1. \tag{3.40}$$

Assume that

$$\{x_t\}_{t=0}^{\infty} \subset V,\ \{\xi_t\}_{t=0}^{\infty} \subset X,\ \{a_t\}_{t=0}^{\infty} \subset [\tau_0, \tau_1], \tag{3.41}$$

$$\|x_0\| \le M \tag{3.42}$$

and that for each integer $t \ge 0$,

$$\xi_t \in \partial f(x_t) + B_X(0, \delta_f), \tag{3.43}$$

$$g_t(v) = \langle \xi_t, v\rangle + (2a_t)^{-1}\|v - x_t\|^2,\ v \in X \tag{3.44}$$

and

$$B_X(x_{t+1}, \delta_C) \cap argmin(g_t, D) \ne \emptyset. \tag{3.45}$$

Then there exists an integer $q \in [1, n_0]$ *such that*

$$f(x_q) \le \inf(f, D) + \epsilon_0,$$

$$\|x_t\| \le 15M + 1,\ t = 0, \ldots, q.$$

We are interested in the best choice of a_t, $t = 0, 1, \ldots$. Assume for simplicity that $\tau_1 = \tau_0$. In order to meet our goal we need to minimize ϵ_0 which obtains its minimal value when

$$\tau_0 = (8\delta_C(M_0 + 1))^{1/2}(L + 1)^{-1}$$

and the minimal value of ϵ_0 is

$$2\delta_C(L + 1) + 2\delta_f(2M_0 + 1) + 2(8\delta_C(M_0 + 1))^{1/2}(L + 1).$$

Thus ϵ_0 is at the same order as $\max\{\delta_C^{1/2}, \delta_f\}$. By (3.40) and the inequalities above, n_0 is at the same order as $\delta_C^{-1/2}\max\{\delta_C^{1/2}, \delta_f\}^{-1}$.

Theorem 3.3 is a generalization of Theorem 3.3 of [92] proved with $\delta_f = \delta_C$.

In this chapter we will prove the following two theorems which have no prototype in [92].

Theorem 3.4 *Let* $\delta_f, \delta_C \in (0, 1)$, $M > 0$ *satisfy*

$$\delta_f \le (24M + 48)^{-1},\ \delta_C \le 8^{-1}(L + 1)^{-2}(16M + 16)^{-1}, \tag{3.46}$$

$$\{x \in V :\ f(x) \le \inf(f, D) + 4\} \subset B_X(0, M), \tag{3.47}$$

$$M_0 = 12M + 12, \tag{3.48}$$

$L > 1$ satisfy

$$|f(v_1) - f(v_2)| \le L\|v_1 - v_2\|$$

$$\text{for all } v_1, v_2 \in V \cap B_X(0, M_0 + 2) \tag{3.49}$$

and let $\{a_t\}_{t=0}^{\infty} \subset (0, \infty)$ satisfy for all integers $t \ge 0$,

$$4\delta_C(M_0 + 1) \le a_t \le (2L + 2)^{-1}(L + 1)^{-1}. \tag{3.50}$$

Assume that

$$\{x_t\}_{t=0}^{\infty} \subset V, \ \{\xi_t\}_{t=0}^{\infty} \subset X, \tag{3.51}$$

$$\|x_0\| \le M \tag{3.52}$$

and that for each integer $t \ge 0$,

$$\xi_t \in \partial f(x_t) + B_X(0, \delta_f), \tag{3.53}$$

$$g_t(v) = \langle \xi_t, v\rangle + (2a_t)^{-1}\|v - x_t\|^2, \ v \in X \tag{3.54}$$

and

$$B_X(x_{t+1}, \delta_C) \cap argmin(g_t, D) \ne \emptyset. \tag{3.55}$$

Then

$$\|x_t\| \le 5M + 3 \text{ for all integers } t \ge 0$$

and for each natural number T,

$$\sum_{t=0}^{T} a_t(f(x_t) - \inf(f, D))$$

$$\le 2M^2 + (\delta_f(2M_0 + 1) + \delta_C(L + 1)) \sum_{t=0}^{T} a_t$$

$$+ 4\delta_C(T + 1)(M_0 + 1) + 2^{-1}(L + 1)^2 \sum_{t=0}^{T} a_t^2. \tag{3.56}$$

Moreover, for each natural number T,

$$f((\sum_{t=0}^{T} a_t)^{-1} \sum_{t=0}^{T} a_t x_t) - \inf(f, D),$$

$$\min\{f(x_t) : t = 0, \dots, T\} - \inf(f, D)$$

$$\leq 2M^2 (\sum_{t=0}^{T} a_t)^{-1} + \delta_f (2M_0 + 1) + \delta_C (L + 1)$$

$$+ 4\delta_C (T + 1)(M_0 + 1)(\sum_{t=0}^{T} a_t)^{-1} + 2^{-1}(L + 1)^2 \sum_{t=0}^{T} a_t^2 (\sum_{t=0}^{T} a_t)^{-1}. \tag{3.57}$$

We are interested in the best choice of a_t, $t = 0, 1, \dots$. Let T be a natural number and $A_T = \sum_{t=0}^{T} a_t$ be given. By Theorem 3.4, in order to make the best choice of a_t, $t = 0, \dots, T$, we need to minimize the function $\sum_{t=0}^{T} a_t^2$ on the set

$$\{a = (a_0, \dots, a_T) \in R^{T+1} : a_i \geq 0, \ i = 0, \dots, T, \ \sum_{i=0}^{T} a_i = A_T\}.$$

By Lemma 2.3, this function has a unique minimizer $a^* = (a_0^*, \dots, a_T^*)$ where $a_i^* = (T + 1)^{-1} A_T$, $i = 0, \dots, T$. This is the best choice of a_t, $t = 0, 1, \dots, T$.

Let T be a natural number and $a_t = a$, $t = 0, \dots, T$. Now we will find the best $a > 0$. By Theorem 3.4, we need to choose a which is a minimizer of the function

$$2((T + 1)a)^{-1} M^2 + \delta_f (2M_0 + 1) + \delta_C (L + 1)$$

$$+4a^{-1} \delta_C (M_0 + 1) + 2^{-1}(L + 1)^2 a.$$

Since T can be arbitrarily large, we need to find a minimizer of the function

$$\phi(a) := 4a^{-1} \delta_C (M_0 + 1) + 2^{-1}(L + 1)^2 a, \ a \in (0, \infty).$$

Clearly, the minimizer is

$$a = (8\delta_C (M_0 + 1))^{1/2} (L + 1)^{-1}$$

and the minimal value of ϕ is

$$(8\delta_C (M_0 + 1))^{1/2} (L + 1).$$

Now we can think about the best choice of T. It is clear that it should be at the same order as $\lfloor \delta_C^{-1} \rfloor$. As a result, we obtain a point $\xi \in U$ such that

$$B_X(\xi, \delta_C) \cap C \neq \emptyset$$

and

$$f(\xi) \leq \inf(f, D) + c_1\delta_C^{1/2} + \delta_f(2M_0 + 1),$$

where the constant $c_1 > 0$ depends only on L and M_0.

Theorem 3.5 *Let*

$$\delta_f, \delta_C \in (0, 1), \ M > 8 \tag{3.58}$$

satisfy

$$D_{min} \cap B_X(0, M) \neq \emptyset, \tag{3.59}$$

$L > 0$ *satisfy*

$$|f(v_1) - f(v_2)| \leq L\|v_1 - v_2\|$$

$$\text{for all } v_1, v_2 \in V \cap B_X(0, 8M + 8), \tag{3.60}$$

$$a = (4L + 4)^{-1}\delta_C^{1/2}, \tag{3.61}$$

$$T = \lfloor 8^{-1}\delta_C^{-1/2} \max\{\delta_f, \delta_C^{1/2}\}^{-1} \rfloor. \tag{3.62}$$

Assume that

$$\{x_t\}_{t=0}^{\infty} \subset V, \ \{\xi_t\}_{t=0}^{\infty} \subset X,$$

$$\|x_0\| \leq M, \ B_X(x_0, \delta_C) \cap D \neq \emptyset \tag{3.63}$$

and that for each integer $t \geq 0$,

$$\xi_t \in \partial f(x_t) + B_X(0, \delta_f), \tag{3.64}$$

$$g_t(v) = \langle \xi_t, v \rangle + (2a_t)^{-1}\|v - x_t\|^2, \ v \in X \tag{3.65}$$

and

$$B_X(x_{t+1}, \delta_C) \cap argmin(g_t, D) \neq \emptyset. \tag{3.66}$$

Then

$$\|x_t\| \leq 3M + 2 \text{ for all integers } t \in [0, T] \tag{3.67}$$

and

$$\sum_{t=0}^{T}(T+1)^{-1}f(x_t) - \inf(f, D),$$

$$f((T+1)^{-1}\sum_{t=0}^{T}x_t) - \inf(f, D), \ \min\{f(x_t) : \ t = 0, \dots, T\} - \inf(f, D)$$

$$\leq 64M^2(L+1)\max\{\delta_f, \delta_C^{1/2}\}. \tag{3.68}$$

3.5 Proof of Theorem 3.3

By (3.35) there exists

$$z \in D_{min} \cap B_X(0, M). \tag{3.69}$$

Assume that T is a natural number and that

$$f(x_t) - f(z) > \epsilon_0, \ t = 1, \dots, T. \tag{3.70}$$

In view of (3.45), there exists

$$\eta \in B_X(x_1, \delta_C) \cap \mathrm{argmin}(g_0, D). \tag{3.71}$$

Relations (3.44), (3.69), and (3.71) imply that

$$\langle \xi_0, \eta \rangle + (2a_0)^{-1}\|\eta - x_0\|^2$$

$$\leq \langle \xi_0, z \rangle + (2a_0)^{-1}\|z - x_0\|^2. \tag{3.72}$$

It follows from (3.37), (3.42), and (3.43) that

$$\|\xi_0\| \leq L + 1. \tag{3.73}$$

In view of (3.8) and (3.41),

$$a_0^{-1} \geq \tau_1^{-1} \geq 4(L+1). \tag{3.74}$$

By (3.42), (3.69), (3.72), and (3.73),

$$(L+1)M + (2a_0)^{-1}(2M+1)^2$$

$$\geq \langle \xi_0, z \rangle + (2a_0)^{-1}\|z - x_0\|^2$$

$$\geq (2a_0)^{-1}\|\eta - x_0\|^2 + \langle \xi_0, \eta - x_0 \rangle + \langle \xi_0, x_0 \rangle$$

$$\geq (2a_0)^{-1}\|\eta - x_0\|^2 - (L+1)\|\eta - x_0\| - (L+1)M.$$

Together with (3.38) and (3.41) this implies that

$$M + (2M+1)^2 \geq \|\eta - x_0\|^2 - 2^{-1}\|\eta - x_0\|,$$

$$(\|\eta - x_0\| - 4^{-1})^2 \leq (4M+1)^2,$$

$$\|\eta - x_0\| \leq 8M. \tag{3.75}$$

Together with (3.48), (3.69), and (3.71) this implies that

$$\|\eta\| \leq 9M, \ \|x_1\| \leq 9M + 1,$$

$$\|\eta - z\| \leq 10M,$$

$$\|x_1 - z\| \leq 10M + 1. \tag{3.76}$$

By induction we show that for every integer $t \in [1, T]$,

$$\|x_t - z\| \leq 14M + 1, \tag{3.77}$$

$$f(x_t) - f(z)$$

$$\leq \delta_f(2M_0 + 1) + \delta_C(L+1)$$

$$+4\tau_0^{-1}\delta_C(M_0+1) + 2^{-1}\tau_1(L+1)^2$$

$$+ (2\tau_0)^{-1}(\|z - x_t\|^2 - \|z - x_{t+1}\|^2). \tag{3.78}$$

Set

$$U = V \cap \{v \in X : \ \|v\| < M_0 + 2\} \tag{3.79}$$

and

$$C = D \cap B_X(0, M_0). \tag{3.80}$$

In view of (3.69) and (3.76), (3.77) holds for $t = 1$.

Assume that an integer $t \in [1, T]$ and that (3.77) holds. It follows from (3.36), (3.69), (3.79), and (3.80) that

$$z \in C \subset B_X(0, M_0). \tag{3.81}$$

In view of (3.36), (3.69), (3.77), and (3.79),

$$x_t \in U \cap B_X(0, M_0 + 1). \tag{3.82}$$

Relations (3.37), (3.43), and (3.82) imply that

$$\xi_t \in \partial f(x_t) + B_X(0, \delta_f) \subset B_X(0, L + 1). \tag{3.83}$$

In view of (3.45), there exists

$$h \in B_X(x_{t+1}, \delta_C) \cap \mathrm{argmin}(g_t, D). \tag{3.84}$$

By (3.44), (3.69), and (3.84),

$$\begin{aligned}&\langle \xi_t, h\rangle + (2a_t)^{-1}\|h - x_t\|^2\\&\le \langle \xi_t, z\rangle + (2a_t)^{-1}\|z - x_t\|^2.\end{aligned} \tag{3.85}$$

In view of (3.85),

$$\begin{aligned}&\langle \xi_t, z\rangle + (2a_t)^{-1}\|z - x_t\|^2\\&\ge (2a_t)^{-1}\|h - x_t\|^2 + \langle \xi_t, h - x_t\rangle + \langle \xi_t, x_t\rangle.\end{aligned}$$

It follows from the inequality above, (3.36), (3.38), (3.41), (3.69), (3.77), and (3.83) that

$$\begin{aligned}&(L + 1)M + (2a_t)^{-1}(14M + 1)^2\\&\ge (2a_t)^{-1}\|h - x_t\|^2 - (L + 1)\|h - x_t\| - (L + 1)(15M + 1)\\&\ge (2a_t)^{-1}(\|h - x_t\|^2 - \|h - x_t\|) - (L + 1)(15M + 1)\\&\ge (2a_t)^{-1}(\|h - x_t\| - 1)^2 - (2a_t)^{-1} - (L + 1)(15M + 1),\\&4(14M + 1)^2 \ge 2 + 16M + (14M + 1)^2 \ge (\|h - x_t\| - 1)^2,\\&\|h - x_t\| \le 28M + 4,\\&\|h\| \le 44M + 5 < M_0.\end{aligned} \tag{3.86}$$

By (3.80), (3.84), and (3.86),

$$h \in C. \tag{3.87}$$

Relations (3.80), (3.84), and (3.87) imply that

$$h \in \mathrm{argmin}(g_t, C)$$

and

$$h \in B_X(x_{t+1}, \delta_C) \cap \mathrm{argmin}(g_t, C). \tag{3.88}$$

It follows from (3.37), (3.38), (3.43), (3.44), (3.69), (3.79), (3.82), (3.88), and Lemma 3.2 which holds with

$$x = x_t,\ a = a_t,\ \xi = \xi_t,\ u = x_{t+1}$$

that

$$a_t(f(x_t) - f(z)) \le \delta_f a_t(2M_0 + 1) + \delta_C a_t(L + 1)$$

$$+4\delta_C(M_0 + 1) + 2^{-1}a_t^2(L + 1)^2$$

$$+2^{-1}\|z - x_t\|^2 - 2^{-1}\|z - x_{t+1}\|^2.$$

Together with the inclusion $a_t \in [\tau_0, \tau_1]$ this implies that

$$f(x_t) - f(z) \le 4\tau_0^{-1}\delta_C(M_0 + 1)$$

$$+(2M_0 + 1)\delta_f + (L + 1)\delta_C + 2^{-1}\tau_1(L + 1)^2$$

$$+ (2\tau_0)^{-1}\|z - x_t\|^2 - (2\tau_0)^{-1}\|z - x_{t+1}\|^2 \tag{3.89}$$

and (3.78) holds.

In view of (3.39), (3.70), (3.77), and (3.89),

$$\|z - x_t\|^2 - \|z - x_{t+1}\|^2 \ge 0,$$

$$\|z - x_{t+1}\| \le \|z - x_t\| \le 14M + 1.$$

Hence by induction we showed that (3.78) holds for all $t = 1, \ldots, T$ and (3.77) holds for all $t = 1, \ldots, T + 1$.

It follows from (3.39), (3.40), (3.42), (3.69), (3.70), and (3.78) that

$$T\epsilon_0 < T(\min\{f(x_t) :\ t = 1, \dots, T\} - f(z))$$

$$\leq \sum_{t=1}^{T}(f(x_t) - f(z))$$

$$\leq (2\tau_0)^{-1}\sum_{t=1}^{T}(\|z - x_t\|^2 - \|z - x_{t+1}\|^2)$$

$$+T(\delta_f(2M_0 + 1) + \delta_C(L + 1)$$

$$+4\delta_C(M_0 + 1)\tau_0^{-1} + 2^{-1}\tau_1(L + 1)^2)$$

$$\leq (2\tau_0)^{-1}(2M + 1)^2 + 4T\tau_0^{-1}\delta_C(M_0 + 1)$$

$$+T(\delta_f(2M_0 + 1) + \delta_C(L + 1) + 2^{-1}\tau_1(L + 1)^2),$$

$$2^{-1}\epsilon_0 < (2\tau_0 T)^{-1}(2M + 1)^2$$

and

$$T < \tau_0^{-1}(2M + 1)^2\epsilon_0^{-1} < n_0.$$

Thus we have shown that if an integer $T \geq 1$ satisfies

$$f(x_t) - f(z) > \epsilon_0,\ t = 1, \dots, T,$$

then $T < n_0$ and (3.77) holds for all $t = 1, \dots, T + 1$. This implies that there exists an integer $q \in \{1, \dots, n_0\}$ such that

$$f(x_q) - f(z) \leq \epsilon_0,$$

$$\|x_t\| \leq 15M + 1,\ t = 0, \dots, q.$$

Theorem 3.3 is proved. □

3.6 Proof of Theorem 3.4

Fix

$$z \in D_{min}. \tag{3.90}$$

Set

$$U = V \cap \{v \in X : \|v\| < M_0 + 2\} \tag{3.91}$$

and

$$C = D \cap B_X(0, M_0), \tag{3.92}$$

$$M_1 = 4M + 3. \tag{3.93}$$

In view of (3.47) and (3.90),

$$\|z\| \le M. \tag{3.94}$$

By (3.52) and (3.94),

$$\|x_0 - z\| \le 2M. \tag{3.95}$$

Assume that $t \ge 0$ is an integer and that

$$\|x_t - z\| \le M_1. \tag{3.96}$$

It follows from (3.48), (3.90), (3.91), (3.93), (3.94), and (3.96) that

$$z \in C \cap B_X(0, M), \tag{3.97}$$

$$\|x_t\| \le M_1 + M, \tag{3.98}$$

$$x_t \in U \cap B_X(0, M_1 + M). \tag{3.99}$$

Relations (3.48), (3.49), (3.53), (3.93), and (3.98) imply that

$$\xi_t \in \partial f(x_t) + B_X(0, \delta_f) \subset B_X(0, L + 1). \tag{3.100}$$

In view of (3.55), there exists

$$h \in B_X(x_{t+1}, \delta_C) \cap \operatorname{argmin}(g_t, D). \tag{3.101}$$

By (3.54), (3.90), and (3.101),

$$\begin{aligned} &\langle \xi_t, h \rangle + (2a_t)^{-1} \|h - x_t\|^2 \\ &\le \langle \xi_t, z \rangle + (2a_t)^{-1} \|z - x_t\|^2. \end{aligned} \tag{3.102}$$

In view of (3.102),

$$\langle \xi_t, z\rangle + (2a_t)^{-1}\|z - x_t\|^2$$
$$\geq (2a_t)^{-1}\|h - x_t\|^2 + \langle \xi_t, h - x_t\rangle + \langle \xi_t, x_t\rangle. \tag{3.103}$$

It follows from (3.50), (3.100), and (3.103) that

$$(2a_t)^{-1}(\|h - x_t\| - 1)^2 - (2a_t)^{-1}$$
$$\leq (2a_t)^{-1}(\|h - x_t\|^2 - \|h - x_t\|)$$
$$\leq (2a_t)^{-1}\|h - x_t\|^2 - (L+1)\|h - x_t\|$$
$$\leq (2a_t)^{-1}\|h - x_t\|^2 + \langle \xi_t, h - x_t\rangle$$
$$\leq (2a_t)^{-1}\|h - x_t\|^2 + \langle \xi_t, h - x_t\rangle + \langle \xi_t, x_t\rangle + \|\xi_t\|\|x_t\|. \tag{3.104}$$

By (3.94), (3.96), (3.98), (3.100), and (3.104),

$$(2a_t)^{-1}(\|h - x_t\| - 1)^2 - (2a_t)^{-1}$$
$$\leq (2a_t)^{-1}\|h - x_t\|^2 + \langle \xi_t, h - x_t\rangle$$
$$+\langle \xi_t, x_t\rangle + (L+1)(M_1 + M)$$
$$\leq (2a_t)^{-1}\|z - x_t\|^2 + \langle \xi_t, z\rangle + (L+1)(M_1 + M)$$
$$\leq (L+1)(M_1 + 2M) + (2a_t)^{-1}M_1^2. \tag{3.105}$$

In view of (3.48), (3.50), (3.93), (3.98), and (3.105),

$$(\|h - x_t\| - 1)^2 \leq 1 + 2M + M_1^2 + M_1,$$
$$\|h - x_t\| \leq M_1 + 2,$$
$$\|h\| \leq M_1 + 2 + M_1 + M$$
$$\leq 3M_1 + 2 = 12M + 11 < M_0. \tag{3.106}$$

By (3.48), (3.92), (3.101), and (3.106),

$$h \in C. \tag{3.107}$$

Relations (3.93), (3.101), and (3.107) imply that

$$h \in \mathrm{argmin}(g_t, C). \tag{3.108}$$

By (3.101) and (3.106),

$$x_{t+1} \in U \cap B_X(0, M_0 + 1). \tag{3.109}$$

It follows from (3.48), (3.53), (3.54), (3.93), (3.97), (3.99), (3.101), (3.108), (3.109), and Lemma 3.2 which holds with

$$x = x_t,\ a = a_t,\ \xi = \xi_t,\ u = x_{t+1}$$

that

$$\begin{aligned}
&a_t(f(x_t) - f(z)) \\
&\le \delta_f a_t(2M_0 + 1) + \delta_C a_t(L + 1) \\
&+4\delta_C(M_0 + 1) + 2^{-1}a_t^2(L + 1)^2 \\
&+ 2^{-1}\|z - x_t\|^2 - 2^{-1}\|z - x_{t+1}\|^2.
\end{aligned} \tag{3.110}$$

There are two cases:

$$\|z - x_{t+1}\| \le \|z - x_t\|; \tag{3.111}$$

$$\|z - x_{t+1}\| > \|z - x_t\|. \tag{3.112}$$

Assume that (3.111) is true. In view of (3.96) and (3.111),

$$\|z - x_{t+1}\| \le \|z - x_t\| \le M_1.$$

Assume that (3.112) holds. It follows from (3.46), (3.48), (3.50), (3.110), and (3.112) that

$$\begin{aligned}
&f(x_t) - f(z) \\
&\le 4a_t^{-1}\delta_C(M_0 + 1) + \delta_f(2M_0 + 1) \\
&+\delta_C(L + 1) + 2^{-1}a_t(L + 1)^2 \le 4.
\end{aligned}$$

By the inequality above, (3.47) and (3.90),

$$\|x_t\| \le M. \tag{3.113}$$

It follows from (3.97), (3.100), (3.103), and (3.113) that

$$(2a_t)^{-1}(\|h - x_t\| - 1)^2 - (2a_t)^{-1}$$

$$\le (2a_t)^{-1}\|h - x_t\|^2 + \langle \xi_t, h - x_t\rangle + \langle \xi_t, x_t\rangle + (L+1)M$$

$$\le (2a_t)^{-1}\|z - x_t\|^2 + \langle \xi_t, z\rangle + (L+1)M$$

$$\le 2(L+1)M + 4(2a_t)^{-1}M^2.$$

By the relation above, (3.50) and (3.113),

$$(\|h - x_t\| - 1)^2 \le 1 + 4M^2 + 2M \le (2M+1)^2,$$

$$\|h - x_t\| \le 2M + 2,$$

$$\|h\| \le 3M + 2. \tag{3.114}$$

In view of (3.101) and (3.114),

$$\|x_{t+1}\| \le 3M + 3. \tag{3.115}$$

By (3.97) and (3.115),

$$\|x_{t+1} - z\| \le 4M + 3.$$

Together with (3.93) this implies that

$$\|x_{t+1} - z\| \le M_1.$$

Hence by induction we showed that for all integers $t \ge 0$,

$$\|x_t - z\| \le M_1 \tag{3.116}$$

and (3.110) is true. It follows from (3.93), (3.97), and (3.116) that

$$\|x_t\| \le 5M + 3 \text{ for all integers } t \ge 0.$$

By (3.90), (3.95), and (3.110), for each natural number T,

$$\sum_{t=0}^{T} a_t(f(x_t) - \inf(f, D))$$

$$\leq 2^{-1}\|z - x_0\|^2 + (\delta_f(2M_0 + 1) + \delta_C(L + 1))\sum_{t=0}^{T} a_t$$

$$+4\delta_C(T + 1)(M_0 + 1) + 2^{-1}(L + 1)^2 \sum_{t=0}^{T} a_t^2$$

$$\leq 2M^2 + (\delta_f(2M_0 + 1) + \delta_C(L + 1))\sum_{t=0}^{T} a_t$$

$$+4\delta_C(T + 1)(M_0 + 1) + 2^{-1}(L + 1)^2 \sum_{t=0}^{T} a_t^2.$$

Theorem 3.4 is proved. □

3.7 Proof of Theorem 3.5

By (3.59), there exists

$$z \in D_{min} \cap B_X(0, M). \tag{3.117}$$

Set

$$U = V \cap \{v \in X : \|v\| < 8M + 8\} \tag{3.118}$$

and

$$C = D \cap B_X(0, 5M + 6). \tag{3.119}$$

In view of (3.63) and (3.116),

$$\|x_0 - z\| \leq 2M. \tag{3.120}$$

Assume that $t \geq 0$ is an integer and that

$$\|x_t - z\| \leq 2M + 2. \tag{3.121}$$

It follows from (3.116)–(3.119), (3.121) that

$$z \in C \cap B_X(0, M), \tag{3.122}$$

$$\|x_t\| \le 3M + 2, \tag{3.123}$$

$$x_t \in U \cap B_X(0, 3M + 2). \tag{3.124}$$

Relations (3.60) and (3.123) imply that

$$\xi_t \in \partial f(x_t) + B_X(0, \delta_f) \subset B_X(0, L + 1). \tag{3.125}$$

In view of (3.66), there exists

$$h \in B_X(x_{t+1}, \delta_C) \cap \operatorname{argmin}(g_t, D). \tag{3.126}$$

By (3.65), (3.116), and (3.126),

$$\begin{aligned} &\langle \xi_t, h \rangle + (2a)^{-1}\|h - x_t\|^2 \\ &\le \langle \xi_t, z \rangle + (2a)^{-1}\|z - x_t\|^2. \end{aligned} \tag{3.127}$$

In view of (3.127),

$$\begin{aligned} &\langle \xi_t, z \rangle + (2a)^{-1}\|z - x_t\|^2 \\ &\ge (2a)^{-1}\|h - x_t\|^2 + \langle \xi_t, h - x_t \rangle + \langle \xi_t, x_t \rangle. \end{aligned} \tag{3.128}$$

It follows from (3.58), (3.61), (3.125), and (3.128) that

$$\begin{aligned} &(2a)^{-1}(\|h - x_t\| - 1)^2 - (2a)^{-1} \\ &\le (2a)^{-1}(\|h - x_t\|^2 - \|h - x_t\|) \\ &\le (2a)^{-1}\|h - x_t\|^2 - (L + 1)\|h - x_t\| \\ &\le (2a)^{-1}\|h - x_t\|^2 + \langle \xi_t, h - x_t \rangle \\ &\le (2a)^{-1}\|h - x_t\|^2 + \langle \xi_t, h - x_t \rangle + \langle \xi_t, x_t \rangle + \|\xi_t\|\|x_t\|. \end{aligned} \tag{3.129}$$

By (3.116), (3.125), (3.128), and (3.129),

$$\begin{aligned} &(2a)^{-1}(\|h - x_t\| - 1)^2 - (2a)^{-1} \\ &\le (2a)^{-1}\|h - x_t\|^2 + \langle \xi_t, h - x_t \rangle + \langle \xi_t, x_t \rangle + (L + 1)\|x_t\| \\ &\le (2a)^{-1}\|z - x_t\|^2 + \langle \xi_t, z \rangle + (L + 1)\|x_t\| \\ &\le (L + 1)(M + \|x_t\|) + (2a)^{-1}\|z - x_t\|^2. \end{aligned} \tag{3.130}$$

In view of (3.61), (3.121), (3.123), and (3.130),

$$(\|h - x_t\| - 1)^2 \leq 1 + \|z - x_t\|^2 + M + \|x_t\|$$

$$1 + 4M^2 + 8M + 4 + 4M + 2 = 4M^2 + 12M + 7 \leq (2M + 3)^2$$

and

$$\|h - x_t\| - 1 \leq 2M + 3, \ \|h\| \leq 5M + 6. \tag{3.131}$$

By (3.119)) and (3.131),

$$h \in C.$$

Together with relations (3.119) and (3.126) this implies that

$$h \in \operatorname{argmin}(g_t, C). \tag{3.132}$$

By (3.126) and (3.131),

$$\|x_{t+1}\| \leq 5M + 7. \tag{3.133}$$

It follows from (3.64), (3.65), (3.116)–(3.119), (3.124), (3.126), (3.132), (3.133), and Lemma 3.2 applied with

$$M_0 = 5M + 6, \ x = x_t, \ \xi = \xi_t, \ u = x_{t+1},$$

that

$$a(f(x_t) - f(z))$$
$$\leq 2^{-1}\|z - x_t\|^2 - 2^{-1}\|z - x_{t+1}\|^2$$
$$+ a\delta_f(10M + 13) + a\delta_C(L + 1)$$
$$+ 4\delta_C(5M + 7) + 2^{-1}a^2(L + 1)^2. \tag{3.134}$$

In view of (3.60), (3.63), (3.67), (3.116), and (3.123),

$$f(x_t) - f(z) \geq -L\delta_C.$$

It follows from (3.61), (3.116), (3.134), and the inequality above that

$$\|z - x_{t+1}\|^2 - \|z - x_t\|^2$$

$$\leq 8\delta_C(5M + 7) + a^2(L + 1)^2$$

$$+2a\delta_f(10M + 3) + 2\delta_C a(L + 1) + 2aL\delta_C$$

$$\leq 8\delta_C(5M + 8) + a^2(L + 1)^2 + 2a\delta_f(10M + 3). \tag{3.135}$$

Thus we have shown that the following property holds:

(a) if for an integer $t \geq 0$ relation (3.121) holds, then (3.134) and (3.135) are true.

Let us show that for all integers $t = 0, \ldots, T$,

$$\|z - x_t\|^2$$

$$\leq \|z - x_0\|^2 + t[8\delta_C(5M + 8) + a^2(L + 1)^2 + 2a\delta_f(10M + 3)]$$

$$\leq \|z - x_0\|^2 + T[8\delta_C(5M + 8) + a^2(L + 1)^2 + 2a\delta_f(10M + 3)]$$

$$\leq 4M^2 + 8M + 4. \tag{3.136}$$

First, note that by (3.61) and (3.62),

$$T[8\delta_C(5M + 8) + a^2(L + 1)^2 + 2a\delta_f(10M + 3)]$$

$$\leq 8T\delta_C(5M + 8) + T\delta_C + 2Ta\delta_f(10M + 3)$$

$$\leq 8^{-1}\min\{\delta_C^{-1}, \delta_C^{-1/2}\delta_f^{-1}\}(8\delta_C(5M + 9) + \delta_f(10M + 3)\delta_C^{1/2}(L + 1)^{-1})$$

$$\leq 5M + 9 + 2M + 1 = 7M + 10 \leq 8M + 4. \tag{3.137}$$

It follows from (3.120) and (3.137) that

$$\|z - x_0\|^2 + T[8\delta_C(5M + 8) + a^2(L + 1)^2 + 2a\delta_f(10M + 3)]$$

$$\leq (2M + 2)^2. \tag{3.138}$$

Assume that $t \in \{0, \ldots, T\} \setminus \{T\}$ and that (3.136) holds. Then

$$\|x_t - z\| \leq 2M + 2$$

(see (3.121)) and by property (a), (3.134) and (3.135) hold.

By (3.135), (3.136), and (3.138),

$$\|z - x_{t+1}\|^2$$

$$\leq \|z - x_t\|^2 + 8\delta_C(5M + 8) + a^2(L + 1)^2 + 2a\delta_f(10M + 3)$$

$$\leq \|z - x_0\|^2$$

$$+(t + 1)[8\delta_C(5M + 8) + a^2(L + 1)^2 + 2a\delta_f(10M + 3)]$$

$$\leq \|z - x_0\|^2$$

$$+T[8\delta_C(5M + 8) + a^2(L + 1)^2 + 2a\delta_f(10M + 3)]$$

$$\leq 4M^2 + 8M + 4$$

and

$$\|z - x_{t+1}\| \leq 2M + 2.$$

Thus we have shown by induction that (3.136) holds for all $t = 0, \ldots, T$. Together with property (a) this implies that (3.134) holds for all $t = 0, \ldots, T$. By (3.116) and (3.136),

$$\|x_t\| \leq 3M + 2,\ t = 0, \ldots, T.$$

It follows from (3.134) that

$$\sum_{t=0}^{T} a(f(x_t) - f(z))$$

$$\leq 2^{-1}\|z - x_0\|^2 + (T + 1)(4\delta_C(5M + 7) + a\delta_f(10M + 13)$$

$$+2^{-1}a^2(L + 1)^2 + a\delta_C(L + 1)).$$

By the relation above, (3.61), (3.62), (3.116), and (3.120),

$$\sum_{t=0}^{T}(T + 1)^{-1}f(x_t) - \inf(f, D),$$

$$\leq \delta_f(10M + 13) + \delta_C(L + 1)$$

$$+4\delta_C(5M + 7)2(L + 1)\delta_C^{-1/2} + 2^{-2}(L + 1)\delta_C^{1/2}$$

$$+4M^2(L+1)8\max\{\delta_C^{1/2},\ \delta_f\}$$

$$\leq 32M^2(L+1)\max\{\delta_C^{1/2},\delta_f\}+44M(L+1)\max\{\delta_C^{1/2},\delta_f\}$$

$$\leq 64M^2(L+1)\max\{\delta_C^{1/2},\delta_f\}.$$

This implies (3.68). Theorem 3.5 is proved. □

3.8 Zero-Sum Games on Bounded Sets

Let $(X, \langle\cdot,\cdot\rangle)$, $(Y, \langle\cdot,\cdot\rangle)$ be Hilbert spaces equipped with the complete norms $\|\cdot\|$ which are induced by their inner products. Let C be a nonempty closed convex subset of X, D be a nonempty closed convex subset of Y, U be an open convex subset of X, and V be an open convex subset of Y such that

$$C \subset U,\ D \subset V. \tag{3.139}$$

Suppose that there exist $L_1, L_2 > 0$, $M_1, M_2 > 0$ such that

$$C \subset B_X(0, M_1),\ D \subset B_Y(0, M_2), \tag{3.140}$$

a function $f : U \times V \to R^1$ possesses the following properties:

(i) for each $v \in V$, the function $f(\cdot, v) : U \to R^1$ is convex;
(ii) for each $u \in U$, the function $f(u, \cdot) : V \to R^1$ is concave,

for each $u \in U$,

$$|f(u, v_1) - f(u, v_2)| \leq L_2\|v_1 - v_2\|$$

$$\text{for all } v_1, v_2 \in V, \tag{3.141}$$

and that for each $v \in V$,

$$|f(u_1, v) - f(u_2, v)| \leq L_1\|u_1 - u_2\|$$

$$\text{for all } u_1, u_2 \in U. \tag{3.142}$$

Recall that for each $(\xi, \eta) \in U \times V$,

$$\partial_x f(\xi,\eta) = \{l \in X :\ f(y,\eta) - f(\xi,\eta) \geq \langle l, y-\xi\rangle \text{ for all } y \in U\},$$

$$\partial_y f(\xi,\eta) = \{l \in Y :\ \langle l, y-\eta\rangle \geq f(\xi,y) - f(\xi,\eta) \text{ for all } y \in V\}. \tag{3.143}$$

In view of properties (i) and (ii) and (3.141)–(3.143), for each $\xi \in U$ and each $\eta \in V$,

$$\emptyset \neq \partial_x f(\xi, \eta) \subset B_X(0, L_1), \tag{3.144}$$

$$\emptyset \neq \partial_y f(\xi, \eta) \subset B_Y(0, L_2). \tag{3.145}$$

Assume that

$$x_* \in C \text{ and } y_* \in D \tag{3.146}$$

satisfy

$$f(x_*, y) \leq f(x_*, y_*) \leq f(x, y_*) \tag{3.147}$$

for each $x \in C$ and each $y \in D$.

Let $\delta_{f,1}, \delta_{f,2}, \delta_C, \delta_D \in (0, 1]$, and $\{a_k\}_{k=0}^{\infty} \subset (0, \infty)$.

Let us describe our algorithm.

Mirror descent algorithm for zero-sum games

Initialization: select arbitrary $x_0 \in U$ and $y_0 \in V$.

Iterative Step: given current iteration vectors $x_t \in U$ and $y_t \in V$ calculate

$$\xi_t \in \partial_x f(x_t, y_t) + B_X(0, \delta_{f,1}),$$

$$\eta_t \in \partial_y f(x_t, y_t) + B_Y(0, \delta_{f,2})$$

and the next pair of iteration vectors $x_{t+1} \in U$, $y_{t+1} \in V$ such that

$$B_X(x_{t+1}, \delta_C) \cap \operatorname{argmin}\{\langle \xi_t, v\rangle + (2a_t)^{-1}\|v - x_t\|^2 : v \in C\} \neq \emptyset,$$

$$B_Y(y_{t+1}, \delta_D) \cap \operatorname{argmin}\{\langle -\eta_t, u\rangle + (2a_t)^{-1}\|u - y_t\|^2 : u \in D\} \neq \emptyset.$$

In this chapter we prove the following result.

Theorem 3.6 *Let $\delta_{f,1}, \delta_{f,2}, \delta_C, \delta_D \in (0, 1]$, and $\{a_k\}_{k=0}^{\infty} \subset (0, \infty)$. Assume that $\{x_t\}_{t=0}^{\infty} \subset U$, $\{y_t\}_{t=0}^{\infty} \subset V$, $\{\xi_t\}_{t=0}^{\infty} \subset X$, $\{\eta_t\}_{t=0}^{\infty} \subset Y$,*

$$B_X(x_0, \delta_C) \cap C \neq \emptyset, \; B_Y(y_0, \delta_D) \cap D \neq \emptyset \tag{3.148}$$

and that for each integer $t \geq 0$,

$$\xi_t \in \partial_x f(x_t, y_t) + B_X(0, \delta_{f,1}),$$

$$\eta_t \in \partial_y f(x_t, y_t) + B_Y(0, \delta_{f,2}),$$

$$B_X(x_{t+1}, \delta_C) \cap argmin\{\langle \xi_t, v\rangle + (2a_t)^{-1}\|v - x_t\|^2 : v \in C\} \neq \emptyset,$$

$$B_Y(y_{t+1}, \delta_D) \cap argmin\{\langle -\eta_t, u\rangle + (2a_t)^{-1}\|u - y_t\|^2 : u \in D\} \neq \emptyset.$$

Let for each natural number T,

$$\widehat{x}_T = (\sum_{i=0}^{T} a_t)^{-1} \sum_{t=0}^{T} a_t x_t,$$

$$\widehat{y}_T = (\sum_{i=0}^{T} a_t)^{-1} \sum_{t=0}^{T} a_t y_t.$$

Then for each natural number T,

$$B_X(\widehat{x}_T, \delta_C) \cap C \neq \emptyset,$$

$$B_Y(\widehat{y}_T, \delta_D) \cap D \neq \emptyset, \tag{3.149}$$

$$|(\sum_{t=0}^{T} a_t)^{-1} \sum_{t=0}^{T} a_t f(x_t, y_t) - f(x_*, y_*)|$$

$$\leq \max\{\delta_{f,1}(2M_1 + 1), \delta_{f,2}(2M_2 + 1)\}$$

$$+2\max\{\delta_C(L_1 + 1), \delta_D(L_2 + 1)\}$$

$$+(T + 1)(\sum_{t=0}^{T} a_t)^{-1} \max\{4\delta_C(M_1 + 1), 4\delta_D(M_2 + 1)\}$$

$$+2^{-1}(\sum_{t=0}^{T} a_t)^{-1} \sum_{t=0}^{T} a_t^2 \max\{(L_1 + 1)^2, (L_2 + 1)^2\}$$

$$+2^{-1}(\sum_{t=0}^{T} a_t)^{-1}(\max\{2M_1, 2M_2\} + 1)^2,$$

$$|f(\widehat{x}_T, \widehat{y}_T) - (\sum_{t=0}^{T} a_t)^{-1} \sum_{t=0}^{T} a_t f(x_t, y_t)|$$

$$\leq \max\{\delta_{f,1}(2M_1 + 1), \delta_{f,2}(2M_2 + 1)\}$$

$$+2\max\{\delta_C(L_1 + 1), \delta_D(L_2 + 1)\}$$

$$+(T+1)(\sum_{t=0}^{T} a_t)^{-1} \max\{4\delta_C(M_1+1), 4\delta_D(M_2+1)\}$$

$$+2^{-1}(\sum_{t=0}^{T} a_t)^{-1} \sum_{t=0}^{T} a_t^2 \max\{(L_1+1)^2, (L_2+1)^2\}$$

$$+2^{-1}(\sum_{t=0}^{T} a_t)^{-1}(\max\{2M_1, 2M_2\}+1)^2,$$

and for each natural number T, each $z \in C$, and each $u \in D$,

$$f(z, \widehat{y}_T) \geq f(\widehat{x}_T, \widehat{y}_T)$$

$$-2\max\{\delta_{f,1}(2M_1+1), \delta_{f,2}(2M_2+1)\}$$

$$-3\max\{\delta_C(L_1+1), \delta_D(L_2+1)\}$$

$$-2(T+1)(\sum_{t=0}^{T} a_t)^{-1} \max\{4\delta_C(M_1+1), 4\delta_D(M_2+1)\}$$

$$-2(\sum_{t=0}^{T} a_t)^{-1} \sum_{t=0}^{T} a_t^2 \max\{(L_1+1)^2, (L_2+1)^2\}$$

$$-(\sum_{t=0}^{T} a_t)^{-1}(\max\{2M_1, 2M_2\}+1)^2,$$

$$f(\widehat{x}_T, u) \leq f(\widehat{x}_T, \widehat{y}_T)$$

$$+2\max\{\delta_{f,1}(2M_1+1), \delta_{f,2}(2M_2+1)\}$$

$$+3\max\{\delta_C(L_1+1), \delta_D(L_2+1)\}$$

$$+2(T+1)(\sum_{t=0}^{T} a_t)^{-1} \max\{4\delta_C(M_1+1), 4\delta_D(M_2+1)\}$$

$$+2(\sum_{t=0}^{T} a_t)^{-1} \sum_{t=0}^{T} a_t^2 \max\{(L_1+1)^2, (L_2+1)^2\}$$

$$+(\sum_{t=0}^{T} a_t)^{-1}(\max\{2M_1, 2M_2\}+1)^2.$$

Proof Evidently, (3.149) holds. It is not difficult to see that

$$\|x_t\| \le M_1 + 1,\ \|y_t\| \le M_2 + 1,\ t = 0, 1, \ldots.$$

Let $t \ge 0$ be an integer. Applying Lemma 3.2 with

$$a = a_t,\ x = x_t,\ f = f(\cdot, y_t),\ \xi = \xi_t,\ u = x_{t+1}$$

we obtain that for each $z \in C$,

$$a_t(f(x_t, y_t) - f(z, y_t))$$

$$\le 2^{-1}\|z - x_t\|^2 - 2^{-1}\|z - x_{t+1}\|^2$$

$$+a_t\delta_{f,1}(2M_1 + 1) + a_t\delta_C(L_1 + 1)$$

$$+4\delta_C(M_1 + 1) + 2^{-1}a_t^2(L_1 + 1)^2.$$

Applying Lemma 3.2 with

$$a = a_t,\ x = y_t,\ f = -f(x_t, \cdot),\ \xi = -\eta_t,\ u = y_{t+1}$$

we obtain that for each $v \in D$,

$$a_t(f(x_t, v) - f(x_t, y_t))$$

$$\le 2^{-1}\|v - y_t\|^2 - 2^{-1}\|v - y_{t+1}\|^2$$

$$+a_t\delta_{f,2}(2M_2 + 1) + a_t\delta_D(L_2 + 1)$$

$$+4\delta_D(M_2 + 1) + 2^{-1}a_t^2(L_2 + 1)^2.$$

For all integers $t \ge 0$ set

$$b_{t,1} = a_t\delta_{f,1}(2M_1 + 1) + a_t\delta_C(L_1 + 1)$$

$$+4\delta_C(M_1 + 1) + 2^{-1}a_t^2(L_1 + 1)^2,$$

$$b_{t,2} = a_t\delta_{f,2}(2M_2 + 1) + a_t\delta_D(L_2 + 1)$$

$$+4\delta_D(M_2 + 1) + 2^{-1}a_t^2(L_2 + 1)^2$$

and define

$$\phi(s) = 2^{-1}s^2, \ s \in R^1.$$

It is easy to see that all the assumptions of Proposition 2.13 hold and it implies Theorem 3.6. □

We are interested in the optimal choice of a_t, $t = 0, 1, \ldots$. Let T be a natural number and $A_T = \sum_{t=0}^{T} a_t$ be given. By Theorem 3.6, in order to make the best choice of a_t, $t = 0, \ldots, T$, we need to minimize the function $\sum_{t=0}^{T} a_t^2$ on the set

$$\{a = (a_0, \ldots, a_T) \in R^{T+1} : a_i \geq 0, \ i = 0, \ldots, T, \ \sum_{i=0}^{T} a_i = A_T\}.$$

By Lemma 2.3, this function has a unique minimizer $a^* = (a_0^*, \ldots, a_T^*)$ where $a_i^* = (T+1)^{-1}A_T$, $i = 0, \ldots, T$ which is the best choice of a_t, $t = 0, 1, \ldots, T$.

Let T be a natural number and $a_t = a$ for all $t = 0, \ldots, T$. Now we will find the best $a > 0$. Since T can be arbitrarily large we need to choice a which is a minimizer of the function

$$\Psi(a) = a^{-1}\max\{4\delta_C(M_1+1), \ 4\delta_D(M_2+1)\} + a\max\{(L_1+1)^2, \ (L_2+1)^2\}.$$

This function has a minimizer

$$a = (\max\{4\delta_C(M_1+1), \ 4\delta_D(M_2+1)\})^{1/2}\max\{(L_1+1)^2, \ (L_2+1)^2\}^{-1}.$$

Now our goal is to find the best $T > 0$. Since in view of the inequalities above, in order to make the best choice of T, it should be at the same order as $\max\{\delta_C, \delta_D\}^{-1}$. For example,

$$T = \lfloor \max\{\delta_C, \delta_D\}^{-1} \rfloor.$$

In this case, we obtain a pair of points $\widehat{x} \in U$, $\widehat{y} \in V$ such that

$$B_X(\widehat{x}, \delta_C) \cap C \neq \emptyset,$$

$$B_Y(\widehat{y}, \delta_C) \cap D \neq \emptyset$$

and for each $z \in C$ and each $v \in D$,

$$f(z, \widehat{y}) \geq f(\widehat{x}, \widehat{y})$$

$$-c\max\{\delta_C, \delta_D\}^{1/2} - 2\max\{\delta_{f,1}, \delta_{f,2}\}(2\max\{M_1, M_2\}+1)$$

and

$$f(\widehat{x}, v) \le f(\widehat{x}, \widehat{y})$$

$$+c \max\{\delta_C, \delta_D\}^{1/2} + 2 \max\{\delta_{f,1}, \delta_{f,2}\}(2 \max\{M_1, M_2\} + 1)$$

where the constant $c > 0$ depends only on L_1, L_2 and M_1, M_2.

Theorem 3.6 is a generalization of Theorem 3.4 of [92] obtained in the case when

$$\delta_{f,1} = \delta_{f,2} = \delta_C = \delta_D.$$

3.9 Zero-Sum Games on Unbounded Sets

Let $(X, \langle \cdot, \cdot \rangle)$, $(Y, \langle \cdot, \cdot \rangle)$ be Hilbert spaces equipped with the complete norms $\| \cdot \|$ which are induced by their inner products. In this section we prove an extension of Theorem 3.6 for games on unbounded sets. This result has no prototype in [92], where only games on bounded sets are studied.

Let A be a nonempty closed convex subset of X, D be a nonempty closed convex subset of Y, W be an open convex subset of X, and V be an open convex subset of Y such that

$$A \subset W, \; D \subset V$$

and let a function $f : W \times V \to R^1$ possess the following properties:

(i) for each $v \in V$, the function $f(\cdot, v) : W \to R^1$ is convex and continuous;
(ii) for each $w \in W$, the function $f(w, \cdot) : V \to R^1$ is concave and continuous.

Suppose that

$$M_0, M_1 > 0, \; M_2 \ge M_0 \tag{3.150}$$

$$\theta_0 \in A \cap B_X(0, M_0), \tag{3.151}$$

$$D \subset B_Y(0, M_0), \tag{3.152}$$

$$L_1 \ge 1, \; L_2 > 0 \tag{3.153}$$

$$\sup\{f(\theta_0, y) : \; y \in V \cap B_Y(0, M_0 + 1)\} < M_1, \tag{3.154}$$

for each $y \in V \cap B_Y(0, M_0 + 1)$,

$$\{x \in W : \; f(x, y) \le M_1 + 4\} \subset B_X(0, M_2), \tag{3.155}$$

for each $v \in V \cap B_Y(0, M_0 + 4)$,

$$|f(u_1, v) - f(u_2, v)| \le L_1 \|u_1 - u_2\|$$

$$\text{for all } u_1, u_2 \in W \cap B_X(0, 10M_2 + 12), \tag{3.156}$$

and that for each $u \in W \cap B_X(0, 10M_2 + 12)$,

$$|f(u, v_1) - f(u, v_2)| \le L_2\|v_1 - v_2\|$$

$$\text{for all } v_1, v_2 \in V \cap B_Y(0, M_0 + 2). \tag{3.157}$$

Let

$$x_* \in A \text{ and } y_* \in D \tag{3.158}$$

satisfy

$$f(x_*, y) \le f(x_*, y_*) \le f(x, y_*) \tag{3.159}$$

for each $x \in A$ and each $y \in D$. In view of (3.151), (3.152), (3.154), (3.155), (3.158), and (3.159),

$$f(x_*, y_*) \le f(\theta_0, y_*) < M_1, \tag{3.160}$$

$$\|x_*\| \le M_2, \ \|\theta_0\| \le M_2. \tag{3.161}$$

In this section we prove the following result which does not have an analog in [92].

Theorem 3.7 *Let $\delta_{f,1}, \delta_{f,2}, \delta_A, \delta_D \in (0, 1/2)$ and $\{a_k\}_{k=0}^{\infty} \subset (0, \infty)$ satisfy*

$$\delta_A \le (36M_2 + 44)^{-1}(L_1 + 1)^{-2}, \tag{3.162}$$

$$\delta_{f,1} \le (18M_2 + 21)^{-1}, \tag{3.163}$$

$$\delta_A(36M_2 + 44) \le a_t \le (L_1 + 1)^{-2} \text{ for all integers } t \ge 0. \tag{3.164}$$

Assume that $\{x_t\}_{t=0}^{\infty} \subset W$, $\{y_t\}_{t=0}^{\infty} \subset V$, $\{\xi_t\}_{t=0}^{\infty} \subset X$, $\{\eta_t\}_{t=0}^{\infty} \subset Y$,

$$B_X(x_0, \delta_A) \cap A \ne \emptyset, \ B_Y(y_0, \delta_D) \cap D \ne \emptyset, \tag{3.165}$$

$$\|x_0\| \le M_2 + 1 \tag{3.166}$$

and that for each integer $t \ge 0$,

$$\xi_t \in \partial_x f(x_t, y_t) + B_X(0, \delta_{f,1}), \tag{3.167}$$

$$\eta_t \in \partial_y f(x_t, y_t) + B_Y(0, \delta_{f,2}), \tag{3.168}$$

$$B_X(x_{t+1}, \delta_A) \cap argmin\{\langle \xi_t, v \rangle + (2a_t)^{-1}\|v - x_t\|^2 : v \in A\} \neq \emptyset, \tag{3.169}$$

$$B_Y(y_{t+1}, \delta_D) \cap argmin\{\langle -\eta_t, u \rangle + (2a_t)^{-1}\|u - y_t\|^2 : u \in D\} \neq \emptyset. \tag{3.170}$$

Let for each natural number T,

$$\widehat{x}_T = (\sum_{i=0}^{T} a_t)^{-1} \sum_{t=0}^{T} a_t x_t, \ \widehat{y}_T = (\sum_{i=0}^{T} a_t)^{-1} \sum_{t=0}^{T} a_t y_t. \tag{3.171}$$

Then for all integers $t \geq 0$,

$$\|x_t\| \leq 5M_2 + 4 \tag{3.172}$$

and for each natural number T,

$$B_X(\widehat{x}_T, \delta_A) \cap A \neq \emptyset, \ B_Y(\widehat{y}_T, \delta_D) \cap D \neq \emptyset, \tag{3.173}$$

$$|(\sum_{t=0}^{T} a_t)^{-1} \sum_{t=0}^{T} a_t f(x_t, y_t) - f(x_*, y_*)|$$

$$\leq \max\{\delta_{f,1}(18M_2 + 21), \delta_{f,2}(2M_0 + 1)\}$$

$$+2\max\{\delta_A(L_1 + 1), \delta_D(L_2 + 1)\}$$

$$+(T + 1)(\sum_{t=0}^{T} a_t)^{-1} \max\{4\delta_A(9M_2 + 11), 4\delta_D(M_0 + 18)\}$$

$$+2^{-1}(\sum_{t=0}^{T} a_t)^{-1} \sum_{t=0}^{T} a_t^2 \max\{(L_1 + 1)^2, (L_2 + 1)^2\}$$

$$+ 2^{-1}(\sum_{t=0}^{T} a_t)^{-1}(10M_2 + 10)^2, \tag{3.174}$$

$$|f(\widehat{x}_T, \widehat{y}_T) - (\sum_{t=0}^{T} a_t)^{-1} \sum_{t=0}^{T} a_t f(x_t, y_t)|$$

$$\leq \max\{\delta_{f,1}(18M_2 + 21), \delta_{f,2}(2M_0 + 1)\}$$

$$+2\max\{\delta_A(L_1 + 1), \delta_D(L_2 + 1)\}$$

$$+(T+1)(\sum_{t=0}^{T} a_t)^{-1} \max\{4\delta_A(9M_2+11), 4\delta_D(M_0+18)\}$$

$$+2^{-1}(\sum_{t=0}^{T} a_t)^{-1} \sum_{t=0}^{T} a_t^2 \max\{(L_1+1)^2, (L_2+1)^2\}$$

$$+2^{-1}(\sum_{t=0}^{T} a_t)^{-1}(10M_2+10)^2 \tag{3.175}$$

and for each natural number T, each $z \in A$, and each $u \in D$,

$$f(z, \widehat{y}_T) \geq f(\widehat{x}_T, \widehat{y}_T)$$

$$-2\max\{\delta_{f,1}(18M_2+21), \delta_{f,2}(2M_0+1)\}$$

$$-3\max\{\delta_A(L_1+1), \delta_D(L_2+1)\}$$

$$-2(T+1)(\sum_{t=0}^{T} a_t)^{-1} \max\{4\delta_A(9M_2+11), 4\delta_D(M_0+18)\}$$

$$-(\sum_{t=0}^{T} a_t)^{-1} \sum_{t=0}^{T} a_t^2 \max\{(L_1+1)^2, (L_2+1)^2\}$$

$$-(\sum_{t=0}^{T} a_t)^{-1}(10M_2+10)^2, \tag{3.176}$$

$$f(\widehat{x}_T, u) \leq f(\widehat{x}_T, \widehat{y}_T)$$

$$+2\max\{\delta_{f,1}(18M_2+21), \delta_{f,2}(2M_0+1)\}$$

$$+3\max\{\delta_A(L_1+1), \delta_D(L_2+1)\}$$

$$+2(T+1)(\sum_{t=0}^{T} a_t)^{-1} \max\{4\delta_A(9M_2+11), 4\delta_D(M_0+18)\}$$

$$+(\sum_{t=0}^{T} a_t)^{-1} \sum_{t=0}^{T} a_t^2 \max\{(L_1+1)^2, (L_2+1)^2\}$$

$$+(\sum_{t=0}^{T} a_t)^{-1}(10M_2+10)^2. \tag{3.177}$$

Proof By (3.152) and (3.165), for all integers $t \geq 0$,

$$\|y_t\| \leq M_0 + 1 \leq M_2 + 1. \tag{3.178}$$

Set

$$C = A \cap B_X(0, 9M_2 + 10) \tag{3.179}$$

and

$$U = W \cap \{x \in X : \|x\| < 10M_2 + 12\}. \tag{3.180}$$

In view of (3.151) and (3.166),

$$\|x_0 - \theta_0\| \leq \|x_0\| + \|\theta_0\| \leq 2M_2 + 1. \tag{3.181}$$

Assume that $t \geq 0$ is an integer and that

$$\|x_t - \theta_0\| \leq 4M_2 + 4. \tag{3.182}$$

It follows from (3.151), (3.180), and (3.182) that

$$\|x_t\| \leq 5M_2 + 4, \tag{3.183}$$

$$x_t \in U \cap B_X(0, 5M_2 + 4). \tag{3.184}$$

By (3.156), (3.167), (3.178), and (3.183),

$$\xi_t \in \partial f(x_t, y_t) + B_X(0, 1) \subset B_X(0, L_1 + 1). \tag{3.185}$$

In view of (3.169), there exists

$$h \in B_X(x_{t+1}, \delta_A) \cap \mathrm{argmin}\{\langle \xi_t, w\rangle + (2a_t)^{-1}\|w - x_t\|^2 : w \in A\}. \tag{3.186}$$

By (3.151) and (3.186),

$$\langle \xi_t, h\rangle + (2a_t)^{-1}\|h - x_t\|^2$$
$$\leq \langle \xi_t, \theta_0\rangle + (2a_t)^{-1}\|\theta_0 - x_t\|^2. \tag{3.187}$$

In view of (3.187),

$$\langle \xi_t, \theta_0 \rangle + (2a_t)^{-1}\|\theta_0 - x_t\|^2$$

$$\geq (2a_t)^{-1}\|h - x_t\|^2 + \langle \xi_t, h - x_t \rangle + \langle \xi_t, x_t \rangle. \tag{3.188}$$

It follows from (3.162), (3.185), and (3.188) that

$$(2a_t)^{-1}(\|h - x_t\| - 1)^2 - (2a_t)^{-1}$$

$$\leq (2a_t)^{-1}(\|h - x_t\|^2 - \|h - x_t\|)$$

$$\leq (2a_t)^{-1}\|h - x_t\|^2 - (L_1 + 1)\|h - x_t\|$$

$$\leq (2a_t)^{-1}\|h - x_t\|^2 + \langle \xi_t, h - x_t \rangle$$

$$\leq (2a_t)^{-1}\|h - x_t\|^2 + \langle \xi_t, h - x_t \rangle + \langle \xi_t, x_t \rangle + \|\xi_t\|\|x_t\|. \tag{3.189}$$

By (3.161), (3.182), (3.184), (3.185), (3.188), and (3.189),

$$(2a_t)^{-1}(\|h - x_t\| - 1)^2 - (2a_t)^{-1}$$

$$\leq (2a_t)^{-1}\|h - x_t\|^2 + \langle \xi_t, h - x_t \rangle + \langle \xi_t, x_t \rangle + (L_1 + 1)(5M_2 + 4)$$

$$\leq (2a_t)^{-1}\|\theta_0 - x_t\|^2 + \langle \xi_t, \theta_0 \rangle + (L_1 + 1)(5M_2 + 4)$$

$$\leq (L_1 + 1)M_2 + (2a_t)^{-1}(4M_2 + 4)^2 + (L_1 + 1)(5M_2 + 4). \tag{3.190}$$

In view of (3.164) and (3.190),

$$(\|h - x_t\| - 1)^2$$

$$\leq 1 + (2a_t)(L_1 + 1)(6M_2 + 4) + (4M_2 + 4)^2$$

$$\leq (4M_2 + 4)^2 + 6M_2 + 5$$

and

$$\|h - x_t\| \leq 4M_2 + 6. \tag{3.191}$$

It follows from (3.183) and (3.191) that

$$\|h\| \leq 9M_2 + 10. \tag{3.192}$$

By (3.179), (3.186), and (3.192),

$$h \in C. \tag{3.193}$$

Relations (3.179), (3.186), and (3.193) imply that

$$h \in \operatorname{argmin}\{\langle \xi_t, w\rangle + (2a_t)^{-1}\|w - x_t\|^2 : \ w \in C\}. \tag{3.194}$$

By (3.180) and (3.192),

$$x_{t+1} \in U. \tag{3.195}$$

It follows from (3.158), (3.167), (3.179), (3.180), (3.184), (3.186), (3.194), (3.195), and Lemma 3.2 applied with

$$M_0 = 9M_2 + 10, \ f = f(\cdot, y_t), \ a = a_t, \ x = x_t, \ \xi = \xi_t, \ u = x_{t+1}$$

that for each $z \in C$,

$$\begin{aligned} & a_t(f(x_t, y_t) - f(z, y_t)) \\ & \le 2^{-1}\|z - x_t\|^2 - 2^{-1}\|z - x_{t+1}\|^2 \\ & + a_t\delta_{f,1}(18M_2 + 21) + a_t\delta_A(L_1 + 1) \\ & + 4\delta_A(9M_2 + 11) + 2^{-1}a_t^2(L_1 + 1)^2. \end{aligned} \tag{3.196}$$

By (3.157), (3.168), (3.170), (3.178)–(3.180), and Lemma 3.2 applied with

$$a = a_t, \ x = y_t, \ f = -f(x_t, \cdot), \ \xi = -\eta_t, \ u = y_{t+1}$$

that for each $v \in D$,

$$\begin{aligned} & a_t(f(x_t, v) - f(x_t, y_t)) \\ & \le 2^{-1}\|v - y_t\|^2 - 2^{-1}\|v - y_{t+1}\|^2 \\ & + a_t\delta_{f,2}(2M_0 + 1) + a_t\delta_D(L_2 + 1) \\ & + 4\delta_D(M_0 + 1) + 2^{-1}a_t^2(L_2 + 1)^2. \end{aligned} \tag{3.197}$$

There are two cases:

$$\|\theta_0 - x_{t+1}\| \le \|\theta_0 - x_t\|; \tag{3.198}$$

$$\|\theta_0 - x_{t+1}\| > \|\theta_0 - x_t\|. \tag{3.199}$$

If (3.198) holds, then in view of (3.182),

$$\|\theta_0 - x_{t+1}\| \le 4M_2 + 4. \tag{3.200}$$

Assume that (3.199) is true. It follows from (3.151), (3.154), (3.163)–(3.165), (3.178), (3.179), and (3.196) with $z = \theta_0$ that

$$\begin{aligned} f(x_t, y_t) &\le f(\theta_0, y_t) \\ &+\delta_{f,1}(18M_2 + 21) + \delta_A(L_1 + 1) \\ &+ 4a_t^{-1}\delta_A(9M_2 + 11) + 1 \le M_1 + 4. \end{aligned} \tag{3.201}$$

By (3.155), (3.178), and (3.201),

$$\|x_t\| \le M_2. \tag{3.202}$$

It follows from (3.151), (3.185), (3.188), (3.189), and (3.202) that

$$\begin{aligned} &(2a_t)^{-1}(\|h - x_t\| - 1)^2 - (2a_t)^{-1} \\ &\le (2a_t)^{-1}\|h - x_t\|^2 + \langle \xi_t, h - x_t\rangle + \langle \xi_t, x_t\rangle + (L_1 + 1)M_2 \\ &\le (2a_t)^{-1}\|\theta_0 - x_t\|^2 + \langle \xi_t, \theta_0\rangle + (L_1 + 1)M_2 \\ &\le (L_1 + 1)M_0 + (2a_t)^{-1}(2M_2 + 1)^2 + (L_1 + 1)M_2. \end{aligned} \tag{3.203}$$

In view of (3.164), (3.202), and (3.203),

$$\begin{aligned} (\|h - x_t\| - 1)^2 &\le 1 + (M_0 + M_2) + (2M_2 + 1)^2 \\ &\le 4M_2^2 + 6M_2 + 2 \le 4(M_2 + 1)^2, \end{aligned}$$

and

$$\|h - x_t\| \le 1 + 2M_2 + 2, \ \|h\| \le 3M_2 + 3. \tag{3.204}$$

By (3.186) and (3.204),

$$\|x_{t+1}\| \le 3M_2 + 4. \tag{3.205}$$

In view of (3.151) and (3.205),

$$\|x_{t+1} - \theta_0\| \le 4M_2 + 4. \tag{3.206}$$

Thus we have shown by induction that for all integers $t \geq 0$, (3.206) holds, (3.196) is true for all

$$z \in C = A \cap B_X(0, 9M_2 + 10),$$

and (3.197) holds for all $v \in D$. It follows from (3.155) and (3.206) that (3.172) is true for all integers $t \geq 0$. For all integers $t \geq 0$ set

$$b_{t,1} = a_t \delta_{f,1}(18M_2 + 21) + a_t \delta_A (L_1 + 1)$$

$$+4\delta_A(9M_2 + 11) + 2^{-1}a_t^2(L_2 + 1)^2,$$

$$b_{t,2} = a_t \delta_{f,2}(2M_0 + 1) + a_t \delta_D (L_2 + 1)$$

$$+4\delta_D(M_0 + 1) + 2^{-1}a_t^2(L_2 + 1)^2$$

and define

$$\phi(s) = 2^{-1}s^2, \ s \in R^1.$$

It is easy to see that all the assumptions of Proposition 2.13 hold and it implies (3.173)–(3.175) for each natural number T and (3.176) and (3.177) for each natural number T, each $u \in D$, and each $z \in A \cap B_X(9M_2 + 10)$. By (3.154), (3.155), and (3.178), for each $z \in A \setminus B_X(9M_2 + 10)$,

$$f(z, \widehat{y}_T) > M_1 + 4 > 4 + f(\theta_0, \widehat{y}_T)$$

and (3.176) holds for all natural numbers T and all $z \in A$. Theorem 3.7 is proved. □

We are interested in the optimal choice of a_t, $t = 0, 1, \ldots$. Let T be a natural number and $A_T = \sum_{t=0}^{T} a_t$ be given. By Theorem 3.7, in order to make the best choice of a_t, $t = 0, \ldots, T$, we need to minimize the function $\sum_{t=0}^{T} a_t^2$ on the set

$$\{a = (a_0, \ldots, a_T) \in R^{T+1} : \ a_i \geq 0, \ i = 0, \ldots, T, \ \sum_{i=0}^{T} a_i = A_T\}.$$

By Lemma 2.3, this function has a unique minimizer $a^* = (a_0^*, \ldots, a_T^*)$ where $a_i^* = (T + 1)^{-1}A_T$, $i = 0, \ldots, T$ which is the best choice of a_t, $t = 0, 1, \ldots, T$.

Let T be a natural number and $a_t = a$ for all $t = 0, \ldots, T$. Now we will find the best $a > 0$. Since T can be arbitrarily large we need to choice a which is a minimizer of the function

$$\Psi(a) = 2a^{-1} \max\{4\delta_A(9M_2 + 1),\ 4\delta_D(M_0 + 18)\}$$

$$+a \max\{(L_1 + 1)^2,\ (L_2 + 1)^2\},\ a > 0.$$

This function has a minimizer

$$a = (2 \max\{4\delta_A(9M_2 + 1),\ 4\delta_D(M_0 + 18)\})^{1/2} \max\{(L_1 + 1),\ (L_2 + 1)\}^{-1}$$

and the minimal value of Ψ is

$$(8 \max\{4\delta_A(9M_2 + 1),\ 4\delta_D(M_0 + 18)\})^{1/2} \max\{(L_1 + 1),\ (L_2 + 1)\}.$$

Now our goal is to find the best $T > 0$. In view of the inequalities above, in order to make the best choice of T, it should be at the same order as $\lfloor \max\{\delta_A, \delta_D\}^{-1} \rfloor$. For example,

$$T = \lfloor \max\{\delta_A, \delta_D\}^{-1} \rfloor + 1.$$

In this case, we obtain a pair of points $\widehat{x} \in U, \widehat{y} \in V$ such that

$$B_X(\widehat{x}, \delta_A) \cap A \neq \emptyset,$$

$$B_Y(\widehat{y}, \delta_D) \cap D \neq \emptyset$$

and for each $z \in A$ and each $v \in D$,

$$f(z, \widehat{y}) \geq f(\widehat{x}, \widehat{y})$$

$$-c \max\{\delta_A, \delta_D\}^{1/2} - 2 \max\{\delta_{f,1}, \delta_{f,2}\} \max\{18M_2 + 21, 2M_0 + 1\}$$

and

$$f(\widehat{x}, v) \leq f(\widehat{x}, \widehat{y})$$

$$+c \max\{\delta_A, \delta_D\}^{1/2} + 2 \max\{\delta_{f,1}, \delta_{f,2}\} \max\{18M_2 + 21, 2M_0 + 1\}$$

where the constant $c > 0$ depends only on L_1, L_2 and M_0, M_1, M_2.

Chapter 4
Gradient Algorithm with a Smooth Objective Function

In this chapter we analyze the convergence of a projected gradient algorithm with a smooth objective function under the presence of computational errors. The problem is described by an objective function and a set of feasible points. For this algorithm each iteration consists of two steps. The first step is a calculation of a gradient of the objective function while in the second one we calculate a projection on the feasible set. In each of these two steps there is a computational error. In general, these two computational errors are different. We show that our algorithm generates a good approximate solution, if all the computational errors are bounded from above by a small positive constant. Moreover, if we know the computational errors for the two steps of our algorithm, we find out what approximate solution can be obtained and how many iterates one needs for this.

4.1 Optimization on Bounded Sets

Let X be a Hilbert space equipped with an inner product $\langle \cdot, \cdot \rangle$ which induces a complete norm $\| \cdot \|$. Let C be a nonempty closed convex subset of X, U be an open convex subset of X such that $C \subset U$, and $f : U \to R^1$ be a convex continuous function.

We suppose that the function f is Fréchet differentiable at every point $x \in U$ and for every $x \in U$ we denote by $f'(x) \in X$ the Fréchet derivative of f at x. It is clear that for any $x \in U$ and any $h \in X$

$$\langle f'(x), h \rangle = \lim_{t \to 0} t^{-1}(f(x + th) - f(x)). \tag{4.1}$$

Recall that for each nonempty set D and each function $g : D \to R^1$,

$$\inf(g, D) = \inf\{g(y) : y \in D\}, \tag{4.2}$$

A. J. Zaslavski, *Convex Optimization with Computational Errors*, Springer Optimization and Its Applications 155, https://doi.org/10.1007/978-3-030-37822-6_4

$$\text{argmin}(g, D) = \text{argmin}\{g(z) : z \in D\}$$

$$= \{z \in D : g(z) = \inf(g, D)\}.$$

We suppose that the mapping $f' : U \to X$ is Lipschitz on all bounded subsets of U. It is well known (see Lemma 2.2) that for each nonempty closed convex set $D \subset X$ and each $x \in X$ there exists a unique point $P_D(x) \in D$ such that

$$\|x - P_D(x)\| = \inf\{\|x - y\| : y \in D\}.$$

In this chapter we study the behavior of a projected gradient algorithm with a smooth objective function which is used for solving convex constrained minimization problems [67, 68, 72].

Suppose that there exist $L > 1$, $M_0 > 0$ such that

$$C \subset B_X(0, M_0), \tag{4.2}$$

$$|f(v_1) - f(v_2)| \le L\|v_1 - v_2\| \text{ for all } v_1, v_2 \in U, \tag{4.3}$$

$$\|f'(v_1) - f'(v_2)\| \le L\|v_1 - v_2\| \text{ for all } v_1, v_2 \in U. \tag{4.4}$$

Let $\delta_f, \delta_C \in (0, 1]$. We describe below our algorithm.

Gradient algorithm
Initialization: select an arbitrary $x_0 \in U \cap B_X(0, M_0)$.
Iterative Step: given a current iteration vector $x_t \in U$ calculate

$$\xi_t \in f'(x_t) + B_X(0, \delta_f)$$

and calculate the next iteration vector $x_{t+1} \in U$ such that

$$\|x_{t+1} - P_C(x_t - L^{-1}\xi_t)\| \le \delta_C.$$

In this chapter we prove the following result.

Theorem 4.1 *Let $\delta_f, \delta_C \in (0, 1]$ and let*

$$x_0 \in U \cap B_X(0, M_0). \tag{4.5}$$

Assume that $\{x_t\}_{t=1}^{\infty} \subset U$, $\{\xi_t\}_{t=0}^{\infty} \subset X$ and that for each integer $t \ge 0$,

$$\|\xi_t - f'(x_t)\| \le \delta \tag{4.6}$$

and

$$\|x_{t+1} - P_C(x_t - L^{-1}\xi_t)\| \le \delta_C. \tag{4.7}$$

Then for each natural number T,

$$\min\{f(x_t) : t = 2, \ldots, T+1\} - \inf(f, C),$$

$$f(\sum_{t=2}^{T+1} T^{-1} x_t) - \inf(f, C)$$

$$\leq (2T)^{-1} L(2M_0+1)^2 + L\delta_C(6M_0+7) + \delta_f(4M_0+4). \tag{4.8}$$

Theorem 4.1 is a generalization of Theorem 4.2 of [92] proved in the case when $\delta_f = \delta_C$.

We are interested in an optimal choice of T. If we choose T in order to minimize the right-hand side of (4.8) we obtain that T should be at the same order as $\max\{\delta_f, \delta_C\}^{-1}$. In this case the right-hand side of (4.8) is at the same order as $\max\{\delta_f, \delta_C\}$. For example, if $T = \lfloor \max\{\delta_f, \delta_C\}^{-1} \rfloor + 1$, then the right-hand side of (4.8) does not exceed

$$2^{-1}(2M_0+1)^2 L \max\{\delta_f, \delta_C\} + \delta_C(6M_0+7)L + (4M_0+5)\delta_f.$$

4.2 Auxiliary Results

Proposition 4.2 *Let D be a nonempty closed convex subset of X, $x \in X$, and $y \in D$. Assume that for each $z \in D$,*

$$\langle z - y, x - y \rangle \leq 0. \tag{4.9}$$

Then $y = P_D(x)$.

Proof Let $z \in D$. By (4.9),

$$\langle z - x, z - x \rangle = \langle z - y + (y - x), z - y + (y - x) \rangle$$

$$= \langle y - x, y - x \rangle + 2\langle z - y, y - x \rangle + \langle z - y, z - y \rangle$$

$$\geq \langle y - x, y - x \rangle + \langle z - y, z - y \rangle$$

$$= \|y - x\|^2 + \|z - y\|^2.$$

Thus $y = P_D(x)$. Proposition 4.2 is proved. □

Proposition 4.3 *Assume that $x, u \in U$, $L > 0$ and that for each $v_1, v_2 \in \{tx + (1-t)u : t \in [0, 1]\}$,*

$$\|f'(v_1) - f'(v_2)\| \leq L\|v_1 - v_2\|.$$

Then

$$f(u) \le f(x) + \langle f'(x), u - x\rangle + 2^{-1}L\|u - x\|^2.$$

Proof For each $t \in [0, 1]$ set

$$\phi(t) = f(x + t(u - x)). \tag{4.10}$$

Clearly, ϕ is a differentiable function and for each $t \in [0, 1]$,

$$\phi'(t) = \langle f'(x + t(u - x)), u - x\rangle. \tag{4.11}$$

By (4.10), (4.11) and the proposition assumptions,

$$f(u) - f(x) = \phi(1) - \phi(0)$$

$$= \int_0^1 \phi'(t)dt = \int_0^1 \langle f'(x + t(u - x)), u - x\rangle dt$$

$$= \int_0^1 \langle f'(x), u - x\rangle dt + \int_0^1 \langle f'(x + t(u - x)) - f'(x), u - x\rangle dt$$

$$\le \langle f'(x), u - x\rangle + \int_0^1 Lt\|u - x\|^2 dt$$

$$= \langle f'(x), u - x\rangle + L\|u - x\|^2 \int_0^1 tdt$$

$$= \langle f'(x), u - x\rangle + L\|u - x\|^2/2.$$

Proposition 4.3 is proved. □

4.3 The Main Lemma

Lemma 4.4 *Let* $\delta_f, \delta_C \in (0, 1]$,

$$u \in B_X(0, M_0 + 1) \cap U, \tag{4.12}$$

$\xi \in X$ *satisfy*

$$\|\xi - f'(u)\| \le \delta_f \tag{4.13}$$

and let $v \in U$ *satisfy*

$$\|v - P_C(u - L^{-1}\xi)\| \le \delta_C. \tag{4.14}$$

Then for each $x \in U$ *satisfying*

$$B(x, \delta_C) \cap C \neq \emptyset \tag{4.15}$$

the following inequalities hold:

$$\begin{aligned} & f(x) - f(v) \\ & \ge 2^{-1}L\|x - v\|^2 - 2^{-1}L\|x - u\|^2 \\ & - \delta_C L(6M_0 + 7) - \delta_f(4M_0 + 5), \end{aligned} \tag{4.16}$$

$$\begin{aligned} & f(x) - f(v) \\ & \ge 2^{-1}L\|u - v\|^2 + L\langle v - u, u - x\rangle \\ & - \delta_C L(4M_0 + 5) - \delta_f(4M_0 + 5). \end{aligned} \tag{4.17}$$

Proof For each $x \in U$ define

$$g(x) = f(u) + \langle f'(u), x - u\rangle + 2^{-1}L\|x - u\|^2. \tag{4.18}$$

Clearly, $g : U \to R^1$ is a convex Fréchet differentiable function, for each $x \in U$,

$$g'(x) = f'(u) + L(x - u), \tag{4.19}$$

$$\lim_{\|x\| \to \infty} g(x) = \infty \tag{4.20}$$

and there exists

$$x_0 \in C \tag{4.21}$$

such that

$$g(x_0) \le g(x) \text{ for all } x \in C. \tag{4.22}$$

By (4.21) and (4.22), for all $z \in C$,

$$\langle g'(x_0), z - x_0\rangle \ge 0. \tag{4.23}$$

In view of (4.19) and (4.23),

$$\langle L^{-1} f'(u) + x_0 - u, z - x_0 \rangle \geq 0 \text{ for all } z \in C. \tag{4.24}$$

Proposition 4.2, (4.21), and (4.24) imply that

$$x_0 = P_C(u - L^{-1} f'(u)). \tag{4.25}$$

It follows from (4.13), (4.14), (4.25), and Lemma 2.2 that

$$\begin{aligned} &\|v - x_0\| \\ &\leq \|v - P_C(u - L^{-1}\xi)\| \\ &+\|P_C(u - L^{-1}\xi) - P_C(u - L^{-1} f'(u))\| \\ &\leq \delta_C + L^{-1}\|\xi - f'(u)\| \leq \delta_C + L^{-1}\delta_f. \end{aligned} \tag{4.26}$$

In view of (4.2) and (4.25),

$$\|x_0\| \leq M_0. \tag{4.27}$$

Relations (4.2) and (4.14) imply that

$$\|v\| \leq M_0 + 1. \tag{4.28}$$

By (4.4), (4.12), (4.18), and Proposition 4.3, for all $x \in U$,

$$f(x) \leq f(u) + \langle f'(u), x - u \rangle + 2^{-1} L \|u - x\|^2 = g(x). \tag{4.29}$$

Let

$$x \in U \tag{4.30}$$

satisfy

$$B(x, \delta_C) \cap C \neq \emptyset. \tag{4.31}$$

It follows from (4.3), (4.26), and (4.30) that

$$|f(x_0) - f(v)| \leq L\|v - x_0\| \leq \delta_C L + \delta_f. \tag{4.32}$$

In view of (4.21) and (4.29),

$$g(x_0) \geq f(x_0). \tag{4.33}$$

By (4.18), (4.33), and convexity of f,

$$f(x) - f(x_0) \geq f(x) - g(x_0)$$

$$= f(x) - f(u) - \langle f'(u), x_0 - u \rangle - 2^{-1}L\|u - x_0\|^2$$

$$\geq f(u) + \langle f'(u), x - u \rangle$$

$$-f(u) - \langle f'(u), x_0 - u \rangle - 2^{-1}L\|u - x_0\|^2$$

$$= \langle f'(u), x - x_0 \rangle - 2^{-1}L\|u - x_0\|^2. \tag{4.34}$$

Relation (4.31) implies that there exists

$$x_1 \in C \tag{4.35}$$

such that

$$\|x_1 - x\| \leq \delta_C. \tag{4.36}$$

By (4.3), (4.30), (4.35), and (4.36),

$$|f(x_1) - f(x)| \leq L\delta_C. \tag{4.37}$$

It follows from (4.23) (with $z = x_1$) and (4.35) that

$$0 \leq \langle g'(x_0), x_1 - x_0 \rangle$$

$$= \langle g'(x_0), x_1 - x \rangle + \langle g'(x_0), x - x_0 \rangle. \tag{4.38}$$

By (4.3), (4.12), (4.20), (4.27), (4.35), and (4.36),

$$\langle g'(x_0), x_1 - x \rangle$$

$$= \langle f'(u) + L(x_0 - u), x_1 - x \rangle$$

$$= \langle f'(u), x_1 - x \rangle + L\langle x_0 - u, x_1 - x \rangle$$

$$\leq L\delta_C + L\delta_C(2M_0 + 1). \tag{4.39}$$

In view of (4.19) and (4.21),

$$\langle g'(x_0), x - x_0 \rangle = \langle f'(u) + L(x_0 - u), x - x_0 \rangle. \tag{4.40}$$

Relations (4.38)–(4.40) imply that

$$\begin{aligned}&\langle f'(u), x - x_0\rangle\\&= \langle g'(x_0), x - x_0\rangle - L\langle x_0 - u, x - x_0\rangle\\&\geq -\langle g'(x_0), x_1 - x\rangle - L\langle x_0 - u, x - x_0\rangle\\&\geq -L\langle x_0 - u, x - x_0\rangle - L\delta_C(2M_0 + 2).\end{aligned} \tag{4.41}$$

It follows from (4.34) and (4.41) that

$$\begin{aligned}f(x) - f(x_0) &\geq \langle f'(u), x - x_0\rangle - 2^{-1}L\|x_0 - u\|^2\\&\geq -L_C\delta(2M_0 + 2) - L\langle x_0 - u, x - x_0\rangle - 2^{-1}L\|x_0 - u\|^2.\end{aligned} \tag{4.42}$$

In view of (4.42) and Lemma 2.1,

$$\begin{aligned}&f(x) - f(x_0)\\&\geq -L\delta_C(2M_0 + 2) - 2^{-1}L\|x_0 - u\|^2\\&-2^{-1}L[\|x - u\|^2 - \|x - x_0\|^2 - \|u - x_0\|^2]\\&= 2^{-1}L\|x - x_0\|^2 - 2^{-1}L\|x - u\|^2 - L\delta_C(2M_0 + 2).\end{aligned} \tag{4.43}$$

By (4.32),

$$f(x) - f(v) \geq f(x) - f(x_0) - \delta_C L - \delta_f. \tag{4.44}$$

It follows from (4.12), (4.14), (4.25), (4.26), (4.30), and (4.31) that

$$\begin{aligned}&|\|x - x_0\|^2 - \|x - v\|^2|\\&= |\|x - x_0\| - \|x - v\||(\|x - x_0\| + \|x - v\|)\\&\leq (\delta_C + L^{-1}\delta_f)(4M_0 + 4)\end{aligned} \tag{4.45}$$

and

$$\begin{aligned}&|\|u - x_0\|^2 - \|u - v\|^2|\\&= |\|u - x_0\| - \|u - v\||(\|u - x_0\| + \|u - v\|)\\&\leq (\delta_C + L^{-1}\delta_f)(4M_0 + 4).\end{aligned} \tag{4.46}$$

In view of (4.42),

$$
\begin{aligned}
& f(x) - f(x_0) \\
& \geq -L\delta_C(2M_0+2) + 2^{-1}L\|x_0 - u\|^2 \\
& -L\langle x_0 - u, x_0 - u\rangle - L\langle x_0 - u, x - x_0\rangle \\
& \geq -L\delta_C(2M_0+2) + 2^{-1}L\|x_0 - u\|^2 \\
& - L\langle x_0 - u, x - u\rangle. \qquad (4.47)
\end{aligned}
$$

By (4.32), (4.43), and (4.45),

$$
\begin{aligned}
& f(x) - f(v) \\
& \geq f(x) - f(x_0) - \delta_C L - \delta_f \\
& \geq 2^{-1}L\|x - x_0\|^2 - 2^{-1}L\|x - u\|^2 \\
& -L\delta_C(2M_0+2) - L\delta_C - \delta_f \\
& \geq 2^{-1}L\|x - v\|^2 - 2^{-1}L\|x - u\|^2 \\
& -L(\delta_C + L^{-1}\delta_f)(4M_0+4) \\
& -L\delta_C(2M_0+2) - L\delta_C - \delta_f
\end{aligned}
$$

and (4.16) holds. It follows from (4.12), (4.26), (4.31), (4.32), (4.46), and (4.47) that

$$
\begin{aligned}
& f(x) - f(v) \\
& \geq f(x) - f(x_0) - \delta_C L - \delta_f \\
& \geq -L\delta_C(2M_0+2) + 2^{-1}L\|x_0 - u\|^2 \\
& -L\langle x_0 - u, x - u\rangle - L\delta_C - \delta_f \\
& \geq -L\delta_C - \delta_f - L(\delta_C + L^{-1}\delta_f)(2M_0+2) + 2^{-1}L\|u - v\|^2 \\
& -L\langle v - u, x - u\rangle - L(\delta_C + L^{-1}\delta_f)(2M_0+2)
\end{aligned}
$$

and (4.17) holds. Lemma 4.4 is proved. □

4.4 Proof of Theorem 4.1

Clearly, the function f has a minimizer on the set C. Fix

$$z \in C \tag{4.48}$$

such that

$$f(z) = \inf(f, C). \tag{4.49}$$

It is easy to see that

$$\|x_t\| \le M_0 + 1, \ t = 0, 1, \dots. \tag{4.50}$$

Let T be a natural number and $t \ge 0$ be an integer. Applying Lemma 4.4 with

$$u = x_t, \ \xi = \xi_t, \ v = x_{t+1}, \ x = z$$

we obtain that

$$f(z) - f(x_{t+1})$$

$$\ge 2^{-1}L\|z - x_{t+1}\|^2 - 2^{-1}L\|z - x_t\|^2$$

$$-\delta_C L(6M_0 + 7) - \delta_f(4M_0 + 5).$$

This implies that

$$Tf(z) - \sum_{t=1}^{T} f(x_{t+1})$$

$$\ge \sum_{t=1}^{T}(2^{-1}L\|z - x_{t+1}\|^2 - 2^{-1}L\|z - x_t\|^2)$$

$$-\delta_C LT(6M_0 + 7) - T\delta_f(4M_0 + 5)$$

$$= 2^{-1}L(\|z - x_{T+1}\|^2 - \|z - x_1\|^2)$$

$$-\delta_C LT(6M_0 + 7) - T\delta_f(4M_0 + 5). \tag{4.51}$$

Relations (4.2), (4.48), (4.50), and (4.51) imply that

$$T\min\{f(x_t) : \ t = 2, \dots, T + 1\} - \inf(f, C),$$

$$Tf(\sum_{t=2}^{T+1} T^{-1}x_t) - \inf(f, C)$$

$$\leq 2^{-1}(2M_0 + 1)^2 L + \delta_C LT(6M_0 + 7) + T\delta_f(4M_0 + 5).$$

Together with (4.49) this implies that

$$\min\{f(x_t) : t = 2, \ldots, T + 1\} - \inf(f, C),$$

$$f(\sum_{t=2}^{T+1} T^{-1}x_t) - \inf(f, C)$$

$$\leq 2^{-1}(2M_0 + 1)^2 T^{-1} L + \delta_C L(6M_0 + 7) + \delta_f(4M_0 + 5).$$

This completes the proof of Theorem 4.1. □

4.5 Optimization on Unbounded Sets

Let X be a Hilbert space with an inner product $\langle \cdot, \cdot \rangle$ which induces a complete norm $\| \cdot \|$.

Let D be a nonempty closed convex subset of X, V be an open convex subset of X such that

$$D \subset V,$$

and $f : V \to R^1$ be a convex Fréchet differentiable function which is Lipschitz on all bounded subsets of V. Set

$$D_{min} = \{x \in D : f(x) \leq f(y) \text{ for all } y \in D\}. \quad (4.52)$$

We suppose that

$$D_{min} \neq \emptyset. \quad (4.53)$$

We will prove the following result.

Theorem 4.5 *Let* $\delta_f, \delta_C \in (0, 1]$, $M > 0$,

$$D_{min} \cap B_X(0, M) \neq \emptyset, \tag{4.54}$$

$$M_0 = 4M + 8, \tag{4.55}$$

$L \geq 1$ *satisfy*

$$|f(v_1) - f(v_2)| \leq L\|v_1 - v_2\|$$

$$\text{for all } v_1, v_2 \in V \cap B_X(0, M_0 + 2), \tag{4.56}$$

$$\|f'(v_1) - f'(v_2)\| \leq L\|v_1 - v_2\|$$

$$\text{for all } v_1, v_2 \in V \cap B_X(0, M_0 + 2), \tag{4.57}$$

$$\epsilon_0 = \delta_f(4M_0 + 6) + \delta_C L(6M_0 + 8) \tag{4.58}$$

and let

$$n_0 = \lfloor 2^{-1}L(2M + 1)^2(\delta_f + L\delta_C)^{-1} \rfloor + 1. \tag{4.59}$$

Assume that $\{x_t\}_{t=0}^{\infty} \subset V$, $\{\xi_t\}_{t=0}^{\infty} \subset X$,

$$\|x_0\| \leq M \tag{4.60}$$

and that for each integer $t \geq 0$,

$$\|\xi_t - f'(x_t)\| \leq \delta_f \tag{4.61}$$

and

$$\|x_{t+1} - P_D(x_t - L^{-1}\xi_t)\| \leq \delta_C. \tag{4.62}$$

Then there exists an integer $q \in [1, n_0 + 1]$ *such that*

$$f(x_q) \leq \inf(f, D) + \epsilon_0,$$

$$\|x_i\| \leq 3M + 3, \ i = 0, \dots, q.$$

Proof By (4.54) there exists

$$z \in D_{min} \cap B_X(0, M). \tag{4.63}$$

By (4.56), (4.60)–(4.63), and Lemma 2.2,

$$\|x_1 - z\| \le \|x_1 - P_D(x_0 - L^{-1}\xi_0)\| + \|P_D(x_0 - L^{-1}\xi_0) - z\|$$

$$\le \delta_C + \|x_0 - z\| + L^{-1}\|\xi_0\|$$

$$\le 1 + 2M + L^{-1}(L+1) \le 2M + 3. \tag{4.64}$$

In view of (4.63) and (4.64),

$$\|x_1\| \le 3M + 3. \tag{4.65}$$

Assume that an integer $T \ge 0$ and that for all $t = 1, \dots, T+1$,

$$f(x_t) - f(z) > \epsilon_0. \tag{4.66}$$

Set

$$U = V \cap \{v \in X : \|v\| < M_0 + 2\} \tag{4.67}$$

and

$$C = D \cap B_X(0, M_0). \tag{4.68}$$

Assume that an integer $t \in [0, T]$ and that

$$\|x_t - z\| \le 2M + 3. \tag{4.69}$$

(In view of (4.60) and (4.63), our assumption is true for $t = 0$.) By (4.55), (4.63), and (4.68),

$$z \in C \subset B_X(0, M_0). \tag{4.70}$$

Relations (4.55), (4.63), (4.67), and (4.69) imply that

$$x_t \in U \cap B_X(0, M_0 + 1). \tag{4.71}$$

It follows from (4.56), (4.61), and (4.71) that

$$\xi_t \in f'(x_t) + B_X(0, 1) \subset B_X(0, L+1). \tag{4.72}$$

By (4.63), (4.69), (4.72), and Lemma 2.2,

$$\|z - P_D(x_t - L^{-1}\xi_t)\|$$

$$\le \|z - x_t + L^{-1}\xi_t\|$$

$$\leq \|z - x_t\| + L^{-1}\|\xi_t\| \leq 2M + 5. \tag{4.73}$$

In view of (4.63) and (4.73),

$$\|P_D(x_t - L^{-1}\xi_t)\| \leq 3M + 5. \tag{4.74}$$

Relations (4.55), (4.68), and (4.74) imply that

$$P_D(x_t - L^{-1}\xi_t) \in C, \tag{4.75}$$

$$P_D(x_t - L^{-1}\xi_t) = P_C(x_t - L^{-1}\xi_t). \tag{4.76}$$

It follows from (4.62), (4.67), and (4.74) that

$$\|x_{t+1}\| \leq 3M + 6, \ x_{t+1} \in U. \tag{4.77}$$

By (4.56), (4.57), (4.61)–(4.63), (4.68), (4.71), (4.77), and Lemma 4.4 applied with

$$u = x_t, \ \xi = \xi_t, \ v = x_{t+1}, \ x = z$$

we obtain that

$$f(z) - f(x_{t+1})$$

$$\geq 2^{-1}L\|z - x_{t+1}\|^2 - 2^{-1}L\|z - x_t\|^2$$

$$- L\delta_C(6M_0 + 7) - \delta_f(4M_0 + 5). \tag{4.78}$$

By (4.58), (4.66), and (4.78),

$$L\delta_C(6M_0 + 8) + \delta_f(4M_0 + 6) = \epsilon_0$$

$$< f(x_{t+1}) - f(z)$$

$$\leq 2^{-1}L(\|z - x_t\|^2 - \|z - x_{t+1}\|^2)$$

$$+ L\delta_C(6M_0 + 7) + \delta_f(4M_0 + 5). \tag{4.79}$$

In view of (4.69) and (4.79),

$$\|z - x_{t+1}\| \leq \|z - x_t\| \leq 2M + 3.$$

Thus by induction we showed that for all $t = 0, \ldots, T + 1$

$$\|z - x_t\| \le 2M + 3, \ \|x_t\| \le 3M + 3$$

and that (4.79) holds for all $t = 0, \dots, T$.

It follows from (4.79) that

$$(1 + T)(L\delta_C(6M_0 + 8) + \delta_f(4M_0 + 6))$$

$$\le (1 + T)(\min\{f(x_t) : t = 1, \dots, T + 1\} - f(z))$$

$$\le \sum_{t=0}^{T}(f(x_{t+1}) - f(z))$$

$$\le 2^{-1}L\sum_{t=0}^{T}(\|z - x_t\|^2 - \|z - x_{t+1}\|^2)$$

$$+(T + 1)(L\delta_C(6M_0 + 7) + \delta_f(4M_0 + 5)).$$

By the relation above, (4.61) and (4.63),

$$(T + 1)(\delta_f + L\delta_C) \le 2^{-1}L\|z - x_0\|^2 \le 2^{-1}L(2M + 1)^2$$

and

$$T < 2^{-1}L(2M + 1)^2(\delta_f + L\delta_C)^{-1} \le n_0.$$

Thus we assumed that an integer $T \ge 0$ satisfies

$$f(x_t) - f(z) > \epsilon_0, \ t = 1, \dots, T + 1$$

and showed that $T \le n_0 - 1$ and

$$\|x_t\| \le 3M + 3, \ t = 0, \dots, T + 1.$$

This implies that there exists a natural number $q \le n_0 + 1$ such that

$$f(x_q) - f(z) \le \epsilon_0,$$

$$\|x_t\| \le 3M + 3, \ t = 0, \dots, q.$$

Theorem 4.5 is proved. □

The next result has no prototype in [92].

Theorem 4.6 *Let $\delta_f, \delta_C \in (0, 1]$, $M > 2$ satisfy*

$$\{x \in V : f(x) \le \inf(f, D) + 3\} \subset B_X(0, M - 2), \tag{4.80}$$

$L \ge 1$ satisfy

$$|f(v_1) - f(v_2)| \le L\|v_1 - v_2\|$$

$$\text{for all } v_1, v_2 \in V \cap B_X(0, 4M + 8), \tag{4.81}$$

$$\|f'(v_1) - f'(v_2)\| \le L\|v_1 - v_2\|$$

$$\text{for all } v_1, v_2 \in V \cap B_X(0, 4M + 8), \tag{4.82}$$

$$\delta_C \le (L(18M + 19))^{-1}, \ \delta_f \le (12M + 3)^{-1}. \tag{4.83}$$

Assume that $\{x_t\}_{t=0}^{\infty} \subset V$, $\{\xi_t\}_{t=0}^{\infty} \subset X$,

$$\|x_0\| \le M \tag{4.84}$$

and that for each integer $t \ge 0$,

$$\|\xi_t - f'(x_t)\| \le \delta_f \tag{4.85}$$

and

$$\|x_{t+1} - P_D(x_t - L^{-1}\xi_t)\| \le \delta_C. \tag{4.86}$$

Then

$$\|x_t\| \le 3M \text{ for all integers } t \ge 0$$

and for each natural number T,

$$\min\{f(x_t) : t = 1, \dots, T + 1\} - \inf(f, D),$$

$$f(\sum_{t=1}^{T+1}(T + 1)^{-1}x_t) - \inf(f, D)$$

$$\le L\delta_C(18M + 19) + \delta_f(12M + 13) + 2LM^2(T + 1)^{-1}.$$

Proof Fix

$$z \in D_{min}. \tag{4.87}$$

Set

$$U = V \cap \{v \in X : \|v\| < 3M + 4\} \tag{4.88}$$

and

$$C = D \cap B_X(0, 3M + 2). \tag{4.89}$$

In view of (4.80) and (4.87),

$$\|z\| \le M - 2. \tag{4.90}$$

By (4.84) and (4.90),

$$\|x_0 - z\| \le 2M. \tag{4.91}$$

Assume that $t \ge 0$ is an integer and that

$$\|x_t - z\| \le 2M. \tag{4.92}$$

By (4.87), (4.89), and (4.90),

$$z \in C \cap B_X(0, M - 1). \tag{4.93}$$

Relations (4.88), (4.90), and (4.92) imply that

$$x_t \in U \cap B_X(0, 3M). \tag{4.94}$$

In view of (4.86),

$$\|x_{t+1} - P_D(x_t - L^{-1}\xi_t)\| \le 1. \tag{4.95}$$

It follows from (4.81), (4.85), and (4.94) that

$$\xi_t \in f'(x_t) + B_X(0, 1) \subset B_X(0, L + 1). \tag{4.96}$$

By (4.88), (4.92), (4.96), and Lemma 2.2,

$$\begin{aligned}
&\|z - P_D(x_t - L^{-1}\xi_t)\| \\
&\le \|z - x_t + L^{-1}\xi_t\| \\
&\le \|z - x_t\| + L^{-1}\|\xi_t\| \\
&\le 2M + L^{-1}(L + 1) \le 2M + 2.
\end{aligned} \tag{4.97}$$

In view of (4.90) and (4.97),

$$\|P_D(x_t - L^{-1}\xi_t)\| \leq 3M + 2. \tag{4.98}$$

Relations (4.89) and (4.98) imply that

$$P_D(x_t - L^{-1}\xi_t) \in C, \tag{4.99}$$

$$P_D(x_t - L^{-1}\xi_t) = P_C(x_t - L^{-1}\xi_t). \tag{4.100}$$

It follows from (4.83), (4.86), (4.88), and (4.98) that

$$\|x_{t+1}\| \leq \|P_D(x_t - L^{-1}\xi_t)\| + \delta_C < 3M + 3, \tag{4.101}$$

$$x_{t+1} \in U. \tag{4.102}$$

By (4.81), (4.82), (4.85), (4.86)–(4.89), (4.94), (4.101), (4.102), and Lemma 4.4 applied with

$$M_0 = 3M + 2,\ u = x_t,\ \xi = \xi_t,\ v = x_{t+1},\ x = z$$

we obtain that

$$\begin{aligned} & f(z) - f(x_{t+1}) \\ & \geq 2^{-1}L\|z - x_{t+1}\|^2 - 2^{-1}L\|z - x_t\|^2 \\ & - L\delta_C(18M + 19) - \delta_f(12M + 13). \end{aligned} \tag{4.103}$$

There are two cases:

$$\|z - x_{t+1}\| \leq \|z - x_t\|; \tag{4.104}$$

$$\|z - x_{t+1}\| > \|z - x_t\|. \tag{4.105}$$

If (4.104) holds, then in view of (4.92),

$$\|z - x_{t+1}\| \leq 2M.$$

Assume that (4.105) holds. It follows from (4.83), (4.103), and (4.105) that

$$\begin{aligned} & f(x_{t+1}) - f(z) \\ & \leq L\delta_C(18M + 19) + \delta_f(12M + 13) \leq 2. \end{aligned} \tag{4.106}$$

By (4.80), (4.87), and (4.106),

$$\|x_{t+1}\| \le M_2. \tag{4.107}$$

In view of (4.93) and (4.107),

$$\|z - x_{t+1}\| \le 2M - 2. \tag{4.108}$$

Thus in both cases

$$\|z - x_{t+1}\| \le 2M$$

and (4.103) holds. Therefore by induction we showed that for all integers $t \ge 0$,

$$\|x_t\| \le 3M$$

and (4.103) holds. Let T be a natural number. By (4.91) and (4.103),

$$\sum_{t=0}^{T}(f(x_{t+1}) - f(z))$$

$$\le (T+1)(L\delta_C(18M+19) + \delta_f(12M+13))$$

$$+2^{-1}L\sum_{t=0}^{T}(\|z - x_t\|^2 - \|z - x_{t+1}\|^2)$$

$$\le (T+1)(L\delta_C(18M+19) + \delta_f(12M+13)) + 2LM^2.$$

This implies that

$$\min\{f(x_t) : t = 1, \dots, T+1\} - f(z),$$

$$f(\sum_{t=1}^{T+1}(T+1)^{-1}x_t) - f(z)$$

$$\le L\delta_C(18M+19) + \delta_f(12M+13) + 2(T+1)^{-1}LM^2.$$

Theorem 4.6 is proved. □

It is clear that a best choice of T should be at the same order as $\max\{\delta_C, \delta_f\}^{-1}$. The next result has no prototype in [92].

Theorem 4.7 *Let $\delta_f, \delta_C \in (0, 1)$, $M > 1$,*

$$\delta_C \le (160M)^{-1}, \ \delta_f \le 120^{-1}, \tag{4.109}$$

$$D_{min} \cap B_X(0, M) \neq \emptyset, \tag{4.110}$$

$L \ge 1$ satisfy

$$|f(v_1) - f(v_2)| \le L\|v_1 - v_2\|$$

$$\text{for all } v_1, v_2 \in V \cap B_X(0, 4M + 10), \tag{4.111}$$

$$\|f'(v_1) - f'(v_2)\| \le L\|v_1 - v_2\|$$

$$\text{for all } v_1, v_2 \in V \cap B_X(0, 4M + 10), \tag{4.112}$$

$$T = \lfloor 36^{-1} \min\{\delta_C^{-1}, \ L\delta_f^{-1}\}\rfloor. \tag{4.113}$$

Assume that $\{x_t\}_{t=0}^{\infty} \subset V$, $\{\xi_t\}_{t=0}^{\infty} \subset X$,

$$\|x_0\| \le M \tag{4.114}$$

and that for each integer $t \ge 0$,

$$\|\xi_t - f'(x_t)\| \le \delta_f \tag{4.115}$$

and

$$\|x_{t+1} - P_D(x_t - L^{-1}\xi_t)\| \le \delta_C. \tag{4.116}$$

Then

$$\|x_t\| \le 3M + 3 \text{ for all integers } t = 0, \dots, T + 1$$

and

$$\sum_{t=0}^{T} (T + 1)^{-1}(f(x_{t+1}) - \inf(f, D)),$$

$$\min\{f(x_t) : \ t = 1, \dots, T + 1\} - \inf(f, D),$$

$$f(\sum_{t=1}^{T+1} (T + 1)^{-1} x_t) - \inf(f, D)$$

$$\leq 72(M+1)^2 \max\{\delta_C L, \delta_f\}.$$

Proof By (4.111) there exists

$$z \in D_{min} \cap B_X(0, M). \tag{4.117}$$

By (4.114) and (4.117),

$$\|x_0 - z\| \leq 2M. \tag{4.118}$$

Set

$$U = V \cap \{v \in X : \|v\| < 4M + 9\} \tag{4.119}$$

and

$$C = D \cap B_X(0, 4M + 8). \tag{4.120}$$

Assume that $t \geq 0$ is an integer and that

$$\|x_t - z\| \leq 2M + 2. \tag{4.121}$$

By (4.117) and (4.121),

$$\|x_t\| \leq 3M + 2. \tag{4.122}$$

In view of (4.119) and (4.122),

$$x_t \in U \cap B_X(0, 3M + 2). \tag{4.123}$$

It follows from (4.111), (4.115), and (4.122) that

$$\|\xi_t\| \leq L + 1. \tag{4.124}$$

By (4.117), (4.121), (4.124), and Lemma 2.2,

$$\|z - P_D(x_t - L^{-1}\xi_t)\|$$

$$\leq \|z - x_t + L^{-1}\xi_t\|$$

$$\leq \|z - x_t\| + L^{-1}\|\xi_t\|$$

$$\leq 2M + 2 + L^{-1}(L + 1) \leq 2M + 4. \tag{4.125}$$

In view of (4.117) and (4.125),

$$\|P_D(x_t - L^{-1}\xi_t)\| \le 3M + 4. \tag{4.126}$$

Relations (4.120) and (4.126) imply that

$$P_D(x_t - L^{-1}\xi_t) \in C, \tag{4.127}$$

$$P_D(x_t - L^{-1}\xi_t) = P_C(x_t - L^{-1}\xi_t). \tag{4.128}$$

It follows from (4.109), (4.116), and (4.126) that

$$\|x_{t+1}\| \le 3M + 5. \tag{4.129}$$

Relations (4.119) and (4.129) imply that

$$x_{t+1} \in U \cap B_X(0, 3M + 5). \tag{4.130}$$

By (4.111), (4.112), (4.115)–(4.117), (4.123), and Lemma 4.4 applied with

$$M_0 = 4M + 8,\ u = x_t,\ \xi = \xi_t,\ v = x_{t+1},\ x = z$$

we obtain that

$$\begin{aligned} &f(z) - f(x_{t+1}) \\ &\ge 2^{-1}L\|z - x_{t+1}\|^2 - 2^{-1}L\|z - x_t\|^2 \\ &- L\delta_C(24M + 55) - \delta_f(16M + 37). \end{aligned} \tag{4.131}$$

It follows from (4.111), (4.116), (4.117), (4.126), and (4.129) that

$$f(x_{t+1}) \ge f(P_D(x_t - L^{-1}\xi_t)) - L\delta_C \ge f(z) - L\delta_C. \tag{4.132}$$

By (4.131) and (4.132),

$$\begin{aligned} &\|z - x_{t+1}\|^2 - \|z - x_t\|^2 \\ &\le 2\delta_C + 2\delta_C(24M + 55) + 2L^{-1}\delta_f(16M + 37). \end{aligned} \tag{4.133}$$

Thus we have shown that the following property holds:

(a) if for an integer $t \ge 0$ relation $\|x_t - z\| \le 2M + 2$ holds, then (4.131) and (4.133) are true.

Let us show that for all integers $t = 0, \dots, T$,

$$\|z - x_t\|^2 \le \|z - x_0\|^2$$

$$+t[2\delta_C(24M + 55) + 2L^{-1}\delta_f(16M + 37)]$$

$$\le \|z - x_0\|^2 + T[2\delta_C(24M + 55) + 2L^{-1}\delta_f(16M + 37)]$$

$$\le 4M^2 + 8M + 4. \tag{4.134}$$

First, note that by (4.113),

$$T[2\delta_C(24M + 55) + 2L^{-1}\delta_f(16M + 37)]$$

$$= 2T\delta_C(24M + 55) + 2TL^{-1}\delta_f(16M + 37)$$

$$\le 18^{-1}(24M + 55) + 18^{-1}(16M + 37) \le 5M + 4. \tag{4.135}$$

It follows from (4.121) and (4.135) that

$$\|z - x_0\|^2 + T[2\delta_C(24M + 55) + 2L^{-1}\delta_f(16M + 13)]$$

$$\le 4M^2 + 8M + 4 = (2M + 2)^2. \tag{4.136}$$

Assume that $t \in \{0, \dots, T\} \setminus \{T\}$ and that (4.134) holds. Then

$$\|x_t - z\| \le 2M + 2$$

and by property (a), relations (4.131) and (4.133) hold. By (4.133) and (4.134),

$$\|z - x_{t+1}\|^2 \le \|z - x_t\|^2$$

$$+2\delta_C(24M + 55) + 2L^{-1}\delta_f(16M + 37)$$

$$\le \|z - x_0\|^2 + (t + 1)[2\delta_C(24M + 55) + 2L^{-1}\delta_f(16M + 37)]$$

$$\le \|z - x_0\|^2 + T[2\delta_C(24M + 55) + 2L^{-1}\delta_f(16M + 37)]$$

$$\le 4M^2 + 8M + 4.$$

Thus we have shown by induction that for all $t = 0, \dots, T$, (4.134) holds and

$$\|x_t - z\| \le 2M + 2. \tag{4.137}$$

Property (a) implies that (4.131) and (4.137) hold for all $t = 0, \dots, T$. In view of (4.117) and (4.137),

$$\|x_t\| \le 3M + 2,\ t = 0, \dots, T.$$

It follows from (4.117), (4.118), and (4.131) that

$$\sum_{t=0}^{T} (T+1)^{-1} f(x_{t+1}) - \inf(f, D)$$

$$\le 2^{-1} L (T+1)^{-1} \|z - x_0\|^2$$

$$+ \delta_C L(24M + 55) + \delta_f(16M + 37)$$

$$\le 72 L M^2 \max\{\delta_C, L^{-1}\delta_f\}$$

$$+ + \delta_C L(24M + 55) + \delta_f(16M + 37)$$

$$\le \max\{\delta_C, L^{-1}\delta_f\}(72LM^2 + L(24M + 55) + L(16M + 37)$$

$$\le \max\{L\delta_C, \delta_f\}(72M^2 + 40M + 92)$$

$$\le 72 \max\{L\delta_C, \delta_f\}(M+1)^2.$$

Property (a), (4.110), (4.133), and (4.137) with $t = T$ imply that

$$\|z - x_{T+1}\|^2 - \|z - x_T\|^2$$

$$\le 2\delta_C(24M + 56) + 2L^{-1}\delta_f(16M + 37)$$

$$\le 4M^2 + 8M + 4 + 160M\delta_C + 106L^{-1}\delta_f$$

$$\le 4M^2 + 8M + 6 \le (2M + 3)^2$$

and

$$x_{T+1} \le 3M + 3.$$

Theorem 4.7 is proved. □

Chapter 5
An Extension of the Gradient Algorithm

In this chapter we analyze the convergence of a gradient type algorithm, under the presence of computational errors, which was introduced by Beck and Teboulle [18] for solving linear inverse problems arising in signal/image processing. This algorithm is used for minimization of the sum of two given convex functions and each of its iteration consists of two steps. The first step is a calculation of a gradient of the first function while in the second one we solve an auxiliary minimization problem. In each of these two steps there is a computational error. In general, these two computational errors are different. We show that our algorithm generates a good approximate solution, if all the computational errors are bounded from above by a small positive constant. Moreover, if we know the computational errors for the two steps of our algorithm, we find out what approximate solution can be obtained and how many iterates one needs for this.

5.1 Preliminaries and the Main Result

Let X be a Hilbert space equipped with an inner product $\langle \cdot, \cdot \rangle$ which induces a complete norm $\| \cdot \|$.

Suppose that $f : X \to R^1$ is a convex Fréchet differentiable function on X and for every $x \in X$ denote by $f'(x) \in X$ the Fréchet derivative of f at x. It is clear that for any $x \in X$ and any $h \in X$

$$\langle f'(x), h \rangle = \lim_{t \to 0} t^{-1}(f(x + th) - f(x)).$$

Recall that for each function $\phi : X \to R^1$,

$$\inf(\phi) = \inf\{\phi(y) : \; y \in X\},$$

A. J. Zaslavski, *Convex Optimization with Computational Errors*, Springer Optimization and Its Applications 155, https://doi.org/10.1007/978-3-030-37822-6_5

$$\text{argmin}(\phi) = \text{argmin}\{\phi(z) : z \in X\} = \{z \in X : \phi(z) = \inf(\phi)\}.$$

We suppose that the mapping $f' : X \to X$ is Lipschitz on all bounded subsets of X.

Let $g : X \to R^1$ be a convex continuous function which is Lipschitz on all bounded subsets of X. Define

$$F(x) = f(x) + g(x), \ x \in X.$$

We suppose that

$$\text{argmin}(F) \neq \emptyset \tag{5.1}$$

and that there exists $c_* \in R^1$ such that

$$g(x) \geq c_* \text{ for all } x \in X. \tag{5.2}$$

For each $u \in X$, each $\xi \in X$, and each $L > 0$ define a convex function

$$G^{(L)}_{u,\xi}(w) = f(u) + \langle \xi, w - u \rangle + 2^{-1}L\|w - u\|^2 + g(w), \ w \in X \tag{5.3}$$

which has a minimizer.

In this chapter we analyze the gradient type algorithm, which was introduced by Beck and Teboulle in [18] for solving linear inverse problems, and prove the following result.

Theorem 5.1 *Let $\delta_f, \delta_G \in (0, 1]$, $M \geq 1$ satisfy*

$$\textit{argmin}(F) \cap B_X(0, M) \neq \emptyset, \tag{5.4}$$

$L \geq 1$ satisfy

$$|f(w_1) - f(w_2)| \leq L\|w_1 - w_2\|$$

$$\textit{for all } w_1, w_2 \in B_X(0, 3M + 2) \tag{5.5}$$

and

$$\|f'(w_1) - f'(w_2)\| \leq L\|w_1 - w_2\|$$

$$\textit{for all } w_1, w_2 \in X, \tag{5.6}$$

$M_1 \geq 3M$ satisfy

$$|f(w)|, \ F(w) \leq M_1 \textit{ for all } w \in B_X(0, 3M + 2), \tag{5.7}$$

$$M_2 = 18(M_1 + |c_*| + 1 + (L+1)^2)^{1/2} + 3M + 2, \tag{5.8}$$

$L_0 \geq 1$ *satisfy*

$$|g(w_1) - g(w_2)| \leq L_0 \|w_1 - w_2\|$$

$$\text{for all } w_1, w_2 \in B_X(0, M_2), \tag{5.9}$$

$$\delta_f \leq 4^{-1}(M_2 + 9M + 4)^{-1},$$

$$\delta_G \leq 4^{-1}(2 + L_0 + L(M_2 + 3M + 3))^{-1}, \tag{5.10}$$

$$\epsilon_0 = 2\delta_f(M_2 + 9M + 4) + 2\delta_G(2 + L_0 + L(M_2 + 3M + 3)) \tag{5.11}$$

and let

$$n_0 = \lfloor 4LM^2\epsilon_0^{-1} \rfloor + 2. \tag{5.12}$$

Assume that $\{x_t\}_{t=0}^{\infty} \subset X$, $\{\xi_t\}_{t=0}^{\infty} \subset X$,

$$\|x_0\| \leq M \tag{5.13}$$

and that for each integer $t \geq 0$,

$$\|\xi_t - f'(x_t)\| \leq \delta_f \tag{5.14}$$

and

$$B_X(x_{t+1}, \delta_G) \cap \mathit{argmin}(G^{(L)}_{x_t,\xi_t}) \neq \emptyset. \tag{5.15}$$

Then there exists an integer $q \in [0, n_0 + 1]$ *such that*

$$\|x_i\| \leq M_2,\ i = 0, \ldots, q$$

and

$$F(x_q) \leq \inf(F) + \epsilon_0.$$

Note that in the theorem above δ_f, δ_G are the computational errors produced by our computer system. The computational error δ_f is produced when we calculate a gradient of f while the computational error δ_G is produced when we solve the auxiliary minimization problem with the objective functions $G^{(L)}_{x_t,\xi_t}$, $t = 0, 1, \ldots$. By Theorem 5.1, after $n_0 + 1$ iterations we obtain an approximate solution x satisfying

$$F(x) \le \inf(f) + \epsilon_0,$$

where ϵ_0, n_0 are defined by (5.11) and (5.12). Theorem 5.1 is a generalization of Theorem 5.1 of [92] proved in the case when $\delta_f = \delta_G$. In this chapter we also prove two extensions of Theorem 5.1 which do not have prototypes in [92].

5.2 Auxiliary Results

Lemma 5.2 ([18]) *Let $u, \xi \in X$ and $L > 0$. Then the function $G^{(L)}_{u,\xi}$ has a point of minimum and $z \in X$ is a minimizer of $G^{(L)}_{u,\xi}$ if and only if*

$$0 \in \xi + L(z - u) + \partial g(z).$$

Proof By (5.2) and (5.3),

$$\lim_{\|w\| \to \infty} G^{(L)}_{u,\xi}(w) = \infty.$$

This implies that the function $G^{(L)}_{u,\xi}$ has a minimizer. Clearly, z is a minimizer of $G^{(L)}_{u,\xi}$ if and only if

$$0 \in \partial G^{(L)}_{u,\xi}(z) = \xi + L(z - u) + \partial g(z).$$

Lemma 5.2 is proved. □

Lemma 5.3 ([92]) *Let $M_0 \ge 1$, $L \ge 1$ satisfy*

$$|f(w_1) - f(w_2)| \le L\|w_1 - w_2\|$$

$$\textit{for all } w_1, w_2 \in B_X(0, M_0 + 2) \tag{5.16}$$

and $M_1 \ge M_0$ satisfy

$$|f(w)|,\ F(w) \le M_1 \textit{ for all } w \in B_X(0, M_0 + 2). \tag{5.17}$$

Assume that

$$u \in B_X(0, M_0 + 1), \tag{5.18}$$

$\xi \in X$ satisfies

$$\|\xi - f'(u)\| \le 1 \tag{5.19}$$

and that $v \in X$ *satisfies*

$$B_X(v, 1) \cap \{z \in X : \; G^{(L)}_{u,\xi}(z) \le \inf(G^{(L)}_{u,\xi}) + 1\} \ne \emptyset. \tag{5.20}$$

Then

$$\|v\| \le (8(M_1 + |c_*| + (L+1)^2 + 1)^{1/2} + M_0 + 2. \tag{5.21}$$

Proof In view of (5.20), there exists

$$\widehat{v} \in B_X(v, 1) \tag{5.22}$$

such that

$$G^{(L)}_{u,\xi}(\widehat{v}) \le \inf(G^{(L)}_{u,\xi}) + 1. \tag{5.23}$$

By (5.1), (5.3), (5.17), (5.18), and (5.23),

$$f(u) + \langle \xi, \widehat{v} - u \rangle + 2^{-1}L\|\widehat{v} - u\|^2 + g(\widehat{v})$$

$$= G^{(L)}_{u,\xi}(\widehat{v}) \le G^{(L)}_{u,\xi}(u) + 1 = F(u) + 1 \le M_1 + 1. \tag{5.24}$$

It follows from (5.2), (5.17), (5.18), and (5.24) that

$$\langle \xi, \widehat{v} - u \rangle + 2^{-1}L\|\widehat{v} - u\|^2 \le 2M_1 + 1 + |c_*|. \tag{5.25}$$

It is clear that

$$2L^{-1}|\langle \xi, \widehat{v} - u \rangle| \le L^{-1}(4^{-1}\|\widehat{v} - u\|^2 + 4\|\xi\|^2). \tag{5.26}$$

Since the function f is Lipschitz on $B_X(0, M_0 + 2)$ relations (5.16), (5.18), and (5.19) imply that

$$\|\xi\| \le \|f'(u)\| + 1 \le L + 1. \tag{5.27}$$

By (5.25)–(5.27),

$$2L^{-1}(2M_1 + 1 + |c_*|)$$

$$\ge \|\widehat{v} - u\|^2 - 2L^{-1}|\langle \xi, \widehat{v} - u \rangle|$$

$$\ge \|\widehat{v} - u\|^2 - 4^{-1}\|\widehat{v} - u\|^2 - 4\|\xi\|^2$$

$$\ge 2^{-1}\|\widehat{v} - u\|^2 - 4(L+1)^2.$$

This implies that

$$\|\widehat{v} - u\|^2 \le 4(M_1 + 1 + |c_*|) + 8(L+1)^2$$

and

$$\|\widehat{v} - u\| \le (4(M_1 + 1 + |c_*|) + 8(L+1)^2)^{1/2}.$$

Together with (5.18) and (5.22) this implies that

$$\|v\| \le \|\widehat{v}\| + 1 \le \|\widehat{v} - u\| + \|u\| + 1$$

$$\le (4(M_1 + 1 + |c_*|) + 8(L+1)^2)^{1/2} + M_0 + 2.$$

Lemma 5.3 is proved. □

Lemma 5.4 *Let* $\delta_f, \delta_G \in (0, 1]$, $M_0 \ge 1$, $L \ge 1$ *satisfy*

$$|f(w_1) - f(w_2)| \le L\|w_1 - w_2\|$$

$$\text{for all } w_1, w_2 \in B_X(0, M_0 + 2), \tag{5.28}$$

$$|f'(w_1) - f'(w_2)| \le L\|w_1 - w_2\| \text{ for all } w_1, w_2 \in X, \tag{5.29}$$

$M_1 \ge M_0$ *satisfy*

$$|f(w)|,\ F(w) \le M_1 \text{ for all } w \in B_X(0, M_0 + 2), \tag{5.30}$$

$$M_2 = (8(M_1 + |c_*| + (L+1)^2 + 1)^{1/2} + M_0 + 2 \tag{5.31}$$

and let $L_0 \ge 1$ *satisfy*

$$|g(w_1) - g(w_2)| \le L_0\|w_1 - w_2\|$$

$$\text{for all } w_1, w_2 \in B_X(0, M_2). \tag{5.32}$$

Assume that

$$u \in B_X(0, M_0 + 1), \tag{5.33}$$

$\xi \in X$ *satisfies*

$$\|\xi - f'(u)\| \le \delta_f \tag{5.34}$$

and that $v \in X$ *satisfies*

$$B_X(v, \delta_G) \cap argmin(G_{u,\xi}^{(L)}) \neq \emptyset. \tag{5.35}$$

Then for each $x \in B_X(0, M_0 + 1)$,

$$F(x) - F(v)$$

$$\geq 2^{-1}L\|v - x\|^2 - 2^{-1}L\|u - x\|^2$$

$$-\delta_f(M_2 + 3M_0 + 3) - \delta_G(1 + L_0 + L(M_0 + M + 2 + 3)).$$

Proof By (5.35) there exists $\widehat{v} \in X$ such that

$$\widehat{v} \in \text{argmin}(G_{u,\xi}^{(L)}) \tag{5.36}$$

and

$$\|v - \widehat{v}\| \leq \delta_G. \tag{5.37}$$

In view of the assumptions of the lemma, Lemma 5.3, (5.31), (5.35), and (5.36),

$$\|v\|, \|\widehat{v}\| \leq M_2. \tag{5.38}$$

Let

$$x \in B_X(0, M_0 + 1). \tag{5.39}$$

Clearly,

$$F(x) = f(x) + g(x), \ \ F(v) = f(v) + g(v). \tag{5.40}$$

Proposition 4.3 and (5.33) imply that

$$g(v) + f(v)$$

$$\leq g(v) + f(u) + \langle f'(u), v - u\rangle + 2^{-1}L\|v - u\|^2. \tag{5.41}$$

By (5.3), (5.33), (5.34), (5.38), (5.40), and (5.41),

$$F(x) - F(v)$$

$$\geq [f(x) + g(x)] - [f(v) + g(v)]$$

$$\geq f(x) + g(x)$$

$$-[f(u) + \langle f'(u), v - u\rangle + 2^{-1}L\|v - u\|^2 + g(v)]$$

$$= [f(x) + g(x)]$$

$$-[f(u) + \langle \xi, v - u\rangle + 2^{-1}L\|v - u\|^2 + g(v)]$$

$$+\langle \xi - f'(u), v - u\rangle$$

$$\geq [f(x) + g(x)] - G^{(L)}_{u,\xi}(v) - \|\xi - f'(u)\|\|v - u\|$$

$$\geq [f(x) + g(x)] - G^{(L)}_{u,\xi}(v) - \delta_f(M_2 + M_0 + 1). \tag{5.42}$$

It follows from (5.3) that

$$G^{(L)}_{u,\xi}(v) - G^{(L)}_{u,\xi}(\widehat{v})$$

$$= \langle \xi, v - u\rangle + 2^{-1}L\|v - u\|^2 + g(v)$$

$$-[\langle \xi, \widehat{v} - u\rangle + 2^{-1}L\|\widehat{v} - u\|^2 + g(\widehat{v})]$$

$$= \langle \xi, v - \widehat{v}\rangle + 2^{-1}L[\|v - u\|^2 - \|\widehat{v} - u\|^2] + g(v) - g(\widehat{v}). \tag{5.43}$$

Relations (5.28), (5.32), and (5.34) imply that

$$\|\xi\| \leq \|f'(u)\| + 1 \leq L + 1. \tag{5.44}$$

In view of (5.37) and (5.44),

$$|\langle \xi, v - \widehat{v}\rangle| \leq (L + 1)\delta_G. \tag{5.45}$$

By (5.33), (5.37), and (5.38),

$$|\|v - u\|^2 - \|\widehat{v} - u\|^2|$$

$$\leq |\|v - u\| - \|\widehat{v} - u\||(\|v - u\| + \|\widehat{v} - u\|)$$

$$\leq \|v - \widehat{v}\|(2M_2 + 2M_0 + 2)$$

$$\leq \delta_G(2M_2 + 2M_0 + 2). \tag{5.46}$$

In view of (5.32), (5.37), and (5.38),

$$|g(v) - g(\widehat{v})| \leq L_0\|v - \widehat{v}\| \leq L_0\delta_G. \tag{5.47}$$

It follows from (5.43) and (5.45–(5.47) that

$$|G_{u,\xi}^{(L)}(v) - G_{u,\xi}^{(L)}(\widehat{v})|$$

$$\leq (L+1)\delta_G + 2^{-1}L\delta_G(2M_0 + 2M_2 + 2) + L_0\delta_G$$

$$= \delta_G(L + 1 + L_0 + L(M_0 + M_2 + 1)). \tag{5.48}$$

Relations (5.42) and (5.48) imply that

$$F(x) - F(v)$$

$$\geq f(x) + g(x) - G_{u,\xi}^{(L)}(v) - \delta_f(M_2 + M_0 + 1)$$

$$\geq f(x) + g(x) - G_{u,\xi}^{(L)}(\widehat{v})$$

$$- \delta_f(M_2 + M_0 + 1) - \delta_G(L_0 + L + 1 + L(M_2 + M_0 + 1)). \tag{5.49}$$

By the convexity of f, (5.33), (5.34), and (5.39),

$$f(x) \geq f(u) + \langle f'(u), x - u\rangle$$

$$\geq f(u) + \langle \xi, x - u\rangle - |\langle f'(u) - \xi, x - u\rangle|$$

$$\geq f(u) + \langle \xi, x - u\rangle - \|f'(u) - \xi\|\|x - u\|$$

$$\geq f(u) + \langle \xi, x - u\rangle - \delta_f(2M_0 + 2). \tag{5.50}$$

Lemma 5.2 and (5.36) imply that there exists

$$l \in \partial g(\widehat{v}) \tag{5.51}$$

such that

$$\xi + L(\widehat{v} - u) + l = 0. \tag{5.52}$$

In view of (5.51) and the convexity of g,

$$g(x) \geq g(\widehat{v}) + \langle l, x - \widehat{v}\rangle. \tag{5.53}$$

It follows from (5.50) and (5.53) that

$$f(x) + g(x)$$

$$\geq f(u) + \langle \xi, x - u \rangle - \delta_f(2M_0 + 2) + g(\widehat{v}) + \langle l, x - \widehat{v} \rangle. \tag{5.54}$$

In view of (5.3),

$$G_{u,\xi}^{(L)}(\widehat{v}) = f(u) + \langle \xi, \widehat{v} - u \rangle + 2^{-1}L\|\widehat{v} - u\|^2 + g(\widehat{v}). \tag{5.55}$$

By (5.52), (5.54), and (5.55),

$$f(x) + g(x) - G_{u,\xi}^{(L)}(\widehat{v})$$

$$\geq \langle \xi, x - \widehat{v} \rangle + \langle l, x - \widehat{v} \rangle - 2^{-1}L\|\widehat{v} - u\|^2 - \delta_f(2M_0 + 2)$$

$$= -\delta_f(2M_0 + 2) + \langle \xi + l, x - \widehat{v} \rangle - 2^{-1}L\|\widehat{v} - u\|^2$$

$$= -2^{-1}L\|\widehat{v} - u\|^2 - L\langle \widehat{v} - u, x - \widehat{v} \rangle - \delta_f(2M_0 + 2)$$

$$= 2^{-1}L\|\widehat{v} - u\|^2 + L\langle \widehat{v} - u, u - x \rangle - \delta_f(2M_0 + 2). \tag{5.56}$$

In view of (5.37)–(5.39),

$$|\|\widehat{v} - x\|^2 - \|v - x\|^2|$$

$$\leq |\|\widehat{v} - x\| - \|v - x\||(\|\widehat{v} - x\| + \|v - x\|)$$

$$\leq \|\widehat{v} - v\|(2M_2 + 2M_0 + 2)$$

$$\leq \delta_G(2M_2 + 2M_0 + 2). \tag{5.57}$$

Lemma 2.1 implies that

$$\langle \widehat{v} - u, u - x \rangle = 2^{-1}[\|\widehat{v} - x\|^2 - \|\widehat{v} - u\|^2 - \|u - x\|^2]. \tag{5.58}$$

By (5.56)–(5.58),

$$f(x) + g(x) - G_{u,\xi}^{(L)}(\widehat{v})$$

$$\geq 2^{-1}L\|\widehat{v} - u\|^2 + 2^{-1}L\|\widehat{v} - x\|^2$$

$$-2^{-1}L\|\widehat{v} - u\|^2 - 2^{-1}L\|u - x\|^2 - \delta_f(2M_0 + 2)$$

$$\geq 2^{-1}L\|\widehat{v} - x\|^2 - 2^{-1}L\|u - x\|^2 - \delta_f(2M_0 + 2)$$

$$\geq 2^{-1}L\|v - x\|^2 - 2^{-1}L\|u - x\|^2$$

$$-2^{-1}L\delta_G(2M_2+2M_0+2)-\delta_f(2M_0+2). \tag{5.59}$$

It follows from (5.49) and (5.59) that

$$F(x)-F(v)$$

$$\geq 2^{-1}L\|v-x\|^2-2^{-1}L\|u-x\|^2$$

$$-\delta_G L(M_2+M_0+1)-\delta_f(2M_0+2)$$

$$-\delta_f(M_2+M_0+1)-\delta_G(1+L_0+L(M_2+M_0+2)).$$

Lemma 5.4 is proved. □

5.3 Proof of Theorem 5.1

By (5.4), there exists

$$z\in \mathrm{argmin}(F)\cap B_X(0,M). \tag{5.60}$$

In view of (5.13) and (5.60),

$$\|x_0-z\|\leq 2M. \tag{5.61}$$

If $F(x_0)\leq F(z)+\epsilon_0$, then the assertion of the theorem holds. Let

$$F(x_0)>F(z)+\epsilon_0. \tag{5.62}$$

If $F(x_1)\leq F(z)+\epsilon_0$, then in view of Lemma 5.3, $\|x_1\|\leq M_2$ and the assertion of the theorem holds. Let

$$F(x_1)>F(z)+\epsilon_0.$$

Assume that $T\geq 0$ is an integer and that for all integers $t=0,\ldots,T$,

$$F(x_{t+1})-F(z)>\epsilon_0. \tag{5.63}$$

We show that for all $t\in\{0,\ldots,T\}$,

$$\|x_t-z\|\leq 2M \tag{5.64}$$

and

$$F(z) - F(x_{t+1})$$

$$\geq 2^{-1}L\|z - x_{t+1}\|^2 - 2^{-1}L\|z - x_t\|^2$$

$$-\delta_f(M_2 + 9M + 3) - \delta_G(1 + L_0 + L(3M + M_2 + 3)). \quad (5.65)$$

In view of (5.61), (5.64) is true for $t = 0$.

Assume that $t \in \{0, \ldots, T\}$ and (5.64) holds. Relations (5.60) and (5.64) imply that

$$\|x_t\| \leq 3M. \quad (5.66)$$

Set

$$M_0 = 3M. \quad (5.67)$$

By (5.14), (5.15), (5.60), (5.66), (5.67), and Lemma 5.4 applied with

$$x = z,\ u = x_t,\ \xi = \xi_t,\ v = x_{t+1},$$

we have

$$F(z) - F(x_{t+1})$$

$$\geq 2^{-1}L\|z - x_{t+1}\|^2 - 2^{-1}L\|z - x_t\|^2$$

$$-\delta_f(M_2 + 3M_0 + 3) - \delta_G(1 + L_0 + L(M_0 + M_2 + 3)). \quad (5.68)$$

It follows from (5.63), (5.67), and (5.68) that

$$\epsilon_0 < F(x_{t+1}) - F(z)$$

$$\leq 2^{-1}L\|z - x_t\|^2 - 2^{-1}L\|z - x_{t+1}\|^2$$

$$+\delta_f(M_2 + 9M + 3) - \delta_G(1 + L_0 + L(M_2 + 3M + 3)). \quad (5.69)$$

In view of (5.11), (5.64), and (5.69),

$$2M \geq \|z - x_t\| \geq \|z - x_{t+1}\|.$$

Thus we have shown by induction that (5.65) holds for all $t = 0, \ldots, T$ and that (5.64) holds for all $t = 0, \ldots, T + 1$. By (5.63) and (5.65),

$$(T+1)\epsilon_0 < \sum_{t=0}^{T}(F(x_{t+1}) - F(z))$$

$$\leq \sum_{t=0}^{T}[2^{-1}L\|z - x_t\|^2 - 2^{-1}L\|z - x_{t+1}\|^2]$$

$$+(T+1)\delta_f(M_2 + 9M + 3)$$

$$+(T+1)\delta_G(1 + L_0 + L(M_2 + 3M + 3))$$

$$\leq 2^{-1}L\|z - x_0\|^2$$

$$+(T+1)\delta_f(M_2 + 9M + 3)$$

$$+(T+1)\delta_G(1 + L_0 + L(M_2 + 3M + 3)).$$

It follows from (5.11) and (5.61) that

$$T\epsilon_0/2 \leq 2^{-1}L(4M^2),$$

$$T \leq 4LM^2\epsilon_0^{-1} < n_0.$$

Thus we have shown that if $T \geq 0$ is an integer and (5.63) holds for all $t = 0, \ldots, T$, then (5.64) holds for all $t = 0, \ldots, T+1$ and $T < n_0$. This implies that there exists an integer $q \in \{1, \ldots, n_0 + 1\}$ such that

$$\|x_i\| \leq 3M, \; i = 0, \ldots, q,$$

$$F(x_q) \leq F(z) + \epsilon_0.$$

Theorem 5.1 is proved. □

5.4 The First Extension of Theorem 5.1

Theorem 5.5 *Let $\delta_f, \delta_G \in (0, 1]$, $M \geq 1$ satisfy*

$$\{x \in X : \; F(x) \leq \inf(F) + 4\} \subset \cap B_X(0, M), \tag{5.70}$$

$L \geq 1$ satisfy

$$|f(w_1) - f(w_2)| \leq L\|w_1 - w_2\|$$

$$\textit{for all } w_1, w_2 \in B_X(0, 3M+2) \tag{5.71}$$

and

$$\|f'(w_1) - f'(w_2)\| \le L\|w_1 - w_2\| \textit{ for all } w_1, w_2 \in X, \tag{5.72}$$

$M_1 \geq 3M + 2$ *satisfy*

$$|f(w)|,\ F(w) \le M_1 \textit{ for all } w \in B_X(0, 3M+2), \tag{5.73}$$

$$M_2 = (8(M_1 + |c_*| + 1 + (L+1)^2)^{1/2} + 3M + 2, \tag{5.74}$$

$L_0 \geq 1$ *satisfy*

$$|g(w_1) - g(w_2)| \le L_0\|w_1 - w_2\|$$

$$\textit{for all } w_1, w_2 \in B_X(0, M_2), \tag{5.75}$$

$$\delta_f \le 4^{-1}(M_2 + 9M + 4)^{-1} \tag{5.76}$$

and

$$\delta_G \le 4^{-1}(2 + L_0 + L(M_2 + 3M + 3))^{-1}. \tag{5.77}$$

Assume that $\{x_t\}_{t=0}^{\infty} \subset X$, $\{\xi_t\}_{t=0}^{\infty} \subset X$,

$$\|x_0\| \le M \tag{5.78}$$

and that for each integer $t \geq 0$,

$$\|\xi_t - f'(x_t)\| \le \delta_f \tag{5.79}$$

and

$$B_X(x_{t+1}, \delta_G) \cap \textit{argmin}(G^{(L)}_{x_t, \xi_t}) \neq \emptyset. \tag{5.80}$$

Then for each integer $t \geq 0$,

$$\|x_t\| \le 3M \tag{5.81}$$

and for each natural number T,

$$F(\sum_{t=1}^{T+1} (T+1)^{-1} x_t) - \inf(F),$$

$$\min\{F(x_t) : t = 1, \dots, T+1\} - \inf(F)$$

$$\leq \delta_f(2M_2 + 9M + 3)$$

$$+\delta_G(1 + L_0 + L(3M + M_2 + 3)) + 2LM^2(T+1)^{-1}.$$

Proof Set

$$M_0 = 3M. \tag{5.82}$$

Fix

$$z \in \text{argmin}(F). \tag{5.83}$$

In view of (5.70) and (5.82),

$$\|z\| \leq M. \tag{5.84}$$

By (5.78) and (5.84),

$$\|x_0 - z\| \leq 2M. \tag{5.85}$$

Assume that $t \geq 0$ is an integer and that

$$\|x_t - z\| \leq 2M \tag{5.86}$$

Relations (5.84) and (5.86) imply that

$$\|x_t\| \leq 3M. \tag{5.87}$$

By (5.79), (5.82), (5.84), (5.86), (5.87), and Lemma 5.4 applied with

$$x = z,\ u = x_t,\ \xi = \xi_t,\ v = x_{t+1},$$

we have

$$F(z) - F(x_{t+1})$$

$$\geq 2^{-1}L\|z - x_{t+1}\|^2 - 2^{-1}L\|z - x_t\|^2$$

$$- \delta_f(2M_2 + 9M + 3) - \delta_G(1 + L_0 + L(3M + M_2 + 3)). \tag{5.88}$$

There are two cases:

$$\|z - x_{t+1}\| \leq \|z - x_t\|; \tag{5.89}$$

$$\|z - x_{t+1}\| > \|z - x_t\|. \tag{5.90}$$

Assume that (5.89) is true. Then in view of (5.86) and (5.89),

$$\|z - x_{t+1}\| \le 2M.$$

Assume that (5.90) holds. It follows from (5.76), (5.77), (5.83), (5.88), and (5.90) that

$$F(x_{t+1}) \le F(z)$$

$$+\delta_f(2M_2 + 9M + 3) + \delta_G(1 + L_0 + L(3M + M_2 + 3))$$

$$\le \inf(F) + 2.$$

By the inequality above, (5.70) and (5.84),

$$\|x_{t+1}\| \le M$$

and

$$\|x_{t+1} - z\| \le 2M.$$

Therefore by induction we showed that for all integers $t \ge 0$, (3.86)–(3.88) hold.

Let T be a natural number. By (5.83), (5.84) and (5.88),

$$\sum_{t=0}^{T} (T+1)^{-1} F(x_{t+1}) - \inf(F)$$

$$\le \delta_f(2M_2 + 9M + 3) + \delta_G(1 + L_0 + L(3M + M_2 + 3))$$

$$+2^{-1} L (T+1)^{-1} \sum_{t=0}^{T} (\|z - x_t\|^2 - \|z - x_{t+1}\|^2)$$

$$\le \delta_f(2M_2 + 9M + 3)$$

$$+\delta_G(1 + L_0 + L(3M + M_2 + 3)) + 2(T+1)^{-1} L M^2.$$

This implies that

$$F(\sum_{t=1}^{T+1} (T+1)^{-1} x_t) - \inf(F),$$

$$\min\{F(x_t) : t = 1, \dots, T+1\} - \inf(F)$$

$$\leq \delta_f(2M_2 + 9M + 3)$$

$$+\delta_G(1 + L_0 + L(3M + M_2 + 3)) + 2(T+1)^{-1}LM^2.$$

This completes the proof of Theorem 5.5. □

By Theorem 5.5, the best choice of T is as the same order as $\max\{\delta_f, \delta_G\}^{-1}$.

5.5 The Second Extension of Theorem 5.1

Theorem 5.6 *Let* $\delta_f, \delta_G \in (0, 1)$, $M > 1$

$$argmin(F) \cap B_X(0, M) \neq \emptyset, \tag{5.91}$$

$L \geq 1$ *satisfy*

$$|f(w_1) - f(w_2)| \leq L\|w_1 - w_2\|$$

$$\text{for all } w_1, w_2 \in B_X(0, 4M + 10) \tag{5.92}$$

and

$$\|f'(w_1) - f'(w_2)\| \leq L\|w_1 - w_2\| \text{ for all } w_1, w_2 \in X, \tag{5.93}$$

$M_1 \geq 4M + 8$ *satisfy*

$$|f(w)|,\ F(w) \leq M_1 \text{ for all } w \in B_X(0, 4M + 10), \tag{5.94}$$

$$M_2 = (8(M_1 + |c_*| + 1 + (L+1)^2)^{1/2} + 4M + 10, \tag{5.95}$$

$L_0 \geq 1$ *satisfy*

$$|g(w_1) - g(w_2)| \leq L_0\|w_1 - w_2\|$$

$$\text{for all } w_1, w_2 \in B_X(0, M_2), \tag{5.96}$$

$$\delta_f, \delta_G < 4^{-1}((M_2 + 12M + 27)L + 1 + L_0)^{-1}, \tag{5.97}$$

$$T = \lfloor 2^{-1}L \min\{\delta_f^{-1}, \delta_G^{-1}\}((M_2 + 12M + 27)L + 1 + L_0)^{-1} \rfloor. \tag{5.98}$$

Assume that $\{x_t\}_{t=0}^{\infty} \subset X$, $\{\xi_t\}_{t=0}^{\infty} \subset X$,

$$\|x_0\| \le M \tag{5.99}$$

and that for each integer $t \ge 0$,

$$\|\xi_t - f'(x_t)\| \le \delta_f \tag{5.100}$$

and

$$B_X(x_{t+1}, \delta_G) \cap \mathit{argmin}(G^{(L)}_{x_t,\xi_t}) \ne \emptyset. \tag{5.101}$$

Then for each integer $t \in [0, T]$,

$$\|x_t\| \le 3M + 2$$

and

$$\sum_{t=0}^{T-1} T^{-1} F(x_{t+1}) - \inf(F),$$

$$F(\sum_{t=0}^{T-1} T^{-1} x_{t+1}) - \inf(F),$$

$$\min\{F(x_t) :\ t = 1, \dots, T\} - \inf(F)$$

$$\le \max\{\delta_G,\ \delta_f\}(8M^2 + 1)((M_2 + 12M + 27)(L + 1) + L_0 + 1).$$

Proof In view of (5.91), there exists

$$z \in \operatorname{argmin}(F) \cap B_X(0, M). \tag{5.102}$$

By (5.99) and (5.102),

$$\|x_0 - z\| \le 2M. \tag{5.103}$$

Assume that $t \ge 0$ is an integer and that

$$\|x_t - z\| \le 2M + 2 \tag{5.104}$$

Relations (5.102) and (5.104) imply that

$$\|x_t\| \le 3M + 2. \tag{5.105}$$

Set

$$M_0 = 4M + 8. \tag{5.106}$$

By (5.100), (5.102), (5.105), (5.106) and Lemma 5.4 applied with

$$x = z,\ u = x_t,\ \xi = \xi_t,\ v = x_{t+1},$$

we have

$$\begin{aligned} &F(z) - F(x_{t+1}) \\ &\geq 2^{-1}L\|z - x_{t+1}\|^2 - 2^{-1}L\|z - x_t\|^2 \\ &-\delta_f(M_2 + 3M_0 + 3) - \delta_G(1 + L_0 + L(M_0 + M_2 + 3)) \\ &\geq 2^{-1}L\|z - x_{t+1}\|^2 - 2^{-1}L\|z - x_t\|^2 \\ &- \delta_f(M_2 + 12M + 27) - \delta_G(1 + L_0 + L(4M + M_2 + 11)). \end{aligned} \tag{5.107}$$

By (5.102) and (5.107),

$$\begin{aligned} &\|z - x_{t+1}\|^2 - \|z - x_t\|^2 \\ &\leq 2L^{-1}\delta_f(M_2 + 12M + 27) \\ &+ 2L^{-1}\delta_G(1 + L_0 + L(4M + M_2 + 11)). \end{aligned} \tag{5.108}$$

Therefore we have shown that the following property holds:

(a) for each integer $t \geq 0$ satisfying (5.104), relations (5.107) and (5.108) hold.

Let us show that for all integers $t = 0, \dots, T$,

$$\begin{aligned} \|z - x_t\|^2 &\leq \|z - x_0\|^2 + t[2L^{-1}\delta_f(M_2 + 12M + 27) \\ &+2L^{-1}\delta_G(1 + L_0 + L(4M + M_2 + 11))] \\ &\leq \|z - x_0\|^2 + T[2L^{-1}\delta_f(M_2 + 12M + 27) \\ &+2L^{-1}\delta_G(1 + L_0 + L(4M + M_2 + 11))] \\ &\leq 4M^2 + 8M + 4. \end{aligned} \tag{5.109}$$

First, note that in view of (5.98),

$$T[2L^{-1}\delta_f(M_2 + 12M + 27)$$

$$+2L^{-1}\delta_G(1 + L_0 + L(4M + M_2 + 11))]$$

$$= 2L^{-1}T\delta_f(M_2 + 12M + 27)$$

$$+ 2L^{-1}T\delta_G(1 + L_0 + L(4M + M_2 + 11))] \le 2. \tag{5.110}$$

By (5.103) and (5.110),

$$\|z - x_0\|^2 + T[2L^{-1}\delta_f(M_2 + 12M + 27)$$

$$+2L^{-1}\delta_G(1 + L_0 + L(4M + M_2 + 11))]$$

$$\le 4M^2 + 2 \le (2M + 2)^2. \tag{5.111}$$

Assume that $t \in \{0, \dots, T\} \setminus \{T\}$ and that (5.109) holds. Then

$$\|x_t - z\| \le 2M + 2$$

and by property (a), relations (5.107) and (5.108) hold. By (5.108), (5.109), and (5.111),

$$\|z - x_{t+1}\|^2 \le \|z - x_t\|^2$$

$$+2L^{-1}\delta_f(M_2 + 12M + 27)$$

$$+2L^{-1}\delta_G(1 + L_0 + L(4M + M_2 + 11))$$

$$\le \|z - x_0\|^2 + (t + 1)[2L^{-1}\delta_f(M_2 + 12M + 27)$$

$$+2L^{-1}\delta_G(1 + L_0 + L(4M + M_2 + 11))]$$

$$\le \|z - x_0\|^2 + T[2L^{-1}\delta_f(M_2 + 12M + 27)$$

$$+2L^{-1}\delta_G(1 + L_0 + L(4M + M_2 + 11))] \le (2M + 2)^2$$

and

$$\|z - x_{t+1}\| \le 2M + 2.$$

Thus we have shown by induction that for all $t = 0, \ldots, T$.

$$\|z - x_t\| \leq 2M + 2.$$

and

$$\|x_t\| \leq 3M + 2.$$

Property (a) implies that for all $t = 0, \ldots, T$, (5.107) holds. It follows from (5.97), (5.98), (5.102), (5.103), (5.106), and (5.107) which holds for all $t = 0, \ldots, T$ that

$$\sum_{t=0}^{T-1} T^{-1} F(x_{t+1}) - \inf(F)$$

$$\leq 2^{-1} T^{-1} L \|z - x_0\|^2 + \delta_f (M_2 + 12M + 27)$$

$$+\delta_G (1 + L_0 + L(4M + M_2 + 11))$$

$$\leq 2LM^2 (T/2)^{-1} + \delta_f (M_2 + 12M + 27)$$

$$+\delta_G (1 + L_0 + L(4M + M_2 + 11))$$

$$\leq \max\{\delta_C, \delta_f\}(8((M_2 + 12M + 27)L + 1 + L_0)M^2$$

$$+M_2 + 12M + 27 + (1 + L_0 + L(4M + M_2 + 8))).$$

Theorem 5.6 is proved. □

Chapter 6
Continuous Subgradient Method

In this chapter we study the continuous subgradient algorithm for minimization of convex nonsmooth functions and for computing the saddle points of convex–concave functions, under the presence of computational errors. The problem is described by an objective function and a set of feasible points. For this algorithm we need a calculation of a subgradient of the objective function and a calculation of a projection on the feasible set. In each of these two calculations there is a computational error produced by our computer system. In general, these two computational errors are different. We show that our algorithm generates a good approximate solution, if all the computational errors are bounded from above by a small positive constant. Moreover, if we know the computational errors for the two calculations of our algorithm, we find out what approximate solution can be obtained and how much time one needs for this.

6.1 Bochner Integrable Functions

Let $(Y, \|\cdot\|)$ be a Banach space and $-\infty < a < b < \infty$. A function $x : [a, b] \to Y$ is strongly measurable on $[a, b]$ if there exists a sequence of functions

$$x_n : [a, b] \to Y,\ n = 1, 2, \ldots$$

such that for any integer $n \geq 1$ the set $x_n([a, b])$ is countable and the set

$$\{t \in [a, b] :\ x_n(t) = y\}$$

is Lebesgue measurable for any $y \in Y$, and

A. J. Zaslavski, *Convex Optimization with Computational Errors*, Springer Optimization and Its Applications 155, https://doi.org/10.1007/978-3-030-37822-6_6

$$x_n(t) \to x(t) \text{ as } n \to \infty$$

in $(Y, \|\cdot\|)$ for almost every (a. e.) $t \in [a, b]$.

Denote by $\mathrm{mes}(E)$ the Lebesgue measure of a Lebesgue measurable set $E \subset R^1$.

The function $x : [a, b] \to Y$ is Bochner integrable if it is strongly measurable and there exists a finite $\int_a^b \|x(t)\|dt$.

If $x : [a, b] \to Y$ is a Bochner integrable function, then for almost every (a. e.) $t \in [a, b]$,

$$\lim_{\Delta t \to 0} (\Delta t)^{-1} \int_t^{t+\Delta t} \|x(\tau) - x(t)\|d\tau = 0$$

and the function

$$y(t) = \int_a^t x(s)ds, \ t \in [a, b]$$

is continuous and a. e. differentiable on $[a, b]$.

Let $-\infty < \tau_1 < \tau_2 < \infty$. Denote by $W^{1,1}(\tau_1, \tau_2; Y)$ the set of all functions $x : [\tau_1, \tau_2] \to Y$ for which there exists a Bochner integrable function $u : [\tau_1, \tau_2] \to Y$ such that

$$x(t) = x(\tau_1) + \int_{\tau_1}^t u(s)ds, \ t \in (\tau_1, \tau_2]$$

(see, e.g., [10, 21]). It is known that if $x \in W^{1,1}(\tau_1, \tau_2; Y)$, then this equation defines a unique Bochner integrable function u which is called the derivative of x and is denoted by x'.

6.2 Convergence Analysis for Continuous Subgradient Method

The study of continuous subgradient algorithms is an important topic in optimization theory. See, for example, [6, 9, 20, 21] and the references mentioned therein. In this chapter we analyze its convergence under the presence of computational errors.

We suppose that X is a Hilbert space equipped with an inner product denoted by $\langle \cdot, \cdot \rangle$ which induces a complete norm $\|\cdot\|$.

Suppose that $f : X \to R^1 \cup \{\infty\}$ is a convex, lower semicontinuous and bounded from below function such that

$$\mathrm{dom}(f) := \{x \in X : \ f(x) < \infty\} \neq \emptyset$$

and that f possesses a minimizer. Set

$$\inf(f) = \inf(f(x) : x \in X\}$$

and

$$\operatorname{argmin}(f) = \{x \in X : f(x) = \inf(f)\}.$$

For each set $D \subset X$ put

$$\inf(f, D) = \inf\{f(z) : z \in D\},$$

$$\sup(f, D) = \sup\{f(z) : z \in D\}.$$

In Sect. 6.4 we will prove the following result which has no prototype in [92].

Theorem 6.1 *Let* $\delta \in (0, 1]$, $0 < \mu_1 < \mu_2$, $M \geq 1$,

$$z \in B_X(0, M) \cap argmin(f), \tag{6.1}$$

and let $\mu : [0, \infty) \to R^1$ *be a Lebesgue measurable function such that*

$$\mu_1 \leq \mu(t) \leq \mu_2 \text{ for all } t \in [0, \infty). \tag{6.2}$$

Assume that

$$T_0 \in [(2\delta)^{-1}, 5(4\delta)^{-1}], \tag{6.3}$$

$x \in W^{1,1}(0, T_0; X)$,

$$x(0) \in dom(f) \cap B_X(0, M) \tag{6.4}$$

and that for almost every $t \in [0, T_0]$,

$$x(t) \in dom(f) \tag{6.5}$$

and

$$B_X(x'(t), \delta) \cap (-\mu(t)\partial f(x(t))) \neq \emptyset. \tag{6.6}$$

Then the following two assertions hold.

1.

$$T_0^{-1} \int_0^{T_0} f(x(t)dt - \inf(f),$$

$$f(T_0^{-1}\int_0^{T_0} x(t)dt) - \inf(f),$$

$$\min\{f(x(t)) : t \in [0, T_0]\} - \leq \inf(f)$$

$$\leq (2M+2)^2\mu_1^{-1}\delta,$$

$$\|z - x(t)\| \leq 2M + 2, \ t \in [0, T_0].$$

and for almost every $t \in [0, T_0]$,

$$2\langle x(t) - z, x'(t)\rangle \leq 2\mu_1(f(z) - f(x(t)) + 2\delta(2M+2).$$

2. *Assume that* $M \geq 4$,

$$\delta \leq \min\{\mu_1(2M+2)^{-1}, \ 16^{-1}\} \tag{6.7}$$

and that

$$\{x \in X : \ f(x) \leq \inf(f) + 4\} \subset B_X(0, 4^{-1}M).$$

Then

$$\|x(T_0)\| \leq M.$$

Theorem 6.1 easily implies the following result.

Theorem 6.2 *Let* $\delta \in (0, 16^{-1}]$, $0 < \mu_1 < 1 \leq \mu_2$, $M \geq 1$, $\mu : [0, \infty) \to R^1$ *be a Lebesgue measurable function such that*

$$\mu_1 \leq \mu(t) \leq \mu_2 \text{ for all } t \in [0, \infty),$$

$$\delta \leq \mu_1(2M+2)^{-1},$$

$$\{x \in X : \ f(x) \leq \inf(f) + 4\} \subset B_X(0, 4^{-1}M),$$

$$T \geq 5(4\delta)^{-1},$$

$x \in W^{1,1}(0, T; X)$,

$$x(0) \in dom(f) \cap B_X(0, M)$$

and that for almost every $t \in [0, T]$,

$$x(t) \in dom(f)$$

and

$$B(x'(t), \delta) \cap (-\mu(t)\partial f(x(t))) \neq \emptyset.$$

Assume that $k \geq 1$ *is an integer,*

$$(2\delta)^{-1}k < T \leq (2\delta)^{-1}(k+1),$$

$$T_i = i(2\delta)^{-1},\ i = 0, \dots, k-1,\ T_k = T.$$

Then

$$\|x(t)\| \leq 3M + 2 \textit{ for all } t \in [0, T],$$

$$\|x(T_i)\| \leq M,\ i = 0, \dots, k$$

and for all integers $i = 0, \dots, k-1$,

$$f((T_{i+1} - T_i)^{-1} \int_{T_i}^{T_{i+1}} x(t)dt) - \inf(f),$$

$$\min\{f(x(t)) :\ t \in [T_i, T_{i+1}]\} - \inf(f)$$

$$\leq (T_{i+1} - T_i)^{-1} \int_{T_i}^{T_{i+1}} f(x(t))dt - \inf(f) \leq (2M+2)^2\delta\mu_1^{-1}.$$

6.3 An Auxiliary Result

Let $V \subset X$ be an open convex set and $g : V \to R^1$ be a convex locally Lipschitzian function.

Let $T > 0$, $x_0 \in X$ and let $u : [0, T] \to X$ be a Bochner integrable function. Set

$$x(t) = x_0 + \int_0^t u(s)ds,\ t \in [0, T].$$

Then $x : [0, T] \to X$ is differentiable and $x'(t) = u(t)$ for almost every $t \in [0, T]$.

Assume that

$$x(t) \in V \text{ for all } t \in [0, T].$$

We claim that the restriction of g to the set

$$\{x(t) :\ t \in [0, T]\}$$

is Lipschitzian. Indeed, since the set

$$\{x(t) : t \in [0, T]\}$$

is compact, the closure of its convex hull C is both compact and convex, and so the restriction of g to C is Lipschitzian. Hence the function

$$(g \cdot x)(t) := g(x(t)), \ t \in [0, T],$$

is absolutely continuous. It follows that for almost every $t \in [0, T]$, both the derivatives $x'(t)$ and $(g \cdot x)'(t)$ exist:

$$x'(t) = \lim_{h \to 0} h^{-1}[x(t+h) - x(t)], \tag{6.8}$$

$$(g \cdot x)'(t) = \lim_{h \to 0} h^{-1}[g(x(t+h)) - g(x(t))]. \tag{6.9}$$

We continue with the following fact (see Proposition 14.2 of [92]).

Proposition 6.3 *Assume that $t \in [0, T]$ and that both the derivatives $x'(t)$ and $(g \cdot x)'(t)$ exist. Then*

$$(g \cdot x)'(t) = \lim_{h \to 0} h^{-1}[g(x(t) + hx'(t)) - g(x(t))]. \tag{6.10}$$

Corollary 6.4 *Let $z \in X$ and $g(y) = \|z - y\|^2$ for all $y \in X$. Then for almost every $t \in [0, T]$, the derivative $(g \cdot x)'(t)$ exists and*

$$(g \cdot x)'(t) = 2\langle x'(t), x(t) - z\rangle.$$

6.4 Proof of Theorem 6.1

Set

$$\phi(t) = \|z - x(t)\|^2, \ t \in [0, T_0]. \tag{6.11}$$

In view of Corollary 6.4, for a. e. $t \in [0, T_0]$, there exist derivatives $x'(t)$, $\phi'(t)$, and

$$\phi'(t) = 2\langle x'(t), x(t) - z\rangle. \tag{6.12}$$

By (6.6), for a. e. $t \in [0, T_0]$, there exist

$$\xi(t) \in \partial f(x(t)) \tag{6.13}$$

such that

$$\|x'(t) + \mu(t)\xi(t)\| \le \delta. \tag{6.14}$$

It follows from (6.12) that for almost every $t \in [0, T_0]$,

$$\phi'(t) = 2\langle x(t) - z, x'(t)\rangle$$

$$= 2\langle x(t) - z, -\mu(t)\xi(t)\rangle + 2\langle x(t) - z, x'(t) + \mu(t)\xi(t)\rangle. \tag{6.15}$$

In view of (6.14), for almost every $t \in [0, T_0]$,

$$|\langle z - x(t), x'(t) + \mu(t)\xi(t)\rangle| \le \delta\|z - x(t)\|. \tag{6.16}$$

By (6.1), (6.13) and (6.17), for almost every $t \in [0, T_0]$,

$$\langle z - x(t), \xi(t)\rangle \le -f(x(t)) + f(z). \tag{6.17}$$

It follows from (6.1), (6.2), (6.15), and (6.16) that for almost every $t \in [0, T_0]$,

$$\phi'(t) \le 2\mu(t)(-f(x(t)) + f(z)) + 2\delta\|z - x(t)\|$$

$$\le 2\mu_1(-f(x(t)) + f(z)) + 2\delta\|z - x(t)\|. \tag{6.18}$$

We show that for all $t \in [0, T_0]$,

$$\|z - x(t)\| \le 2M + 2. \tag{6.19}$$

Assume the contrary. Then there exists

$$\tau \in (0, T_0]$$

such that

$$\|z - x(t)\| < 2M + 2, \ t \in [0, \tau), \tag{6.20}$$

$$\|z - x(\tau)\| = 2M + 2. \tag{6.21}$$

By (6.1), (6.4), (6.11), (6.18), (6.20), and (6.21),

$$8M + 4 = (2M + 2)^2 - 4M^2$$

$$\le \|z - x(\tau)\|^2 - \|z - x(0)\|^2$$

$$= \phi(\tau) - \phi(0) = \int_0^\tau 2\delta \|z - x(t)\| dt$$

$$\leq 2\tau\delta(2M + 2). \tag{6.22}$$

In view of (6.3) and (6.22),

$$5(4\delta)^{-1} \geq T_0 \geq \tau \geq (8M + 4)(4M + 4)^{-1}\delta^{-1} \geq 3(2\delta)^{-1}.$$

The contradiction we have reached proves that

$$\|z - x(t)\| \leq 2M + 2, \ t \in [0, T_0]. \tag{6.23}$$

By (6.18), for almost every $t \in [0, T_0]$

$$\phi'(t) \leq 2\mu_1(f(z) - f(x(t)) + 2\delta(2M + 2). \tag{6.24}$$

It follows from (6.1), (6.4), (6.11), (6.12), and (6.14) that

$$4M^2 \geq \phi(0) - \phi(T_0)$$

$$= -\int_0^{T_0} \phi'(t)dt \geq 2\mu_1 \int_0^{T_0} (f(x(t)) - f(z))dt - 2\delta T_0(2M + 2). \tag{6.25}$$

By (6.3) and (6.25),

$$T_0^{-1} \int_0^{T_0} (f(x(t)) - f(z))dt$$

$$\leq (2\mu_1)^{-1}[4M^2 T_0^{-1} + 2\delta(M + 2)]$$

$$\leq 2\delta(2\mu_1)^{-1}(4M^2 + 4M + 4) = (2M + 2)^2 \mu_1^{-1}\delta. \tag{6.26}$$

Therefore assertion 1 is proved.

Let us complete the proof of assertion 2. By (6.7) and (6.16),

$$T_0^{-1} \int_0^{T_0} (f(x(t)) - f(z))dt \leq 1.$$

This implies that

$$4T_0^{-1}\text{mes}\{t \in [0, T_0] : \ f(x(t)) > f(z) + 4\} < 1,$$

$$\text{mes}\{t \in [0, T_0] : \ f(x(t)) > f(z) + 4\} < 4^{-1}T_0.$$

This implies that there exists

$$t_0 \in [3T_0/4, T_0] \tag{6.27}$$

such that

$$f(x(t_0)) \le f(z) + 4. \tag{6.28}$$

By (6.1), (6.28), and the assumptions of the assertion,

$$\|x(t_0)| \le M/4, \ \|z\| \le M/4. \tag{6.29}$$

It follows from (6.1), (6.3), (6.11), (6.12), (6.18), (6.23), and (6.27) that

$$\|z - x(T_0)\|^2 - \|z - x(t_0)\|^2$$

$$= \int_{t_0}^{T_0} \phi'(t)dt$$

$$\le 28(T_0 - t_0)(2M + 2) = (3/2)T_0\delta(2M + 2) \le 4M + 4.$$

In view of the relation above, (6.1) and (6.29),

$$\|z - x(T_0)\|^2 \le M^2/4 + 4M + 4 \le (M/2 + 4)^2$$

and

$$\|x(T_0)\| \le \|z\| + M/2 + 4 \le 3M/4 + 4 \le M.$$

Assertion 2 is proved. This completes the proof of Theorem 6.1. □

6.5 Continuous Subgradient Method for Zero-Sum Games

In this section we study continuous subgradient method for zero-sum games with two players. This topic was not considered in [92].

Let $(X, \langle\cdot,\cdot\rangle)$, $(Y, \langle\cdot,\cdot\rangle)$ be Hilbert spaces equipped with the complete norms $\|\cdot\|$ which are induced by their inner products. Let a borelian function $f : X \times Y \to R^1 \cup \{-\infty\}$ possesses the following property:

$$\{(x, y) \in X \times Y : \ f(x, y) > -\infty\} = X \times C,$$

where C is a nonempty bounded convex subset of Y.

Recall that for for each concave function $g : Y \to R^1 \cup \{-\infty\}$ and each $x \in Y$ satisfying $g(x) > -\infty$,

$$\partial g(x) = \{l \in Y : \langle l, y - x \rangle \geq g(y) - g(x) \text{ for all } y \in Y\}.$$

Clearly, for each $x \in Y$ satisfying $g(x) > -\infty$,

$$\partial g(x) = -(\partial(-g)(x)).$$

Suppose that the following properties hold:

(i) for each $y \in C$, the function $f(\cdot, y) : X \to R^1$ is convex and lower semicontinuous;
(ii) for each $x \in X$, the function $f(x, \cdot) : Y \to R^1 \cup \{-\infty\}$ is concave and upper semicontinuous.

Let

$$x_* \in X \text{ and } y_* \in C \tag{6.30}$$

satisfy

$$f(x_*, y) \leq f(x_*, y_*) \leq f(x, y_*) \tag{6.31}$$

for each $x \in X$ and each $y \in C$.

Let $M > 0$,

$$\|x_*\| < M/8, \tag{6.32}$$

$$C \subset B_Y(0, M), \tag{6.33}$$

f is bounded from above on

$$B_X(0, M/3) \times C$$

and let for every $y \in C$,

$$\{x \in X : f(x, y) \leq f(x_*, y_*) + 4\} \subset B_X(0, M/3). \tag{6.34}$$

In Sect. 6.7 we prove the following result.

Theorem 6.5 *Let*

$$0 < \mu_1 < \mu_2, \tag{6.35}$$

$$\mu \in [\mu_1, \mu_2], \tag{6.36}$$

$$\delta_{f,1}, \delta_{f,2} \in (0, 1),\ \delta_{f,1} < \mu_1/M. \tag{6.37}$$

Assume that $T > 0$, $x \in W^{1,1}(0, T; X)$, $y \in W^{1,1}(0, T; Y)$,

$$y(0) \in C, \tag{6.38}$$

for almost every $t \in [0, T]$,

$$y(t) \in C, \tag{6.39}$$

$$B_Y(y'(t), \delta_{f,2}) \cap (\mu \partial_y f(x(t), y(t))) \neq \emptyset, \tag{6.40}$$

$$B_X(x'(t), \delta_{f,1}) \cap (-\mu \partial_x f(x(t), y(t))) \neq \emptyset \tag{6.41}$$

and that

$$\|x(0)\| \leq M/8. \tag{6.42}$$

Then

$$\|x(t)\| \leq M \textit{ for all } t \in [0, T], \tag{6.43}$$

$$|T^{-1} \int_0^T f(x(t), y(t))dt - f(x_*, y_*)|$$

$$\leq 2M^2T^{-1}\mu_1^{-1} + 2M\mu_1^{-1} \max\{\delta_{f,1}, \delta_{f,2}\}, \tag{6.44}$$

$$|f(T^{-1} \int_0^T x(t)dt, T^{-1} \int_0^T y(t)dt) - T^{-1} \int_0^T f(x(t), y(t))dt|$$

$$\leq 2M^2T^{-1}\mu_1^{-1} + 2M\mu_1^{-1} \max\{\delta_{f,1}, \delta_{f,2}\}, \tag{6.45}$$

for each $v \in C$,

$$f(T^{-1} \int_0^T x(t)dt, v)$$

$$\leq f(T^{-1} \int_0^T x(t)dt, T^{-1} \int_0^T y(t)dt)$$

$$+ 4M^2T^{-1}\mu_1^{-1} + 4M\mu_1^{-1} \max\{\delta_{f,1}, \delta_{f,2}\}, \tag{6.46}$$

for each $z \in X$,

$$f(z, T^{-1}\int_0^T y(t)dt)$$

$$\geq f(T^{-1}\int_0^T x(t)dt, T^{-1}\int_0^T y(t)dt)$$

$$-4M^2T^{-1}\mu_1^{-1} - 3M\mu_1^{-1}\max\{\delta_{f,1}, \delta_{f,2}\}. \quad (6.47)$$

Clearly, the best choice of T should be at the same order as $\max\{\delta_{f,1}, \delta_{f,2}\}^{-1}$.

6.6 An Auxiliary Result

Proposition 6.6 *Let*

$$0 < \mu_1 < \mu_2, \quad (6.48)$$

$$\mu \in [\mu_1, \mu_2], \quad (6.49)$$

$$\delta_{f,1}, \delta_{f,2} \in (0, 1),\ \ \delta_{f,1} < \mu_1/M. \quad (6.50)$$

Assume that $T > 0$, $x \in W^{1,1}(0, T; X)$, $y \in W^{1,1}(0, T; Y)$,

$$y(0) \in C, \quad (6.51)$$

and that for almost every $t \in [0, T]$,

$$y(t) \in C, \quad (6.52)$$

$$B_Y(y'(t), \delta_{f,2}) \cap (\mu\partial_y f(x(t), y(t))) \neq \emptyset, \quad (6.53)$$

$$B_X(x'(t), \delta_{f,1}) \cap (-\mu\partial_x f(x(t), y(t))) \neq \emptyset. \quad (6.54)$$

Then the following assertions hold.

1. Let $z \in X$ *and* $v \in C$. *Then for almost every* $t \in [0, T]$,

$$(d/dt)(\|x(t) - z\|^2)$$

$$\leq 2\delta_{f,1}\|z - x(t)\| + 2\mu(f(z, y(t)) - f(x(t), y(t))), \quad (6.55)$$

$$(d/dt)(\|y(t) - v\|^2)$$

$$\leq 2\delta_{f,2}\|v - y(t)\| + 2\mu(-f(x(t), v) + f(x(t), y(t))). \quad (6.56)$$

2. *Let*

$$\|x(0)\| \le M/8. \tag{6.57}$$

Then for each $t \in [0, T]$,

$$\|x(t)\| \le M \tag{6.58}$$

and for each $z \in X$, *each* $v \in C$ *and almost every* $t \in [0, T]$,

$$(d/dt)(\|x(t) - z\|^2)$$

$$\le 2\delta_{f,1}(\|z\| + M) + 2\mu(f(z, y(t)) - f(x(t), y(t))), \tag{6.59}$$

$$(d/dt)(\|y(t) - v\|^2)$$

$$\le 4\delta_{f,2}M + 2\mu(-f(x(t), v) + f(x(t), y(t))). \tag{6.60}$$

Proof Let $z \in X$ and $v \in C$. For all $t \in [0, T]$ set

$$\phi_1(t) = \|z - x(t)\|^2,$$

$$\phi_2(t) = \|v - y(t)\|^2. \tag{6.61}$$

By (6.53) and (6.54), for almost every $t \in [0, T]$, there exist

$$\xi(t) \in \partial_x f(x(t), y(t)), \tag{6.62}$$

and

$$\eta(t) \in \partial_y f(x(t), y(t)) \tag{6.63}$$

such that

$$\|x'(t) + \mu\xi(t)\| \le \delta_{f,1} \tag{6.64}$$

and

$$\|y'(t) - \mu\eta(t)\| \le \delta_{f,2}. \tag{6.65}$$

Corollary 6.4 and (6.61) imply that for almost every $t \in [0, T]$,

$$\phi_1'(t) = 2\langle x'(t), x(t) - z\rangle, \tag{6.66}$$

$$\phi_2(t) = 2\langle y'(t), y(t) - v\rangle. \tag{6.67}$$

In view of (6.66) and (6.67), for almost every $t \in [0, T]$,

$$\phi_1'(t) = 2\langle x'(t), x(t) - z\rangle$$

$$= 2\langle x(t) - z, -\mu\xi(t)\rangle + 2\langle x(t) - z, x'(t) + \mu\xi(t)\rangle \tag{6.68}$$

and

$$\phi_2'(t) = 2\langle y(t) - v, y'(t)\rangle$$

$$= 2\langle y(t) - v, \mu\eta(t)\rangle + 2\langle y(t) - v, y'(t) - \mu\eta(t)\rangle. \tag{6.69}$$

By (6.64) and (6.65), for almost every $t \in [0, T]$,

$$|\langle x(t) - z, x'(t) + \mu\xi(t)\rangle| \leq \delta_{f,1}\|z - x(t)\| \tag{6.70}$$

and

$$|\langle y(t) - v, y'(t) - \mu\eta(t)\rangle| \leq \delta_{f,2}\|v - y(t)\|. \tag{6.71}$$

It follows from (6.62) and (6.63) that for almost every $t \in [0, T]$,

$$\langle z - x(t), \xi(t)\rangle \leq f(z, y(t)) - f(x(t), y(t)) \tag{6.72}$$

and

$$\langle v - y(t), \eta(t)\rangle \geq f(x(t), v) - f(x(t), y(t)). \tag{6.73}$$

By (6.61) and (6.68)–(6.73), for almost every $t \in [0, T]$,

$$(d/dt)(\|x(t) - z\|^2)$$

$$\leq 2\delta_{f,1}\|z - x(t)\| + 2\mu(f(z, y(t)) - f(x(t), y(t))) \tag{6.74}$$

and

$$(d/dt)(\|y(t) - v\|^2)$$

$$\leq 2\delta_{f,2}\|v - y(t)\| + 2\mu(-f(x(t), v) + f(x(t), y(t))). \tag{6.75}$$

Thus assertion 1 is proved.

Let us complete the proof of assertion 2. We show that

$$\|x(t)\| \leq M \text{ for all } t \in [0, T].$$

In view of (6.32) and (6.57),

$$\|x_* - x(0)\| \le M/4. \tag{6.76}$$

By (6.31) and (6.55) with $z = x_*$, for almost every $t \in [0, T]$,

$$(d/dt)(\|x(t) - x_*\|^2)$$
$$\le 2\delta_{f,1}\|x_* - x(t)\| + 2\mu(f(x_*, y(t)) - f(x(t), y(t)))$$
$$\le 2\delta_{f,1}\|x_* - x(t)\| + 2\mu(f(x_*, y_*) - f(x(t), y(t))). \tag{6.77}$$

Define

$$E = \{\tau \in [0, T] :$$
$$\|x(t) - x_*\| \le (3/4)M \text{ for all } t \in 0, \tau]\}. \tag{6.78}$$

In view of (6.76),

$$E \ne \emptyset.$$

Set

$$\tau_0 = \sup E. \tag{6.79}$$

Since the function x is continuous we have

$$\tau_0 \in E. \tag{6.80}$$

In order to complete the proof of assertion 2 it is sufficient to show that

$$\tau_0 = T_0.$$

Assume the contrary. Then

$$\tau_0 < T_0. \tag{6.81}$$

In view of (6.76),

$$\tau > 0.$$

Relations (6.78)–(6.81) imply that

$$\|x(\tau_0) - x_*\| = (3/4)M. \tag{6.82}$$

There exists

$$\tau_1 \in (0, \tau_0)$$

such that for all $t \in [\tau_1, \tau_0]$ we have

$$\|x(t) - x_*\| \geq (2/3)M.$$

Together with (6.32) this implies that for all $t \in [\tau_1, \tau_0]$,

$$\|x(t)\| \geq (2/3)M - M/8 > M/3.$$

By the relation above, (6.34) and (6.52), for almost every $t \in [\tau_1, \tau_0]$,

$$f(x(t), y(t)) > f(x_*, y_*) + 4. \tag{6.83}$$

It follows from (6.49) and (6.77)–(6.80) that

$$0 \leq \|x(\tau_0) - x_*\|^2 - \|x(\tau_1) - x_*\|^2$$

$$\leq 2(\tau_0 - \tau_1)\delta_{f,1}M - 8\mu_1(\tau_0 - \tau_1)$$

and

$$4\mu_1 \leq \delta_{f,1}M.$$

This contradicts (6.50). The contradiction we have reached proves that

$$\tau_0 = T_0$$

and completes the proof of assertion 2 and Proposition 6.6. □

6.7 Proof of Theorem 6.5

Proposition 6.6 and the assumptions of the theorem imply that for each $t \in [0, T]$,

$$\|x(t)\| \leq M \tag{6.84}$$

and for each $z \in X$, each $v \in C$ and almost every $t \in [0, T]$,

$$(d/dt)(\|x(t) - z\|^2)$$

$$\leq 2\delta_{f,1}(\|z\| + M) + 2\mu(f(z, y(t)) - f(x(t), y(t))), \tag{6.85}$$

$$(d/dt)(\|y(t) - v\|^2)$$

$$\leq 4\delta_{f,2}M + 2\mu(-f(x(t), v) + f(x(t), y(t))). \tag{6.86}$$

By (6.85) and (6.86), for each $z \in X$, each $v \in C$,

$$T^{-1}(\|x(T) - z\|^2 - \|x(0) - z\|^2)$$

$$+2\mu T^{-1} \int_0^T f(x(t), y(t))dt - 2\mu T^{-1} \int_0^T f(z, y(t))dt$$

$$\leq 2\delta_{f,1}(\|z\| + M) \tag{6.87}$$

and

$$T^{-1}(\|y(T) - v\|^2 - \|y(0) - v\|^2)$$

$$-2\mu T^{-1} \int_0^T f(x(t), y(t))dt + 2\mu T^{-1} \int_0^T f(x(t), v)dt$$

$$\leq 4\delta_{f,2}M. \tag{6.88}$$

Let $z \in X$ satisfy

$$\|z\| \leq M \tag{6.89}$$

and $v \in C$. Relations (6.84), (6.85), and (6.89) imply that

$$T^{-1} \int_0^T f(x(t), y(t))dt - T^{-1} \int_0^T f(z, y(t))dt$$

$$\leq (2\mu T)^{-1} 4M^2 + 2M\delta_{f,1}(2\mu)^{-1}. \tag{6.90}$$

By (6.33) and (6.88),

$$T^{-1} \int_0^T f(x(t), v)dt - T^{-1} \int_0^T f(x(t), y(t))dt$$

$$\leq (2\mu T)^{-1} 4M^2 + 4M\delta_{f,2}(2\mu)^{-1}. \tag{6.91}$$

It follows from (6.30), (6.31), (6.90) with $v = y_*$ and (6.91) with $z = x_*$ that

$$T^{-1} \int_0^T f(x(t), y(t))dt - f(x_*, y_*)$$

$$\geq T^{-1}\int_0^T f(x(t),y(t))dt - f(T^{-1}\int_0^T x(t)dt, y_*)$$

$$\geq T^{-1}\int_0^T f(x(t),y(t))dt - T^{-1}\int_0^T f(x(t),y_*)dt$$

$$\geq -(\mu T)^{-1}2M^2 - 2M\delta_{f,2}\mu^{-1} \tag{6.92}$$

and

$$T^{-1}\int_0^T f(x(t),y(t))dt - f(x_*,y_*)$$

$$\leq T^{-1}\int_0^T f(x(t),y(t))dt - f(x_*, T^{-1}\int_0^T y(t)dt)$$

$$\leq T^{-1}\int_0^T f(x(t),y(t))dt - T^{-1}\int_0^T f(x_*,y(t))dt$$

$$\leq (\mu T)^{-1}2M^2 + M\delta_{f,1}\mu^{-1}. \tag{6.93}$$

Set

$$x_T = T^{-1}\int_0^T x(t)dt,$$

$$y_T = T^{-1}\int_0^T y(t)dt. \tag{6.94}$$

Since the set C is convex and closed we have

$$y_T \in C. \tag{6.95}$$

Relations (6.33), (6.84), (6.94), and (6.95) imply that

$$\|x_T\|,\ \|y_T\| \leq M. \tag{6.96}$$

By (6.84), (6.90) with $z = x_T$ and (6.96),

$$T^{-1}\int_0^T f(x(t),y(t))dt - T^{-1}\int_0^T f(x_T,y(t))dt$$

$$\leq 2M^2(\mu T)^{-1} + 2\delta_{f,1}M\mu^{-1}. \tag{6.97}$$

By (6.52), (6.91) with $v = y_T$ and (6.96),

$$T^{-1} \int_0^T f(x(t), y(t))dt - T^{-1} \int_0^T f(x(t), y_T)dt$$

$$\geq -2M^2(\mu T)^{-1} - 2\delta_{f,2} M \mu^{-1}. \tag{6.98}$$

Since the function $f(\cdot, y_T)$ is convex and lower semicontinuous it follows from (6.94) and (6.98) that

$$f(x_T, y_T) - T^{-1} \int_0^T f(x(t), y_T)dt$$

$$= f(T^{-1} \int_0^T x(t)dt, y_T) - T^{-1} \int_0^T f(x(t), y_T)dt$$

$$\leq T^{-1} \int_0^T f(x(t), y_T)dt - T^{-1} \int_0^T f(x(t), y(t))dt$$

$$\leq 2M^2(\mu T)^{-1} + 2\delta_{f,2} M \mu^{-1}. \tag{6.99}$$

Since the function $f(x_T, \cdot)$ is concave and upper semicontinuous it follows from (6.94) and (6.97) that

$$f(x_T, y_T) - T^{-1} \int_0^T f(x(t), y(t))dt$$

$$= f(x_T, T^{-1} \int_0^T y(t)dt) - T^{-1} \int_0^T f(x(t), y(t))dt$$

$$\geq T^{-1} \int_0^T f(x_T, y(t))dt - T^{-1} \int_0^T f(x(t), y(t))dt$$

$$\geq -2M^2(\mu T)^{-1} - 2\delta_{f,1} M \mu^{-1}. \tag{6.100}$$

By (6.99) and (6.100),

$$|f(x_T, y_T) - T^{-1} \int_0^T f(x(t), y(t))dt|$$

$$\leq \max\{2M^2(\mu T)^{-1} + 2\delta_{f,1} M \mu^{-1}, \ 2M^2(\mu T)^{-1} + 2\delta_{f,2} M \mu^{-1}\}$$

$$= 2M^2(\mu T)^{-1} + 2M\mu^{-1} \max\{\delta_{f,1}, \delta_{f,2}\}. \tag{6.101}$$

Let

$$v \in C, \; z \in B_X(0, M). \tag{6.102}$$

Then (6.90) and (6.91) hold. It follows from (6.94) and properties (i) and (ii) that

$$T^{-1} \int_0^T f(z, y(t)dt \le f(z, T^{-1} \int_0^T y(t)dt) = f(z, y_T) \tag{6.103}$$

and

$$T^{-1} \int_0^T f(x(t), v)dt) \ge f(T^{-1} \int_0^T x(t)dt, v) = f(x_T, v). \tag{6.104}$$

It follows from (6.90) and (6.103) that

$$\begin{aligned} &T^{-1} \int_0^T f(x(t), y(t))dt - f(z, y_T) \\ &\le T^{-1} \int_0^T f(x(t), y(t))dt - T^{-1} \int_0^T f(z, y(t))dt \\ &\le 2M^2(\mu T)^{-1} + M\mu^{-1}\delta_{f,1}. \end{aligned} \tag{6.105}$$

In view of (6.101) and (6.105),

$$\begin{aligned} &f(z, y_T) \\ &\ge T^{-1} \int_0^T f(x(t), y(t))dt - 2M^2(\mu T)^{-1} - M\mu^{-1}\delta_{f,1} \\ &\ge f(x_T, y_T) - 2M\mu^{-1} \max\{\delta_{f,1}, \delta_{f,2}\} \\ &\quad - 4M^2(\mu T)^{-1} - M\mu^{-1}\delta_{f,1}. \end{aligned} \tag{6.106}$$

It follows from (6.91) and (6.104) that

$$\begin{aligned} &T^{-1} \int_0^T f(x(t), y(t))dt - f(x_T, v) \\ &\ge T^{-1} \int_0^T f(x(t), y(t))dt - T^{-1} \int_0^T f(x(t), v)dt \\ &\le -2M^2(\mu T)^{-1} - 2M\mu^{-1}\delta_{f,2}. \end{aligned} \tag{6.107}$$

In view of (6.101) and (6.107),

$$f(x_T, v)$$

$$\leq T^{-1} \int_0^T f(x(t), y(t))dt + 2M^2(\mu T)^{-1} + 2M\mu^{-1}\delta_{f,2}$$

$$\leq f(x_T, y_T) + 2M\mu^{-1} \max\{\delta_{f,1}, \delta_{f,2}\}$$

$$+4M^2(\mu T)^{-1} + 2\delta_{f,2}M\mu^{-1}.$$

If $\|z\| > M$, then it follows from (6.34), (6.93), (6.96), and (6.101) that

$$f(z, y_T) \geq f(x_*, y_*) + 4$$

$$\geq T^{-1} \int_0^T f(x(t), y(t))dt - 2M^2(\mu T)^{-1} - M\mu^{-1}\delta_{f,1}$$

$$\geq f(x_T, y_T) - 2M\mu^{-1} \max\{\delta_{f,1}, \delta_{f,2}\} - 4M^2(\mu T)^{-1} - \delta_{f,1}M\mu^{-1}.$$

Theorem 6.5 is proved. □

6.8 Continuous Subgradient Projection Method

Let X be a Hilbert space with an inner product $\langle \cdot, \cdot \rangle$ which induces a complete norm $\| \cdot \|$. Let C be a nonempty, convex, and closed set in the Hilbert space X, U be an open and convex subset of X such that

$$C \subset U,$$

and $f : U \to R^1$ be a convex locally Lipschitzian function.

Let $x \in U$ and $\xi \in X$. Set

$$f^0(x, \xi) = \lim_{t \to 0^+} t^{-1}[f(x + t\xi) - f(x)],$$

$$\partial f(x; \xi) = \{l \in \partial f(x) : \langle l, \xi \rangle = f^0(x, \xi)\}.$$

It is a well-known fact of the convex analysis that

$$\partial f(x; \xi) \neq \emptyset.$$

Let $M > 1$, $L > 0$ and assume that

$$C \subset B_X(0, M-1),\ \{y \in X :\ d(y, C) \leq 1\} \subset U,$$

$$|f(v_1) - f(v_2)| \leq L\|v_1 - v_2\|$$

$$\text{for all } v_1, v_2 \in B_X(0, M+1) \cap U. \tag{6.108}$$

We will prove the following result which also has no prototype in [92].

Theorem 6.7 *Let* $\delta_f, \delta_C \in (0, 1]$,

$$0 < \mu_1 < \mu_2,\ \mu_1 \leq 1, \tag{6.109}$$

$T_0 > 0$,

$$\mu \in [\mu_1, \mu_2]. \tag{6.110}$$

Assume that $x \in W^{1,1}(0, T_0; X)$,

$$d(x(0), C) \leq \delta_C \tag{6.111}$$

and that for almost every $t \in [0, T_0]$*, there exists* $\xi(t) \in X$ *such that*

$$B_X(\xi(t), \delta_f) \cap \partial f(x(t); x'(t)) \neq \emptyset, \tag{6.112}$$

$$P_C(x(t) - \mu\xi(t)) \in B_X(x(t) + x'(t), \delta_C). \tag{6.113}$$

Then

$$f(T_0^{-1} \int_0^T x(s)ds) - \inf(f, C),$$

$$\min\{f(x(t)) :\ t \in [0, T_0]\} - \inf(f, C)$$

$$\leq T_0^{-1} \int_0^T f(x(s))ds - \inf(f, C)$$

$$\leq (2M^2\mu_1^{-1} + 2(M+1))T_0^{-1}$$

$$+6\delta_f M + \mu_1^{-1}\delta_C(4M + \mu_2(L+1)) + 2\delta_C L T_0^{-1}.$$

Clearly, the best choice of T should be at the same order as $\max\{\delta_f, \delta_C\}^{-1}$.

6.9 An Auxiliary Result

Lemma 6.8 *Let D be a nonempty convex closed subset of X, $T, \delta > 0$, $x \in W^{1,1}(0, T; X)$,*

$$d(x(0), D) \leq \delta, \tag{6.114}$$

$$D_\delta = \{x \in X : d(x, D) \leq \delta\}, \tag{6.115}$$

$$B_X(x(t) + x'(t), \delta) \cap D \neq \emptyset \text{ for almost every } t \in [0, T]. \tag{6.116}$$

Then

$$x(t) \in D_\delta \text{ for all } t \in [0, T].$$

Proof For almost every $t \in [0, T]$ set

$$\phi(t) = x(t) + x'(t). \tag{6.117}$$

It is clear that $\phi : [0, T] \to X$ is a Bochner integrable function. In view of (6.116) and (6.117), for almost every $t \in [0, T]$,

$$B(\phi(t), \delta) \cap D \neq \emptyset. \tag{6.118}$$

Clearly, D_δ is a closed convex set, for each $x \in D_\delta$,

$$B(x, \delta) \cap D \neq \emptyset \tag{6.119}$$

and in view of (6.115) and (6.118),

$$\phi(t) \in D_\delta \text{ for almost every } t \in [0, T]. \tag{6.120}$$

Evidently, the function $e^s\phi(s)$, $s \in [0, T]$ is Bochner integrable.

We claim that for all $t \in [0, T]$,

$$x(t) = e^{-t}x(0) + e^{-t}\int_0^t e^s\phi(s)ds. \tag{6.121}$$

Clearly, (6.121) holds for $t = 0$. For every $t \in (0, T]$ we have

$$\int_0^t e^s\phi(s)ds = \int_0^t e^s(x(s) + x'(s))ds$$

$$= \int_0^t (e^s x(s))'ds = e^t x(t) - x(0).$$

This implies (6.121) for all $t \in [0, T]$. By (6.121), for all $t \in [0, T]$,

$$x(t) = e^{-t}x(0) + (1 - e^{-t})(1 - e^{-t})^{-1}e^{-t} \int_0^t e^s \phi(s)ds$$

$$= e^{-t}x(0) + (1 - e^{-t}) \int_0^t e^s (e^t - 1)^{-1} \phi(s)ds. \tag{6.122}$$

In view of (6.120), for all $t \in [0, T]$,

$$\int_0^t e^s (e^t - 1)^{-1} \phi(s)ds \in D_\delta. \tag{6.123}$$

Relations (6.114), (6.122), and (6.123) imply that

$$x(t) \in D_\delta \text{ for all } t \in [0, T].$$

Lemma 6.8 is proved. □

6.10 Proof of Theorem 6.7

Let

$$z \in C. \tag{6.124}$$

Set

$$C_{\delta_C} = \{x \in X : d(x, C) \le \delta_C\}. \tag{6.125}$$

For almost every $t \in [0, T_0]$ set

$$\phi(t) = x(t) + x'(t). \tag{6.126}$$

Lemma 6.8, (6.111), and (6.113) imply that

$$x(t) \in C_{\delta_C} \text{ for all } t \in [0, T_0]. \tag{6.127}$$

It follows from (6.125) and (6.127) that for every $t \in [0, T_0]$, there exists

$$\widehat{x}(t) \in C \tag{6.128}$$

such that

$$\|x(t) - \widehat{x}(t)\| \le \delta_C. \tag{6.129}$$

By (6.128) and Lemma 2.2, for almost every $t \in [0, T_0]$,

$$\langle \widehat{x}(t) - P_C(x(t) - \mu\xi(t)), x(t) - \mu\xi(t) - P_C(x(t) - \mu\xi(t))\rangle \le 0. \tag{6.130}$$

Inequality (6.130) implies that for almost every $t \in [0, T_0]$,

$$\langle x(t) - P_C(x(t) - \mu\xi(t)), x(t) - \mu\xi(t) - P_C(x(t) - \mu\xi(t))\rangle$$

$$\le \langle x(t) - \widehat{x}(t), x(t) - \mu\xi(t) - P_C(x(t) - \mu\xi(t))\rangle. \tag{6.131}$$

It follows from (6.124) and Lemma 2.2 that

$$\langle z - P_C(x(t) - \mu\xi(t)), x(t) - \mu\xi(t) - P_C(x(t) - \mu\xi(t))\rangle \le 0. \tag{6.132}$$

In view of (6.42), for almost every $t \in [0, T_0]$ there exists

$$\widehat{\xi}(t) \in \partial f(x(t); x'(t)) \tag{6.133}$$

such that

$$f^0(x(t), x'(t)) = \langle \widehat{\xi}(t), x'(t)\rangle, \tag{6.134}$$

$$\|\xi(t) - \widehat{\xi}(t)\| \le \delta_f. \tag{6.135}$$

In view of (6.133), for almost every $t \in [0, T_0]$,

$$f(z) \ge f(x(t)) + \langle \widehat{\xi}(t), z - x(t)\rangle. \tag{6.136}$$

By (6.113), for almost every $t \in [0, T_0]$,

$$\|(x(t) + x'(t)) - P_C(x(t) - \mu\xi(t))\| \le \delta_C. \tag{6.137}$$

Relations (6.126) and (6.137) imply that for almost every $t \in [0, T_0]$,

$$\langle z - x(t) - x'(t), x(t) - \mu\xi(t) - x(t) - x'(t)\rangle$$

$$= \langle z - \phi(t), x(t) - \mu\xi(t) - \phi(t)\rangle$$

$$= \langle z - \phi(t), x(t) - \mu\xi(t) - P_C(x(t) - \mu\xi(t))\rangle$$

$$+\langle z - \phi(t), P_C(x(t) - \mu\xi(t)) - \phi(t)\rangle$$

$$\leq \langle z - \phi(t), x(t) - \mu\xi(t) - P_C(x(t) - \mu\xi(t))\rangle$$

$$+ \delta_C \|z - \phi(t)\|. \tag{6.138}$$

In view of (6.110), (6.125), and (6.127), for all $t \in [0, T_0]$,

$$\|x(t)\| \leq M. \tag{6.139}$$

It follows from (6.110), (6.113), (6.126), and (6.139) that for almost every $t \in [0, T_0]$,

$$\|\phi(t)\| = \|x(t) + x'(t)\| \leq M, \tag{6.140}$$

$$\|x'(t)\| \leq 2M. \tag{6.141}$$

By (6.110), (6.124), (6.126), (6.138), and (6.140), for almost every $t \in [0, T_0]$,

$$\langle z - x(t) - x'(t), -\mu\xi(t) - x'(t)\rangle$$

$$\leq \langle z - \phi(t), x(t) - \mu\xi(t) - P_C(x(t) - \mu\xi(t))\rangle + 2M\delta_C. \tag{6.142}$$

By (6.110)–(6.112), (6.126), (6.132), and (6.137),

$$\langle z - \phi(t), x(t) - \mu\xi(t) - P_C(x(t) - \mu\xi(t))\rangle$$

$$\leq \langle z - P_C(x(t) - \mu\xi(t)), x(t) - \mu\xi(t) - P_C(x(t) - \mu\xi(t))\rangle$$

$$+\delta_C \|x(t) - \mu\xi(t) - P_C(x(t) - \mu\xi(t))\|$$

$$\leq \delta_C(2M + \mu_2(L + 1)). \tag{6.143}$$

In view of (6.142) and (6.143), for almost every $t \in [0, T_0]$,

$$\langle x(t) + x'(t) - z, \mu\xi(t) + x'(t)\rangle \leq \delta_C(4M + \mu_2(L + 1)). \tag{6.144}$$

It follows from (6.110), (6.124), (6.135), (6.136), and (6.141) that for almost every $t \in [0, T_0]$,

$$f(x(t)) - f(z)$$

$$\leq \langle \widehat{\xi}(t), x(t) - z\rangle$$

$$= \langle \xi(t), x(t) - z + x'(t)\rangle - \langle \xi(t), x'(t)\rangle$$

$$+\langle \widehat{\xi}(t) - \xi(t), x(t) - z + x'(t)\rangle$$

$$\langle \xi(t) - \widehat{\xi}(t), x'(t) \rangle$$

$$\leq \langle \xi(t), x(t) - z + x'(t) \rangle - \langle \xi(t), x'(t) \rangle + 4M\delta_f. \tag{6.145}$$

Relations (6.144) and (6.145) imply that for almost every $t \in [0, T_0]$,

$$f(x(t)) - f(z)$$

$$\leq \mu^{-1}\langle -x'(t), x(t) + x'(t) - z \rangle + \mu^{-1}(4M + \mu_2(L+1))\delta_C$$

$$- \langle \xi(t), x'(t) \rangle + 4M\delta_f. \tag{6.146}$$

By (6.110) and (6.146), for almost every $t \in [0, T_0]$,

$$f(x(t)) - f(z)$$

$$\leq -\mu^{-1}\|x'(t)\|^2 - \mu^{-1}\langle x'(t), x(t) - z \rangle$$

$$- \langle \xi(t), x'(t) \rangle + \delta_C(4M + \mu_2(L+1))\mu_1^{-1} + 4M\delta_f \tag{6.147}$$

and

$$\mu^{-1}\|x'(t)\|^2 + \mu^{-1}\langle x'(t), x(t) - z \rangle$$

$$+\langle \xi(t), x'(t) \rangle + f(x(t)) - f(z)$$

$$\leq \delta_C(4M + \mu_2(L+1))\mu_1^{-1} + 4M\delta_f. \tag{6.148}$$

In view of (6.135), (6.141), and (6.148),

$$\mu^{-1}\|x'(t)\|^2 + \mu^{-1}\langle x'(t), x(t) - z \rangle$$

$$+\langle \widehat{\xi}(t), x'(t) \rangle + f(x(t)) - f(z)$$

$$\leq \delta_C(4M + \mu_2(L+1))\mu_1^{-1} + 6M\delta_f. \tag{6.149}$$

It follows from (6.134), (6.149), and Corollary 6.4 that for almost every $t \in [0, T_0]$,

$$(2\mu)^{-1}(d/dt)(\|x(t) - z\|^2) + f^0(x(t), x'(t))$$

$$+f(x(t)) - f(z) + \mu^{-1}\|x'(t)\|^2$$

$$\leq 6\delta_f M + \delta_C(4M + \mu_2(L+1))\mu_1^{-1}. \tag{6.150}$$

Using Proposition 6.3, the equality

$$f^0(x(t), x'(t)) = (f \circ x)'(t),$$

and integrating inequality (6.150) over the interval $[0, t]$, we obtain that for all $t \in [0, T_0]$,

$$\begin{aligned}
&(2\mu)^{-1}\|x(t) - z\|^2 - (2\mu)^{-1}\|x(0) - z\|^2 \\
&+f(x(t)) - f(x(0)) \\
&+\int_0^t (f(x(s)) - f(z))ds \\
&\le 6\delta_f Mt + \mu_1^{-1}\delta_C(4M + \mu_2(L+1)). \qquad (6.151)
\end{aligned}$$

Relations (6.108), (6.110), (6.111), and (6.129) imply that for all $t \in [0, T_0]$,

$$\begin{aligned}
&f(x(t)) \ge \inf(f, C) - \delta_C L, \\
&f(x(t)) \le \sup(f, C) + \delta_C L. \qquad (6.152)
\end{aligned}$$

By (6.110), (6.151), and (6.152), for all $t \in [0, T_0]$,

$$\begin{aligned}
&(2\mu_2)^{-1}\|x(t) - z\|^2 - (2\mu_1)^{-1}\|x(0) - z\|^2 \\
&+\inf(f, C) - f(x(0)) - \delta_C L \\
&+\int_0^t (f(x(s)) - f(z))ds \\
&\le 6\delta_f Mt + \delta_C t(4M + \mu_2(L+1))\mu_1^{-1}.
\end{aligned}$$

The relation above with $t = T_0$, (6.110), (6.111), (6.128), (6.129), and (6.152) imply that

$$\begin{aligned}
&\int_0^{T_0} (f(x(s)) - f(z))ds \\
&\le (2\mu_1)^{-1}\|x(0) - z\|^2 + \sup(f, C) - \inf(f, C) \\
&+2\delta_C L + 6\delta_f M T_0 + \delta_C T_0(4M + \mu_2(L+1))\mu_1^{-1} \\
&\le 2M^2\mu_1^{-1} + 2LM + 2\delta_C L + 6\delta_f M T_0 \\
&+\delta_C T_0(4M + \mu_2(L+1))\mu_1^{-1}.
\end{aligned}$$

This implies that

$$f(T_0^{-1}\int_0^{T_0} x(s)ds) - \inf(f, C)$$

$$\min\{f(x(t)) : t \in [0, T_0]\} - \inf(f, C)$$

$$\leq T_0^{-1}\int_0^{T_0} f(x(s))ds - \inf(f, C)$$

$$\leq (2M^2\mu_1^{-1} + 2LM)T_0^{-1}$$

$$+6\delta_f M + \mu_1^{-1}\delta_C(4M + \mu_2(L + 1)) + 2\delta_C LT_0^{-1}.$$

This completes the proof of Theorem 6.7. □

6.11 Continuous Subgradient Projection Method on Unbounded Sets

In Sect. 6.13 of chapter we obtain an extension of Theorem 6.7 for minimization problems on unbounded sets. Note that in [92] we study continuous subgradient projection method only for problems on bounded sets.

Let X be a Hilbert space with an inner product $\langle\cdot,\cdot\rangle$ which induces a complete norm $\|\cdot\|$

Let C be a nonempty, convex, and closed set in the Hilbert space X, U be an open and convex subset of X such that

$$C \subset U,$$

and $f : U \to R^1$ be a convex locally Lipschitzian function.

Let $M > 4$, $L \geq 1$,

$$B_X(0, M - 1) \cap \text{argmin}(f, C) \neq \emptyset, \tag{6.153}$$

$$\{y \in X : d(y, C) \leq 1\} \subset U, \tag{6.154}$$

$$0 < \mu_1 < \mu_2. \tag{6.155}$$

We suppose that

$$\{x \in U : f(x) \leq \inf(f, C) + 1\} \subset B_X(0, M/4), \tag{6.156}$$

$$c_* < 0,$$

$$f(v) \ge c_* \text{ for all } v \in U, \tag{6.157}$$

$$\widehat{M} > 4M + 210\mu_2\mu_1^{-1}, \tag{6.158}$$

$$(8\mu_2)^{-1}\widehat{M}^2$$

$$> 4M^2\mu_1^{-1} - 2c_* + 2|\sup\{f(v) : v \in B_X(0, M)\}|, \tag{6.159}$$

$$|f(v_1) - f(v_2)| \le L\|v_1 - v_2\|$$

$$\text{for all } v_1, v_2 \in B_X(0, \widehat{M} + 1) \cap U, \tag{6.160}$$

$\delta_f, \delta_C \in (0, 1)$ satisfy

$$4\max\{\delta_f, \delta_C\}(16(L + 1)\mu_1^{-1}M^2$$

$$+ \sup\{|f(v)| : v \in B_X(0, M)\} - c_*$$

$$+14\widehat{M} + 6\mu_2(L + 3) + 3$$

$$+ \mu_1^{-1}(5\widehat{M} + 2\mu_2(L + 1) + L + 4)) \le 1. \tag{6.161}$$

6.12 An Auxiliary Result

Proposition 6.9 *Let*

$$z \in B_X(0, M) \cap argmin(f, C), \tag{6.162}$$

$T_0 > 0$, $x \in W^{1,1}(0, T_0; X)$,

$$d(x(0), C) \le \delta_C \tag{6.163}$$

and for almost every $t \in [0, T_0]$ *there exist* $\xi(t) \in X$ *such that*

$$B_X(\xi(t), \delta_f) \cap \partial f(x(t); x'(t)) \ne \emptyset, \tag{6.164}$$

$$P_C(x(t) - \mu\xi(t)) \in B_X(x(t) + x'(t), \delta_C). \tag{6.165}$$

Then for almost every $t \in [0, T_0]$,

$$(2\mu)^{-1}(d/dt)(\|x(t)-z\|^2)$$

$$+(f\circ x)'(t)+f(x(t))-f(z)$$

$$\leq \delta_f(3\|x'(t)\|+\|x(t)-z\|)$$

$$+\delta_C(\mu_1^{-1}\|x(t)-z\|+\mu_1^{-1}\|z-\phi(t)\|+\|\xi(t)\|)$$

and for all $t\in[0,T_0]$,

$$d(x(t),C)\leq\delta_C.$$

Proof In view of (6.156) and (6.162),

$$\|z\|\leq M-1 \tag{6.166}$$

and

$$f(z)=\inf(f,C). \tag{6.167}$$

For almost every $t\in[0,T_0]$ set

$$\phi(t)=x(t)+x'(t). \tag{6.168}$$

It is clear that $\phi:[0,T_0]\to X$ is a Bochner integrable function. In view of (6.105) and (6.168), for almost every $t\in[0,T_0]$,

$$B_X(\phi(t),\delta_C)\cap C\neq\emptyset. \tag{6.169}$$

Define

$$C_{\delta_C}=\{x\in X:\ d(x,C)\leq\delta_C\}. \tag{6.170}$$

Clearly, C_δ is a convex closed set, for each $x\in C_\delta$,

$$B_X(x,\delta_C)\cap C\neq\emptyset \tag{6.171}$$

and in view of (6.169),

$$\phi(t)\in C_{\delta_C}\text{ for almost every }t\in[0,T_0]. \tag{6.172}$$

Lemma 6.8, (6.163), and (6.172) imply that

$$x(t)\in C_{\delta_C}\text{ for all }t\in[0,T_0]. \tag{6.173}$$

It follows from (6.171) and (6.173) that for every $t \in [0, T_0]$ there exists

$$\widehat{x}(t) \in C \tag{6.174}$$

such that

$$\|x(t) - \widehat{x}(t)\| \le \delta_C. \tag{6.175}$$

By (6.174) and Lemma 2.2, for almost every $t \in [0, T_0]$,

$$\langle \widehat{x}(t) - P_C(x(t) - \mu\xi(t)), x(t) - \mu\xi(t) - P_C(x(t) - \mu\xi(t))\rangle \le 0. \tag{6.176}$$

By (6.176),

$$\langle x(t) - P_C(x(t) - \mu\xi(t)), x(t) - \mu\xi(t) - P_C(x(t) - \mu\xi(t))\rangle$$

$$\le \langle x(t) - \widehat{x}(t), x(t) - \mu\xi(t) - P_C(x(t) - \mu\xi(t))\rangle. \tag{6.177}$$

It follows from (6.162) and Lemma 2.2 that

$$\langle z - P_C(x(t) - \mu\xi(t)), x(t) - \mu\xi(t) - P_C(x(t) - \mu\xi(t))\rangle \le 0. \tag{6.178}$$

In view of (6.164), for almost every $t \in [0, T_0]$ there exists

$$\widehat{\xi}(t) \in \partial f(x(t); x'(t)) \tag{6.179}$$

such that

$$f^0(x(t), x'(t)) = \langle \widehat{\xi}(t), x'(t)\rangle, \tag{6.180}$$

$$\|\xi(t) - \widehat{\xi}(t)\| \le \delta_f. \tag{6.181}$$

In view of (6.109) and (6.179), for almost every $t \in [0, T_0]$,

$$f(z) \ge f(x(t)) + \langle \widehat{\xi}(t), z - x(t)\rangle. \tag{6.182}$$

By (6.165), for almost every $t \in [0, T_0]$,

$$\|(x(t) + x'(t)) - P_C(x(t) - \mu\xi(t))\| \le \delta_C. \tag{6.183}$$

Relations (6.168) and (6.183) imply that for almost every $t \in [0, T_0]$,

$$\langle z - x(t) - x'(t), x(t) - \mu\xi(t) - x(t) - x'(t)\rangle$$

$$= \langle z - \phi(t), x(t) - \mu\xi(t) - \phi(t)\rangle$$

$$= \langle z - \phi(t), x(t) - \mu\xi(t) - P_C(x(t) - \mu\xi(t))\rangle$$

$$+\langle z - \phi(t), P_C(x(t) - \mu\xi(t)) - \phi(t)\rangle$$

$$\leq \langle z - \phi(t), x(t) - \mu\xi(t) - P_C(x(t) - \mu\xi(t))\rangle$$

$$+ \delta_C \|z - \phi(t)\|. \tag{6.184}$$

By (6.168), (6.178), (6.183), and the definition of P_C, for a. e. $t \in [0, T_0]$,

$$\langle z - \phi(t), x(t) - \mu\xi(t) - P_C(x(t) - \mu\xi(t))\rangle$$

$$\leq \langle z - P_C(x(t) - \mu\xi(t)), x(t) - \mu\xi(t) - P_C(x(t) - \mu\xi(t))\rangle$$

$$+\delta_C \|x(t) - \mu\xi(t) - P_C(x(t) - \mu\xi(t))\|$$

$$\leq \delta_C \|x(t) - \mu\xi(t) - z\|. \tag{6.185}$$

In view of (6.184) and (6.185), for a. e. $t \in [0, T_0]$,

$$\langle x(t) + x'(t) - z, \mu\xi(t) + x'(t)\rangle$$

$$\leq \delta_C \|z - \phi(t)\| + \delta_C(\|x(t) - z - \mu\xi(t)\|). \tag{6.186}$$

It follows from (6.181) and (6.182) that for almost every $t \in [0, T_0]$,

$$f(x(t)) - f(z)$$

$$\leq \langle \widehat{\xi}(t), x(t) - z\rangle$$

$$= \langle \xi(t), x(t) - z + x'(t)\rangle - \langle \xi(t), x'(t)\rangle$$

$$+\langle \widehat{\xi}(t) - \xi(t), x(t) - z + x'(t)\rangle$$

$$+\langle \xi(t) - \widehat{\xi}(t), x'(t)\rangle$$

$$\leq \langle \xi(t), x(t) - z + x'(t)\rangle - \langle \xi(t), x'(t)\rangle$$

$$+ \delta_f(\|x(t) - z\| + 2\|x'(t)\|). \tag{6.187}$$

In view of (6.184) and (6.185), for almost every $t \in [0, T_0]$,

$$\langle \xi(t), x(t) - z + x'(t)\rangle$$

$$\leq \mu^{-1}\langle -x'(t), x(t) - z + x'(t)\rangle$$

$$+\langle \xi(t) + \mu^{-1}x'(t), x(t) - z + x'(t)\rangle$$

$$\leq \mu^{-1}\langle -x'(t), x(t) - z + x'(t)\rangle$$

$$+\mu^{-1}\delta_C\|z - \phi(t)\|$$

$$+\langle z - \phi(t), x(t) - \mu\xi(t) - P_C(x(t) - \mu\xi(t))\rangle\mu^{-1}$$

$$\leq \mu^{-1}\delta_C\|z - \phi(t)\|$$

$$-\langle \mu^{-1}x'(t), x(t) - z + x'(t)\rangle$$

$$+ \mu^{-1}\delta_C\|x(t) - \mu\xi(t - z\|. \tag{6.188}$$

Relations (6.187) and (6.188) imply that for almost every $t \in [0, T_0]$,

$$f(x(t)) - f(z)$$

$$\leq \mu^{-1}\langle -x'(t), x(t) + x'(t) - z\rangle$$

$$+\mu^{-1}\delta_C(\|z - \phi(t)\| + \|x(t) - z\| + \mu\|\xi(t)\|)$$

$$- \langle \xi(t), x'(t)\rangle + \delta_f(\|x(t) - z\| + 2\|x'(t)\|). \tag{6.189}$$

By (6.189), for almost every $t \in [0, T_0]$,

$$f(x(t)) - f(z)$$

$$\leq -\mu^{-1}\|x'(t)\|^2 - \mu^{-1}\langle x'(t), x(t) - z\rangle$$

$$-\langle \xi(t), x'(t)\rangle$$

$$+\delta_C\mu^{-1}(\|z - \phi(t)\| + \|x(t) - z\| + \mu\|\xi(t)\|)$$

$$+ \delta_f(\|x(t) - z\| + 2\|x'(t)\|). \tag{6.190}$$

In view of (6.155), (6.181), and (6.190),

$$\mu^{-1}\|x'(t)\|^2 + \mu^{-1}\langle x'(t), x(t) - z\rangle$$

$$+\langle \widehat{\xi}(t), x'(t)\rangle + f(x(t)) - f(z)$$

$$\leq \delta_f\|x'(t)\| + \delta_f\|x(t) - z\| + 2\|x'(t)\|$$

$$+\delta_C(\|\xi(t)\| + \mu_1^{-1}\|z - \phi(t)\| + \mu^{-1}\|x(t) - z\|). \tag{6.191}$$

Corollary 6.4, (6.181), and (6.191) imply that for almost every $t \in [0, T_0]$,

$$(2\mu)^{-1}(d/dt)(\|x(t) - z\|^2) + f^0(x(t), x'(t))$$

$$+f(x(t)) - f(z) + \mu^{-1}\|x'(t)\|^2$$

$$\leq \delta_f(\|x(t) - z\| + 3\|x'(t)\|)$$

$$+\delta_C(\|\xi(t)\| + \mu_1^{-1}\|z - \phi(t)\| + \mu^{-1}\|x(t) - z\|).$$

Together with Proposition 6.3 and the equality

$$f^0(x(t), x'(t)) = (f \circ x)'(t)$$

for almost every $t \in [0, T_0]$, this implies that for almost every $t \in [0, T_0]$,

$$(2\mu)^{-1}(d/dt)(\|x(t) - z\|^2)$$

$$+(f \circ x)'(t) + f(x(t)) - f(z)$$

$$\leq \delta_f(\|x(t) - z\| + 3\|x'(t)\|)$$

$$+\delta_C(\|\xi(t)\| + \mu_1^{-1}\|z - \phi(t)\| + \mu^{-1}\|x(t) - z\|).$$

Proposition 6.9 is proved. □

6.13 The Convergence Result

We will prove the following result.

Theorem 6.10 *Let*

$$0 < \mu_1 \leq 1 < \mu_2, \tag{6.192}$$

$$\epsilon_0 = 4\max\{\delta_f, \delta_C\}[8(L+1)(2M^2\mu_1^{-1}$$

$$+\sup\{|f(v)| : v \in B_X(0, M)\} - c_*)$$

$$+14\widehat{M} + 3\mu_2(L+1) + 3 + \mu_1^{-1}(5\widehat{M} + 2\mu_2(L+1) + L + 4)]. \tag{6.193}$$

Then following two assertions hold.

1. *Let*

$$8^{-1}\min\{\delta_f^{-1},\delta_C^{-1}\}(L+1)^{-1}\le T_0\le \min\{\delta_f^{-1},\delta_C^{-1}\}(L+1)^{-1}, \tag{6.194}$$

$$\mu_1\le\mu\le\mu_2.$$

Assume that $x\in W^{1,1}(0,T_0;X)$,

$$x(0)\in B_X(0,M),$$

$$d(x(0),C)\le\delta_C \tag{6.195}$$

and that for almost every $t\in[0,T_0]$, *there exists* $\xi(t)\in X$ *such that*

$$B_X(\xi(t),\delta_f)\cap\partial f(x(t);x'(t))\ne\emptyset, \tag{6.196}$$

$$P_C(x(t)-\mu\xi(t))\in B(x(t)+x'(t),\delta_C). \tag{6.197}$$

Then

$$\|x(t)\|\le\widehat{M}\ \textit{for all}\ t\in[0,T_0],$$

$$\min\{f(x(t)):\ t\in[0,T_0]\}-\inf(f,C),$$

$$f(T_0^{-1}\int_0^{T_0}x(s)ds)-\inf(f,C)$$

$$\le T_0^{-1}\int_0^{T_0}f(x(s))ds-\inf(f,C)\le\epsilon_0/4,$$

$$mes\{t\in[0,T_0]:\ f(x(t))>\inf(f,C)+\epsilon_0\}\le T_0/4,$$

$$d(x(t),C)\le\delta_C,\ t\in[0,T_0]$$

and there exists $t\in[(3/4)T_0,T_0]$ *such that*

$$\|x(t)\|\le M.$$

2. *Assume that*

$$T\ge\max\{\delta_f^{-1},\delta_C^{-1}\}(L+1)^{-1},$$

$x\in W^{1,1}(0,T;X)$,

$$x(0) \in B_X(0, M),$$

$$d(x(0), C) \le \delta_C$$

and that for almost every $t \in [0, T]$, there exists $\xi(t) \in X$ such that (6.196) *and* (6.197) *hold. Then*

$$\|x(t)\| \le \widehat{M},\ t \in [0, T]$$

and there exists a sequence of numbers $\{T_i\}_{i=0}^k$, where k is a natural number, such that

$$T_0 = 0,\ T_k = T$$

and that for all integers $i = 0, \dots, k-1$,

$$4^{-1}\min\{\delta_f^{-1}, \delta_C^{-1}\}(L+1)^{-1} \le T_{i+1} - T_i \le \min\{\delta_f^{-1}, \delta_C^{-1}\}(L+1)^{-1},$$

$$\|x(T_i)\| \le M,\ i = 0, \dots, k-1$$

and that for all $i = 0, \dots, k-1$,

$$\min\{f(x(t)) :\ t \in [T_i, T_{i+1}]\} - \inf(f, C),$$

$$f((T_{i+1} - T_i)^{-1} \int_{T_i}^{T_{i+1}} x(s)ds) - \inf(f, C)$$

$$\le (T_{i+1} - T_i)^{-1} \int_{T_i}^{T_{i+1}} f(x(s))ds - \inf(f, C) \le \epsilon_0.$$

Proof In view of (6.156), there exists

$$z \in B_X(0, M-1) \cap \mathrm{argmin}(f, C). \tag{6.198}$$

Proposition 6.9 implies that for almost every $t \in [0, T_0]$,

$$(2\mu)^{-1}(d/dt)(\|x(t) - z\|^2)$$

$$+(f \circ x)'(t) + f(x(t)) - f(z)$$

$$\le \delta_f(\|x(t) - z\| + 3\|x'(t)\|)$$

$$+ \delta_C(\|\xi(t)\| + 1 + \mu^{-1}\|z - \phi(t)\| + \mu^{-1}\|x(t) - z\|). \tag{6.199}$$

For almost every $t \in [0, T_0]$ set

$$\phi(t) = x(t) + x'(t). \tag{6.200}$$

Define

$$E = \{\tau \in (0, T_0] : \|x(t)\| \le \widehat{M} \text{ for all } t \in [0, \tau]\}. \tag{6.201}$$

Since the function x is continuous it follows from (6.158) and (6.195) that

$$E \neq \emptyset.$$

Set

$$\tau = \sup(E). \tag{6.202}$$

Clearly,

$$\tau \in E \tag{6.203}$$

and

$$\|x(t)\| \le \widehat{M}, \ t \in [0, \tau]. \tag{6.204}$$

By (6.160), (6.196), and (6.204), for almost every $t \in [0, \tau]$,

$$\|\xi(t)\| \le L + 1. \tag{6.205}$$

Lemma 2.2, (6.198), (6.204), and (6.205) imply that for almost every $t \in [0, \tau]$,

$$\begin{aligned} &\|P_C(x(t) - \mu\xi(t))\| \\ &\le \|P_C(x(t) - \mu\xi(t)) - z\| + \|z\| \\ &\le \|x(t) - \mu\xi(t) - z\| + \|z\| \\ &\le \widehat{M} + \mu_2(L + 1) + 2M. \end{aligned} \tag{6.206}$$

It follows from (6.197), (6.200), (6.204), and (6.206) that for almost every $t \in [0, \tau]$,

$$\begin{aligned} \|\phi(t)\| &\le \|P_C(x(t) - \mu\xi(t))\| + 1 \\ &\le \widehat{M} + \mu_2(L + 1) + 2M + 1 \end{aligned} \tag{6.207}$$

and

$$\|x'(t)\| \le \|x(t)\| + \|\phi(t)\|$$

$$\le 2\widehat{M} + \mu_2(L+1) + 2M + 1. \tag{6.208}$$

Thus for a. e. $t \in [0, \tau]$, (6.207) and (6.208) are true. Combined with (6.198), (6.199), and (6.201)–(6.203), this implies that for almost every $t \in [0, \tau]$,

$$(2\mu)^{-1}(d/dt)(\|x(t) - z\|^2)$$

$$+(f \circ x)'(t) + f(x(t)) - f(z)$$

$$\le \delta_f(6\widehat{M} + 3\mu_2(L+1) + 6M + 3 + \widehat{M} + M)$$

$$+\delta_C(\mu_1^{-1}\widehat{M} + \mu_1^{-1}(\widehat{M} + \mu_2(L+1) + 4M + 1) + L + 1).$$

Integrating the inequality above over the interval $[t_1, t_2] \subset [0, \tau]$, we obtain that for all $t_1, t_2 \in [0, \tau]$ satisfying $t_1 < t_2$,

$$(2\mu)^{-1}\|x(t_2) - z\|^2 - (2\mu)^{-1}\|x(t_1) - z\|^2$$

$$+f(x(t_2)) - f(x(t_1))$$

$$+\int_{t_1}^{t_2} (f(x(s)) - f(z))ds$$

$$\le (t_2 - t_1)\delta_f(14\widehat{M} + 3\mu_2(L+1) + 3)$$

$$+(t_2 - t_1)\delta_C\mu_1^{-1}(5\widehat{M} + \mu_2(L+1) + 4)$$

$$+ \delta_C(L+1)(t_2 - t_1). \tag{6.209}$$

Since the function x is continuous it follows from (6.201) to (6.203) that at least one of the following equalities holds:

$$\tau = T_0; \tag{6.210}$$

$$\|x(\tau)\| = \widehat{M}. \tag{6.211}$$

Assume that (6.211) holds. By (6.157), (6.192), (6.195), (6.198), (6.209) with $t_2 = \tau$, $t_1 = 0$ and (6.211),

$$(2\mu_2)^{-1}(\widehat{M} - M)^2 - 2(\mu_1)^{-1}4\widehat{M}^2$$

$$+c_* - \sup\{f(v) : v \in B_X(0, M)\}$$

$$\leq \tau\delta_f(14\widehat{M} + 3\mu_2(L+1) + 3)$$

$$+ \tau\delta_C\mu_1^{-1}(5\widehat{M} + 6\mu_2(L+1) + 1) + \tau\delta_C(L+1). \tag{6.212}$$

By (6.158) and (6.159),

$$(2\mu_2)^{-1}(\widehat{M} - 2M)^2 - 2(\mu_1)^{-1}4\widehat{M}^2$$

$$+c_* - \sup\{f(v) : v \in B_X(0, M)\}$$

$$\geq 2^{-1}\mu_2^{-1}\widehat{M}^2/4 - 2\mu_1^{-1}M^2$$

$$+c_* - \sup\{f(v) : v \in B_X(0, M)\}$$

$$\geq 16^{-1}\mu_2^{-1}\widehat{M}^2. \tag{6.213}$$

It follows from (6.194), (6.212), (6.213), and the inequality $\tau \leq T_0$ that

$$16^{-1}\mu_2^{-1}\widehat{M}^2$$

$$\leq \tau\delta_f(14\widehat{M} + 3\mu_2(L+1) + 3)$$

$$+\tau\delta_C\mu_1^{-1}(5\widehat{M} + 6\mu_2(L+1) + 1)$$

$$+\tau\delta_C(L+1)$$

$$\leq (L+1)^{-1}\min\{\delta_f^{-1}, \delta_C^{-1}\}\delta_f(14\widehat{M} + 3\mu_2(L+1) + 3)$$

$$+(L+1)^{-1}\min\{\delta_f^{-1}, \delta_C^{-1}\}\delta_C[\mu_1^{-1}(5\widehat{M}$$

$$+6\mu_2(L+1) + 1) + L + 1)]$$

$$\leq 7\widehat{M} + 3\mu_2 + 1$$

$$+ 5\mu_1^{-1}\widehat{M} + 6\mu_2\mu_1^{-1} + \mu_1^{-1} + 1. \tag{6.214}$$

In view of (6.158), (6.192), and (6.214),

$$\widehat{M}^2 \leq 16 \cdot 7\mu_2\widehat{M} + 48\mu_2^2$$

$$+16\mu_2 + 90\widehat{M}\mu_2\mu_1^{-1}$$

$$+96\mu_2^2\mu_1^{-1} + 16\mu_2\mu_1^{-1} + 16\mu_2$$

$$\mu_2\mu_1^{-1}(202\widehat{M}+16+160\mu_2+32)$$

$$\leq \mu_2\mu_1^{-1}(202\widehat{M}+208\mu_2) \leq 203\mu_2\mu_1^{-1}\widehat{M}$$

and

$$\widehat{M} \leq 203\mu_2\mu_1^{-1}.$$

This contradicts (6.158). The contradiction we have reached proves that

$$\|x(\tau)\| < \widehat{M}.$$

Therefore

$$\tau = T_0$$

and

$$\|x(t)\| \leq \widehat{M},\ t \in [0, T_0]. \tag{6.215}$$

It follows from (6.157), (6.192), (6.195), (6.198), and (6.209) with $t_2 = T_0$ and $t_1 = 0$ that

$$\int_0^{T_0} (f(x(s)) - f(z))ds$$

$$\leq (2\mu_1)^{-1}\|x(0)-z\|^2 + f(x(0)) - f(x(T_0))$$

$$+T_0\delta_f(14\widehat{M}+3\mu_2(L+1)+3)$$

$$+T_0\delta_C\mu_1^{-1}(5\widehat{M}+\mu_2(L+1)+4)+\delta_C T_0(L+1)$$

$$\leq (2\mu_1)^{-1}4M^2 - c_* + \sup\{f(v) :\ v \in B_X(0,M)\}$$

$$+T_0\delta_f(14\widehat{M}+3\mu_2(L+1)+3)$$

$$+ T_0\delta_C\mu_1^{-1}(5\widehat{M}+\mu_2(L+1)+4)+\mu_1^{-1}\delta_C T_0(L+1). \tag{6.216}$$

By (6.191), (6.192), (6.193), (6.198), and (6.216),

$$T_0^{-1}\int_0^{T_0} (f(x(s)) - \inf(f, C))ds$$

$$\leq ((2\mu_1)^{-1}4M^2 + \sup\{f(v) :\ v \in B_X(0,M)\}$$

$$-c_*)8\max\{\delta_f, \delta_C\}(L+1)$$

$$+\delta_f(14\widehat{M}+3\mu_2(L+1)+3)$$

$$+\delta_C\mu_1^{-1}(5\widehat{M}+\mu_2(L+1)+4) \le \epsilon_0/4. \quad (6.217)$$

By (6.217),

$$f(T_0^{-1}\int_0^T x(s)ds) - \inf(f, C)$$

$$\min\{f(x(t)) : t \in [0, T_0]\} - \inf(f, C)$$

$$\le T_0^{-1}\int_0^T f(x(s))ds - \inf(f, C) \le \epsilon_0/4. \quad (6.218)$$

Define

$$E = \{t \in [0, T_0] : f(x(t)) > \inf(f, C) + \epsilon_0\}. \quad (6.219)$$

In view of (6.217) and (6.219),

$$T_0^{-1}\mathrm{mes}(E)\epsilon_0 \le T_0^{-1}\int_0^{T_0}(f(x(s)) - \inf(f, C))ds \le \epsilon_0/4$$

and

$$\mathrm{mes}(E) \le T_0/4. \quad (6.220)$$

Proposition 6.9 implies that

$$d(x(t), C) \le \delta_C, \ t \in [0, T_0]. \quad (6.221)$$

Set

$$t_0 = \max\{t \in [0, T_0] : f(x(t)) - \inf(f, C) \le \epsilon_0\}.$$

It follows from (6.219) and (6.220) that

$$t_0 > 3T_0/4. \quad (6.222)$$

By the choice of t_0, (6.156), (6.161), (6.193), and the inequalities

$$\epsilon_0 \le 1, \ f(x(t_0)) \le \inf(f, C) + 1$$

we have

$$\|x(t_0)\| \le M. \tag{6.223}$$

Now Assertion 1 follows from (6.215) and (6.217)–(6.223). Assertion 2 follows from Assertion 1 applied by induction. Theorem 6.10 is proved. □

6.14 Subgradient Projection Algorithm for Zero-Sum Games

Let $(X, \langle\cdot,\cdot\rangle)$, $(Y, \langle\cdot,\cdot\rangle)$ be Hilbert spaces equipped with the complete norms $\|\cdot\|$ which are induced by their inner products. Let C be a nonempty closed convex subset of X, D be a nonempty closed convex subset of Y, U be an open convex subset of X, and V be an open convex subset of Y such that

$$C \subset U,\ D \subset V, \tag{6.224}$$

and let a borelian function $f : U \times V \to R^1$ possess the following properties:

(i) for each $v \in V$, the function $f(\cdot, v) : U \to R^1$ is convex and locally Lipschitzian;
(ii) for each $u \in U$, the function $f(u, \cdot) : V \to R^1$ is concave and locally Lipschitzian.

Let

$$\delta_C, \delta_D,\ \delta_{f,1}, \delta_{f,2} \in (0, 1]. \tag{6.225}$$

For each $(\xi, \eta) \in U \times V$, set

$$\partial_x f(\xi, \eta) = \{l \in X : \ f(y, \eta) - f(\xi, \eta) \ge \langle l, y - \xi\rangle \text{ for all } y \in U\},$$

$$\partial_y f(\xi, \eta) = \{l \in Y : \ \langle l, y - \eta\rangle \ge f(\xi, y) - f(\xi, \eta) \text{ for all } y \in V\}. \tag{6.226}$$

Let $x \in U$, $y \in V$, $\xi \in X$, and $\eta \in Y$. Set

$$f^0(x, y, \xi) = \lim_{t\to 0^+} t^{-1}[f(x + t\xi, y) - f(x, y)], \tag{6.227}$$

$$f^0(x, y, \eta) = \lim_{t\to 0^+} t^{-1}[f(x, y + t\eta) - f(x, y)], \tag{6.228}$$

$$\partial_x f(x, y; \xi) = \{l \in \partial_x f(x, y) : \ \langle l, \xi\rangle = f^0(x, y, \xi)\}. \tag{6.229}$$

$$\partial_y f(x, y; \eta) = \{l \in \partial_y f(x, y) : \ \langle l, \eta\rangle = f^0(x, y, \eta)\}. \tag{6.230}$$

We study subgradient projection algorithm for the zero-sum game defined by the triplet (f, U, V) and prove in Sect. 6.16 a convergence result which is based on an auxiliary result of Sect. 5.15.

6.15 An Auxiliary Result

Proposition 6.11 *Let*

$$T > 0,\ x \in W^{1,1}(0, T; X),\ \ y \in W^{1,1}(0, T; Y), \mu > 0, \tag{6.231}$$

$$x(t) \in U,\ y(t) \in V,\ t \in [0, T],$$

$$d(x(0), C) \le \delta_C,\ d(y(0), D) \le \delta_D \tag{6.232}$$

and let for almost every $t \in [0, T]$ there exist $\xi(t) \in X$, $\eta(t) \in Y$ such that

$$B_X(\xi(t), \delta_{f,1}) \cap \partial_x f(x(t), y(t)) \ne \emptyset, \tag{6.233}$$

$$P_C(x(t) - \mu\xi(t)) \in B_X(x(t) + x'(t), \delta_C), \tag{6.234}$$

$$B_Y(\eta(t), \delta_{f,2}) \cap \partial_y f(x(t), y(t)) \ne \emptyset, \tag{6.235}$$

$$P_D(y(t) + \mu\eta(t)) \in B_Y(y(t) + y'(t), \delta_D). \tag{6.236}$$

Then for almost every $t \in [0, T]$, every $z \in C$, and every $v \in D$,

$$(2\mu)^{-1}(d/dt)(\|x(t) - z\|^2)$$

$$+f(x(t), y(t)) - f(z, y(t))$$

$$\le 4^{-1}\mu\|\xi(t)\|^2$$

$$+(\mu^{-1}\delta_C + \delta_{f,1})(2 + 3\|z\| + 3\|x(t)\| + 2\mu\|\xi(t)\|),$$

$$(2\mu)^{-1}(d/dt)(\|y(t) - v\|^2)$$

$$+f(x(t), v) - f(x(t), y(t))$$

$$\le 4^{-1}\mu\|\eta(t)\|^2$$

$$+(\mu^{-1}\delta_D + \delta_{f,2})(2 + 3\|v\| + 3\|y(t)\| + 2\mu\|\eta(t)\|)$$

and for all $t \in [0, T]$,

$$B_X(x(t), \delta_C) \cap C \neq \emptyset,$$

$$B_Y(y(t), \delta_D) \cap D \neq \emptyset.$$

Proof For almost every $t \in [0, T]$ set

$$\phi_X(t) = x(t) + x'(t),$$

$$\phi_Y(t) = y(t) + y'(t). \tag{6.237}$$

It is clear that $\phi_X : [0, T] \to X$ and $\phi_Y : [0, T] \to Y$ are Bochner integrable functions. In view of (6.138), (6.234), and (6.236), for almost every $t \in [0, T]$,

$$B_X(\phi_X(t), \delta_C) \cap C \neq \emptyset, \tag{6.238}$$

$$B_Y(\phi_Y(t), \delta_D) \cap D \neq \emptyset. \tag{6.239}$$

Define

$$C_\delta = \{x \in X : d(x, C) \leq \delta_C\}, \tag{6.240}$$

$$D_\delta = \{y \in Y : d(y, D) \leq \delta_D\}. \tag{6.241}$$

Clearly, C_δ and D_δ are convex closed sets, for each $x \in C_\delta$,

$$B(x, \delta_C) \cap C \neq \emptyset, \tag{6.242}$$

for each $x \in D_\delta$,

$$B(x, \delta_D) \cap D \neq \emptyset, \tag{6.243}$$

In view of (6.238) and (6.239),

$$\phi_X(t) \in C_\delta,\ \phi_Y(t) \in D_\delta$$

$$\text{for almost every } t \in [0, T]. \tag{6.244}$$

Lemma 6.8, (6.232), (6.237), (6.240), (6.241), and (6.244) imply that for every $t \in [0, T]$,

$$x(t) \in C_\delta,\ y(t) \in D_\delta.$$

It follows from the relation above that for every $t \in [0, T]$ there exists

$$\widehat{x}(t) \in C,\ \widehat{y}(t) \in D$$

$$\text{such that } \|x(t) - \widehat{x}(t)\| \le \delta_C, \ \|y(t) - \widehat{y}(t)\| \le \delta_D. \tag{6.245}$$

By (6.245) and Lemma 2.2, for almost every $t \in [0, T]$,

$$\langle \widehat{x}(t) - P_C(x(t) - \mu\xi(t)), x(t) - \mu\xi(t) - P_C(x(t) - \mu\xi(t)) \rangle \le 0 \tag{6.246}$$

and

$$\langle \widehat{y}(t) - P_D(y(t) + \mu\eta(t)), y(t) + \mu\eta(t) - P_D(y(t) + \mu\eta(t)) \rangle \le 0. \tag{6.247}$$

Relations (6.246) and (6.247) imply that for almost every $t \in [0, T]$,

$$\begin{aligned} &\langle x(t) - P_C(x(t) - \mu\xi(t)), x(t) - \mu\xi(t) - P_C(x(t) - \mu\xi(t)) \rangle \\ &\le \langle x(t) - \widehat{x}(t), x(t) - \mu\xi(t) - P_C(x(t) - \mu\xi(t)) \rangle \end{aligned} \tag{6.248}$$

and

$$\begin{aligned} &\langle y(t) - P_D(y(t) + \mu\eta(t)), y(t) + \mu\eta(t) - P_D(y(t) + \mu\eta(t)) \rangle \\ &\le \langle y(t) - \widehat{y}(t), y(t) + \mu\eta(t) - P_D(y(t) + \mu\eta(t)) \rangle. \end{aligned} \tag{6.249}$$

Let

$$z \in C, \ v \in D. \tag{6.250}$$

It follows from (6.250) and Lemma 2.2 that

$$\langle z - P_C(x(t) - \mu\xi(t)), x(t) - \mu\xi(t) - P_C(x(t) - \mu\xi(t)) \rangle \le 0 \tag{6.251}$$

and

$$\langle v - P_D(y(t) + \mu\eta(t)), y(t) + \mu\eta(t) - P_D(y(t) + \mu\eta(t)) \rangle \le 0. \tag{6.252}$$

In view of (6.229), (6.230), (6.233), and (6.235), for almost every $t \in [0, T]$ there exists

$$\widehat{\xi}(t) \in \partial_x f(x(t), y(t)) \tag{6.253}$$

such that

$$f^0(x(t), y(t), x'(t) \ge \langle \widehat{\xi}(t), x'(t) \rangle, \tag{6.254}$$

$$\|\xi(t) - \widehat{\xi}(t)\| \le \delta_{f,1} \tag{6.255}$$

and

$$\widehat{\eta}(t) \in \partial_y f(x(t), y(t)) \tag{6.256}$$

such that

$$f^0(x(t), y(t), y'(t)) \leq \langle \widehat{\eta}(t), y'(t) \rangle, \tag{6.257}$$

$$\|\eta(t) - \widehat{\eta}(t)\| \leq \delta_{f,2}. \tag{6.258}$$

In view of (6.226), (6.253), and (6.256), for almost every $t \in [0, T]$,

$$f(z, y(t)) \geq f(x(t), y(t)) + \langle \widehat{\xi}(t), z - x(t) \rangle,$$

$$f(x(t), v) - f(x(t), y(t)) \leq \langle \widehat{\eta}(t), v - y(t) \rangle$$

and

$$f(x(t), y(t)) - f(z, y(t))$$

$$\leq \langle \widehat{\xi}(t), x(t) - z + x'(t) \rangle - \langle \widehat{\xi}(t), x'(t) \rangle \tag{6.259}$$

$$f(x(t), y(t)) - f(x(t), v)$$

$$\geq \langle \widehat{\eta}(t), y(t) - v + y'(t) \rangle - \langle \widehat{\eta}(t), y'(t) \rangle. \tag{6.260}$$

By (6.234) and (6.236), for almost every $t \in [0, T]$,

$$\|(x(t) + x'(t)) - P_C(x(t) - \mu\xi(t))\| \leq \delta_C \tag{6.261}$$

and

$$\|(y(t) + y'(t)) - P_D(y(t) + \mu\eta(t))\| \leq \delta_D. \tag{6.262}$$

Relations (6.234), (6.237), (6.245), (6.261), and (6.262) imply that for almost every $t \in [0, T]$,

$$\langle z - x(t) - x'(t), x(t) - \mu\xi(t) - x(t) - x'(t) \rangle$$

$$= \langle z - \phi_X(t), x(t) - \mu\xi(t) - \phi_X(t) \rangle$$

$$= \langle z - \phi_X(t), x(t) - \mu\xi(t) - P_C(x(t) - \mu\xi(t)) \rangle$$

$$+ \langle z - \phi_X(t), P_C(x(t) - \mu\xi(t)) - \phi_X(t) \rangle$$

$$\leq \langle z - \phi_X(t), x(t) - \mu\xi(t) - P_C(x(t) - \mu\xi(t))\rangle$$

$$+ \delta_C \|z - \phi_X(t)\| \tag{6.263}$$

and

$$\langle v - y(t) - y'(t), y(t) + \mu\eta(t) - y(t) - y'(t)\rangle$$

$$= \langle v - \phi_Y(t), y(t) + \mu\eta(t) - \phi_Y(t)\rangle$$

$$= \langle v - \phi_Y(t), y(t) + \mu\eta(t) - P_C(y(t) + \mu\eta(t))\rangle$$

$$+\langle v - \phi_Y(t), P_D(y(t) + \mu\eta(t)) - \phi_Y(t)\rangle$$

$$\leq \langle v - \phi_Y(t), y(t) + \mu\eta(t) - P_D(y(t) + \mu\eta(t))\rangle$$

$$+ \delta_D \|v - \phi_Y(t)\|. \tag{6.264}$$

By (6.237), (6.250)–(6.252), (6.261), and (6.262),

$$\langle z - \phi_X(t), x(t) - \mu\xi(t) - P_C(x(t) - \mu\xi(t))\rangle$$

$$\leq \langle z - P_C(x(t) - \mu\xi(t)), x(t) - \mu\xi(t) - P_C(x(t) - \mu\xi(t))\rangle$$

$$+\langle P_C(x(t) - \mu\xi(t)) - \phi_X(t), x(t) - \mu\xi(t) - P_C(x(t) - \mu\xi(t))\rangle$$

$$\leq \langle z - P_C(x(t) - \mu\xi(t)), x(t) - \mu\xi(t) - P_C(x(t) - \mu\xi(t))\rangle$$

$$+\delta_C \|x(t) - \mu\xi(t) - z\|$$

$$\leq \langle z - P_C(x(t) - \mu\xi(t)), x(t) - \mu\xi(t) - P_C(x(t) - \mu\xi(t))\rangle$$

$$+\delta_C(\|x(t)\| + \mu\|\xi(t)\| + \|z\|)$$

$$\leq \delta_C(\|x(t)\| + \mu\|\xi(t)\| + \|z\|) \tag{6.265}$$

and

$$\langle v - \phi_Y(t), y(t) + \mu\eta(t) - P_D(y(t) + \mu\eta(t))\rangle$$

$$\leq \langle v - P_D(y(t) + \mu\eta(t)), y(t) + \mu\eta(t) - P_D(y(t) + \mu\eta(t))\rangle$$

$$+\langle P_D(y(t) + \mu\eta(t)) - \phi_Y(t), y(t) + \mu\eta(t) - P_D(y(t) + \mu\eta(t))\rangle$$

$$\leq \langle v - P_D(y(t) + \mu\eta(t)), y(t) + \mu\eta(t) - P_D(y(t) + \mu\eta(t))\rangle$$

$$+\delta_D\|y(t)+\mu\eta(t)-v\|$$

$$\leq \langle v-P_D(y(t)+\mu\eta(t)), y(t)+\mu\eta(t)-P_D(y(t)+\mu\eta(t))\rangle$$

$$+\delta_D(\|y(t)\|+\mu\|\eta(t)\|+\|v\|)$$

$$\leq \delta_D(\|y(t)\|+\mu\|\eta(t)\|+\|v\|). \tag{6.266}$$

Lemma 2.2, (6.237), (6.250), and (6.261)–(6.266) imply that for almost every $t \in [0, T]$,

$$\langle x(t)+x'(t)-z, \mu\xi(t)+x'(t)\rangle$$

$$\leq \langle z-\phi_X(t), x(t)-\mu\xi(t)-P_C(x(t)-\mu\xi(t))\rangle$$

$$+\delta_C\|z-\phi_X(t)\|$$

$$\leq \delta_C(\|z-x(t)-x'(t)\|+\|x(t)\|+\mu\|\xi(t)\|+\|z\|)$$

$$\leq \delta_C(1+\|z\|+\|x(t)\|+\mu\|\xi(t)\|+\|x(t)\|+\mu\|\xi(t)\|+\|z\|) \tag{6.267}$$

and

$$\langle y(t)+y'(t)-v, -\mu\eta(t)+y'(t)\rangle$$

$$\leq \langle v-\phi_Y(t), y(t)+\mu\eta(t)-P_D(y(t)+\mu\eta(t)))\rangle$$

$$+\delta_C\|v-\phi_Y(t)\|$$

$$\leq \delta_D(\|v-y(t)-y'(t)\|+\|y(t)\|+\mu\|\eta(t)\|+\|v\|)$$

$$\leq \delta_D(1+\|v\|+\|y(t)\|+\mu\|\eta(t)\|+\|y(t)\|+\mu\|\eta(t)\|+\|v\|). \tag{6.268}$$

It follows from (6.163), (6.250), (6.253), (6.254), (6.256), (6.258), (6.261), and Lemma 2.2 that for almost every $t \in [0, T]$,

$$f(x(t), y(t)) - f(z, y(t))$$

$$\leq \langle\widehat{\xi}(t), x(t)-z\rangle$$

$$= \langle\xi(t), x(t)-z+x'(t)\rangle - \langle\xi(t), x'(t)\rangle$$

$$+\langle\widehat{\xi}(t)-\xi(t), x(t)-z+x'(t)\rangle$$

$$+\langle\xi(t)-\widehat{\xi}(t), x'(t)\rangle$$

$$\le \langle \xi(t), x(t) - z + x'(t)\rangle - \langle \xi(t), x'(t)\rangle$$

$$+\delta_{f,1}\|x(t) - z + 2x'(t)\|$$

$$\le \langle \xi(t), x(t) - z + x'(t)\rangle - \langle \xi(t), x'(t)\rangle$$

$$+\delta_{f,1}(2\|x(t) - z + x'(t)\| + \|z\| + \|x(t)\|)$$

$$\le \langle \xi(t), x(t) - z + x'(t)\rangle - \langle \xi(t), x'(t)\rangle$$

$$+ \delta_{f,1}(\|z\| + \|x(t)\| + 2 + 2(\|z\| + \|x(t)\| + \mu\|\xi(t)\|)) \tag{6.269}$$

and

$$f(x(t), y(t)) - f(x(t), v)$$

$$\ge \langle \widehat{\eta}(t), y(t) - v\rangle$$

$$= \langle \eta(t), y(t) - v + y'(t)\rangle - \langle \eta(t), y'(t)\rangle$$

$$+\langle \widehat{\eta}(t) - \eta(t), y(t) - v + y'(t)\rangle$$

$$\langle \eta(t) - \widehat{\eta}(t), y'(t)\rangle$$

$$\ge \langle \eta(t), y(t) - v + y'(t)\rangle - \langle \eta(t), y'(t)\rangle$$

$$-\delta_{f,2}\|y(t) - v + 2y'(t)\|$$

$$\ge \langle \eta(t), y(t) - v + y'(t)\rangle - \langle \eta(t), y'(t)\rangle$$

$$-\delta_{f,2}(2\|y(t) - v + y'(t)\| + \|v\| + \|y(t)\|)$$

$$\ge \langle \eta(t), y(t) - v + y'(t)\rangle - \langle \eta(t), y'(t)\rangle$$

$$- \delta_{f,2}(2 + 2(\|y(t)\| + \mu\|\eta(t)\| + \|v\|) + \|v\| + \|y(t)\|). \tag{6.270}$$

In view of (6.267), for almost every $t \in [0, T]$,

$$\langle \xi(t), x(t) - z + x'(t)\rangle$$

$$\le \mu^{-1}\langle -x'(t), x(t) - z + x'(t)\rangle$$

$$+\langle \xi(t) + \mu^{-1}x'(t), x(t) - z + x'(t)\rangle$$

$$\le \mu^{-1}\langle -x'(t), x(t) - z + x'(t)\rangle$$

$$+ \mu^{-1}\delta_C(1 + 2\|z\| + 2\|x(t)\| + 2\mu\|\xi(t)\|). \tag{6.271}$$

In view of (6.268), for almost every $t \in [0, T]$,

$$\langle \eta(t), y(t) - v + y'(t)\rangle$$
$$= \mu^{-1}\langle y'(t), y(t) - v + y'(t)\rangle$$
$$+\langle \eta(t) - \mu^{-1}y'(t), y(t) - v + y'(t)\rangle$$
$$\geq \mu^{-1}\langle y'(t), y(t) - v + y'(t)\rangle$$
$$- \mu^{-1}\delta_D(1 + 2\|v\| + 2\|y(t)\| + 2\mu\|\eta(t)\|). \tag{6.272}$$

Relations (6.162), (6.163), (6.269), and (6.271) imply that for almost every $t \in [0, T]$,

$$f(x(t), y(t)) - f(z, y(t))$$
$$\leq \mu^{-1}\langle -x'(t), x(t) + x'(t) - z\rangle$$
$$-\langle \xi(t), x'(t)\rangle$$
$$+\mu^{-1}\delta_C(1 + 2\|z\| + 2\|x(t)\| + 2\mu\|\xi(t)\|)$$
$$+ \delta_{f,1}(3\|z\| + \|x(t)\| + 2 + 2\mu\|\xi(t)\|). \tag{6.273}$$

By (6.270) are (6.272),

$$f(x(t), y(t)) - f(x(t), v)$$
$$\leq \mu^{-1}\langle y'(t), y(t) + y'(t) - v\rangle$$
$$-\langle \eta(t), y'(t)\rangle$$
$$-\delta_{f,2}(2 + 3\|v\| + 3\|y(t)\| + 2\mu\|\eta(t)\|)$$
$$- \mu^{-1}\delta_D(1 + 2\|v\| + 2\|y(t)\| + 2\mu\|\eta(t)\|). \tag{6.274}$$

In view of (6.273) and (6.274), for almost every $t \in [0, T]$,

$$\mu^{-1}\|x'(t)\|^2 + \mu^{-1}\langle x'(t), x(t) - z\rangle$$
$$+\langle \xi(t), x'(t)\rangle + f(x(t), y(t)) - f(z, y(t))$$

$$\le (\mu^{-1}\delta_C + \delta_{f,1})(2 + 3\|z\| + 3\|x(t)\| + 2\mu\|\xi(t)\|) \tag{6.275}$$

and

$$\mu^{-1}\|y'(t)\|^2 + \mu^{-1}\langle y'(t), y(t) - v\rangle$$

$$-\langle \eta(t), y'(t)\rangle + f(x(t), v) - f(x(t), y(t))$$

$$\le (\mu^{-1}\delta_D + \delta_{f,2})(2 + 3\|v\| + 3\|y(t)\| + 2\mu\|\eta(t)\|). \tag{6.276}$$

Corollary 6.4, (6.275), and (6.276) imply that for almost every $t \in [0, T]$,

$$(2\mu)^{-1}(d/dt)(\|x(t) - z\|^2)$$

$$+f(x(t), y(t)) - f(z, y(t))$$

$$\le \mu^{-1}\langle x'(t), x(t) - z\rangle$$

$$+f(x(t), y(t)) - f(z, y(t))$$

$$+\mu^{-1}\|x'(t) + 2^{-1}\mu\xi(t)\|^2$$

$$\le \mu^{-1}\langle x'(t), x(t) - z\rangle$$

$$+f(x(t), y(t)) - f(z, y(t))$$

$$+\mu^{-1}\|x'(t)\|^2 + \langle x'(t), \xi(t)\rangle + 4^{-1}\mu\|\xi(t)\|^2$$

$$\le 4^{-1}\mu\|\xi(t)\|^2 + (\mu^{-1}\delta_C + \delta_{f,1})(2 + 3\|z\| + 3\|x(t)\| + 2\mu\|\xi(t)\|)$$

and

$$(2\mu)^{-1}(d/dt)(\|y(t) - v\|^2)$$

$$+f(x(t), v) - f(x(t), y(t))$$

$$\le \mu^{-1}\langle y'(t), y(t) - v\rangle$$

$$+f(x(t), v)) - f(x(t), y(t))$$

$$+\mu^{-1}\|y'(t) - 2^{-1}\mu\eta(t)\|^2$$

$$= \mu^{-1}\langle y'(t), y(t) - v\rangle$$

$$+f(x(t), v) - f(x(t), y(t))$$

$$+\mu^{-1}\|y'(t)\|^2 - \langle y'(t), \eta(t)\rangle + 4^{-1}\mu\|\eta(t)\|^2$$

$$\leq 4^{-1}\mu\|\eta(t)\|^2 + (\mu^{-1}\delta_D + \delta_{f,2})(2 + 3\|v\| + 3\|y(t)\| + 2\mu\|\eta(t)\|).$$

Proposition 6.11 is proved. □

6.16 A Convergence Result for Games on Bounded Sets

Suppose that $L_1, L_2 \geq 1$, $M_1, M_2 > 1$,

$$C \subset B_X(0, M_1 - 1),\ D \subset B_Y(0, M_2 - 1), \tag{6.277}$$

$$\{x \in X :\ d(x, C) \leq 1\} \subset U,$$

$$\{y \in Y :\ d(y, D) \leq 1\} \subset V\}$$

and that the function $f : U \times V \to R^1$ is continuous and has the following properties:

(iii) for each $v \in V \cap B_Y(0, M_2 + 1)$,

$$|f(u_1, v) - f(u_2, v)| \leq L_1\|u_1 - u_2\|$$

$$\text{for all } u_1, u_2 \in U \cap B_X(0, M_1 + 1);$$

(iv) for each $u \in U \cap B_X(0, M_1 + 1)$,

$$|f(u, v_1) - f(u, v_2)| \leq L_2\|v_1 - v_2\|$$

$$\text{for all } v_1, v_2 \in V \cap B_Y(0, M_2 + 1).$$

Let

$$x_* \in C \text{ and } y_* \in D \tag{6.278}$$

satisfy

$$f(x_*, y) \leq f(x_*, y_*) \leq f(x, y_*) \tag{6.279}$$

for each $x \in C$ and each $y \in D$.

In this section we prove the following result.

Theorem 6.12 *Let*

$$M, T > 0,\ x \in W^{1,1}(0, T; X),\ y \in W^{1,1}(0, T; Y),$$

$$x(t) \in U,\ y(t) \in V,\ t \in [0, T],$$

$$d(x(0), C) \le \delta_C,\ d(y(0), D) \le \delta_D \tag{6.280}$$

and for almost every $t \in [0, T]$ there exist $\xi(t) \in X$, $\eta(t) \in Y$ such that

$$B_X(\xi(t), \delta_{f,1}) \cap \partial_x f(x(t), y(t)) \neq \emptyset, \tag{6.281}$$

$$P_C(x(t) - \mu\xi(t)) \in B_X(x(t) + x'(t), \delta_C), \tag{6.282}$$

$$B_Y(\eta(t), \delta_{f,2}) \cap \partial_y f(x(t), y(t)) \neq \emptyset, \tag{6.283}$$

$$P_D(y(t) + \mu\eta(t)) \in B_X(yt) + y'(t), \delta_D). \tag{6.284}$$

Then for almost every $t \in [0, T]$,

$$B_X(x(t), \delta_C) \cap C \neq \emptyset,$$

$$B_Y(y(t), \delta_D) \cap D \neq \emptyset.$$

$$B_X(T^{-1}\int_0^T x(t)dt, \delta_C) \cap C \neq \emptyset,$$

$$B_Y(T^{-1}\int_0^T y(t)dt, \delta_D) \cap D \neq \emptyset,$$

$$|T^{-1}\int_0^T f(x(t), y(t))dt - f(x_*, y_*)|$$

$$\le \max\{6M_1\mu^{-1}\delta_C + 2\delta_C(L_1 + 1) + 6\delta_{f,1}M_1 + 2\mu\delta_{f,1}(L_1 + 1)$$

$$+4^{-1}\mu(L_1 + 1)^2 + (2\mu T)^{-1}4M_1^2,$$

$$4(2T\mu)^{-1}M_2^2 + 6M_2\mu^{-1}\delta_D + 2\delta_D(L_2 + 1)$$

$$+6\delta_{f,2}M_2 + 2\mu\delta_{f,2}(L_2 + 1) + 4^{-1}\mu(L_2 + 1)^2\},$$

$$|f(T^{-1}\int_0^T x(t)dt, T^{-1}\int_0^T y(t)dt) - T^{-1}\int_0^T f(x(t), y(t))dt|$$

$$\leq \max\{6M_1\mu^{-1}\delta_C + \delta_C(3L_1+1) + 6\delta_{f,1}M_1 + 2\mu\delta_{f,1}(L_1+1)$$

$$+4^{-1}\mu(L_1+1)^2 + (2\mu T)^{-1}4M_1^2,$$

$$4(2T\mu)^{-1}M_2^2 + 6M_2\mu^{-1}\delta_D + \delta_D(3L_2+1)$$

$$+6\delta_{f,2}M_2 + 2\mu\delta_{f,2}(L_2+1) + 4^{-1}\mu(L_2+1)^2\},$$

and for every $z \in C$ and every $v \in D$,

$$f(z, T^{-1}\int_0^T y(t)dt) \geq f(T^{-1}\int_0^T x(t)dt, T^{-1}\int_0^T y(t)dt)$$

$$-2\max\{6M_1\mu^{-1}\delta_C + \delta_C(3L_1+1) + 6\delta_{f,1}M_1 + 2\mu\delta_{f,1}(L_1+1)$$

$$+4^{-1}\mu(L_1+1)^2 + (2\mu T)^{-1}4M_1^2,$$

$$(2T\mu)^{-1}4M_2^2 + 6M_2\mu^{-1}\delta_D + \delta_D(3L_2+1)$$

$$+6\delta_{f,2}M_2 + 2\mu\delta_{f,2}(L_2+1) + 4^{-1}\mu(L_2+1)^2\},$$

$$f(T^{-1}\int_0^T x(t)dt, v) \leq f(T^{-1}\int_0^T x(t)dt, T^{-1}\int_0^T y(t)dt)$$

$$+2\max\{6M_1\mu^{-1}\delta_C + \delta_C(3L_1+2) + 6\delta_{f,1}M_1$$

$$+4^{-1}\mu(L_1+1)^2 + (2\mu T)^{-1}4M_1^2,$$

$$4(2T\mu)^{-1}M_2^2 + 6M_2\mu^{-1}\delta_D + \delta_D(3L_2+1)$$

$$+6\delta_{f,2}M_2 + 2\mu\delta_{f,2}(L_2+1) + 4^{-1}\mu(L_2+1)^2\}.$$

Proof Let

$$z \in C,\ v \in D. \tag{6.285}$$

Proposition 6.11 implies that for almost every $t \in [0, T]$,

$$(2\mu)^{-1}(d/dt)(\|x(t) - z\|^2)$$

$$+f(x(t), y(t)) - f(z, y(t))$$

$$\leq 4^{-1}\mu\|\xi(t)\|^2 + (\mu^{-1}\delta_C + \delta_{f,1})(2 + 3\|z\|$$

$$+ 3\|x(t)\| + 2\mu\|\xi(t)\|), \tag{6.286}$$

$$(2\mu)^{-1}(d/dt)(\|y(t) - v\|^2)$$

$$+f(x(t), v) - f(x(t), y(t))$$

$$\le 4^{-1}\mu\|\eta(t)\|^2 + (\mu^{-1}\delta_D + \delta_{f,2})(2 + 3\|v\|$$

$$+ 3\|y(t)\| + 2\mu\|\eta(t)\|) \tag{6.287}$$

and that for all $t \in [0, T]$,

$$B_X(x(t), \delta_C) \cap C \neq \emptyset, \ B_Y(y(t), \delta_D) \cap D \neq \emptyset. \tag{6.288}$$

In view of (6.225), (6.277), and (6.288), for all $t \in [0, T]$,

$$\|x(t)\| \le M_1,$$

$$\|y(t)\| \le M_2. \tag{6.289}$$

Properties (iii) and (iv), (6.281), (6.283), and (6.289) imply that for almost every $t \in [0, T]$,

$$\|\xi(t)\| \le L_1 + 1,$$

$$\|\eta(t)\| \le L_2 + 1. \tag{6.290}$$

By (6.277), (6.285)–(6.287), (6.289), and (6.290), for almost every $t \in [0, T]$,

$$(2\mu)^{-1}(d/dt)(\|x(t) - z\|^2)$$

$$+f(x(t), y(t)) - f(z, y(t))$$

$$\le 4^{-1}\mu(L_1 + 1)^2$$

$$+(\mu^{-1}\delta_C + \delta_{f,1})(2 + 3(M_1 - 1) + 3M_1 + 2\mu(L_1 + 1))$$

$$\le 4^{-1}\mu(L_1 + 1)^2 + (\mu^{-1}\delta_C + \delta_{f,1})(6M_1 + 2\mu(L_1 + 1))$$

$$= 6M_1\mu^{-1}\delta_C + 2\delta_C(L_1 + 1)$$

$$+ 6M_1\delta_{f,1} + 2\mu\delta_{f,1}(L_1 + 1) + 4^{-1}\mu(L_1 + 1)^2 \tag{6.291}$$

and

$$(2\mu)^{-1}(d/dt)(\|y(t)-v\|^2)$$

$$+f(x(t),v)-f(x(t),y(t))$$

$$\leq 4^{-1}\mu(L_2+1)^2$$

$$+(\mu^{-1}\delta_D+\delta_{f,2})(2+3(M_2-1)+3M_2+2\mu(L_2+1))$$

$$\leq 4^{-1}\mu(L_2+1)^2+(\mu^{-1}\delta_D+\delta_{f,2})(6M_2+2\mu(L_2+1))$$

$$=6M_2\mu^{-1}\delta_D+2\delta_D(L_2+1)$$

$$+6M_2\delta_{f,2}+2\mu\delta_{f,2}(L_2+1)+4^{-1}\mu(L_2+1)^2. \tag{6.292}$$

By integrating (6.291) and (6.292) we obtain that

$$T^{-1}((2\mu)^{-1}\|x(T)-z\|^2-(2\mu)^{-1}\|x(0)-z\|^2)$$

$$+T^{-1}\int_0^T f(x(t),y(t))dt-T^{-1}\int_0^T f(z,y(t))dt$$

$$\leq 6M_1\mu^{-1}\delta_C+2\delta_C(L_1+1)$$

$$+6M_1\delta_{f,1}+2\mu\delta_{f,1}(L_1+1)+4^{-1}\mu(L_1+1)^2 \tag{6.293}$$

and

$$T^{-1}((2\mu)^{-1}\|y(T)-v\|^2-(2\mu)^{-1}\|y(0)-z\|^2)$$

$$+T^{-1}\int_0^T f(x(t),v)dt-T^{-1}\int_0^T f(x(t),y(t))dt$$

$$\leq 6M_2\mu^{-1}\delta_D+2\delta_D(L_2+1)$$

$$+6M_2\delta_{f,2}+2\mu\delta_{f,2}(L_2+1)+4^{-1}\mu(L_2+1)^2. \tag{6.294}$$

It follows from (6.277), (6.280), (6.285), (6.293), and (6.294) that

$$T^{-1}\int_0^T f(x(t),y(t))dt-T^{-1}\int_0^T f(z,y(t))dt$$

$$\leq (2\mu T)^{-1}4M_1^2+6M_1\mu^{-1}\delta_C+2\delta_C(L_1+1)$$

$$+6M_1\delta_{f,1}+2\mu\delta_{f,1}(L_1+1)+4^{-1}\mu(L_1+1)^2 \tag{6.295}$$

and

$$T^{-1}\int_0^T f(x(t),v)dt - T^{-1}\int_0^T f(x(t),y(t))dt$$

$$\leq (2\mu T)^{-1}4M_2^2 + 6M_2\mu^{-1}\delta_D + 2\delta_D(L_2+1)$$

$$+ 6M_2\delta_{f,2} + 2\mu\delta_{f,2}(L_2+1) + 4^{-1}\mu(L_2+1)^2. \tag{6.296}$$

Set

$$\Delta_1 = (2\mu T)^{-1}4M_1^2 + 6M_1\mu^{-1}\delta_C + 2\delta_C(L_1+1)$$

$$+ 6M_1\delta_{f,1} + 2\mu\delta_{f,1}(L_1+1) + 4^{-1}\mu(L_1+1)^2 \tag{6.297}$$

and

$$\Delta_2 = (2\mu T)^{-1}4M_2^2 + 6M_2\mu^{-1}\delta_D + 2\delta_D(L_2+1)$$

$$+ 6M_2\delta_{f,2} + 2\mu\delta_{f,2}(L_2+1) + 4^{-1}\mu(L_2+1)^2. \tag{6.298}$$

By (6.278), (6.279), and (6.295)–(6.298),

$$T^{-1}\int_0^T f(x(t),y(t))dt - f(x_*,y_*)$$

$$\geq T^{-1}\int_0^T f(x(t),y(t))dt - T^{-1}\int_0^T f(x(t),y_*)dt \geq -\Delta_2$$

and

$$T^{-1}\int_0^T f(x(t),y(t))dt - f(x_*,y_*)$$

$$\leq T^{-1}\int_0^T f(x(t),y(t))dt - T^{-1}\int_0^T f(x_*,y(t))dt \leq \Delta_1.$$

The relations above imply that

$$|T^{-1}\int_0^T f(x(t),y(t))dt - f(x_*,y_*)| \leq \max\{\Delta_1,\Delta_2\}.$$

Clearly, the functions $P_C(x(t))$, $t \in [0,T]$ and $P_D(y(t))$, $t \in [0,T]$ are Bochner integrable. Set

$$\widehat{x} = T^{-1} \int_0^T P_C(x(t))dt, \ \widehat{y} = T^{-1} \int_0^T P_C(y(t))dt, \tag{6.299}$$

$$x_T = T^{-1} \int_0^T x(s)ds, \ y_T = T^{-1} \int_0^T y(s)ds. \tag{6.300}$$

In view of (6.299),

$$\widehat{x} \in C, \ \widehat{y} \in D. \tag{6.301}$$

By (6.288) and (6.299),

$$\|\widehat{x} - T^{-1} \int_0^T x(s)ds\| \le \delta_C,$$

$$\|\widehat{y} - T^{-1} \int_0^T y(s)ds\| \le \delta_D. \tag{6.302}$$

It follows from (6.295) to (6.298) and (6.301) that

$$T^{-1} \int_0^T f(x(t), y(t))dt - T^{-1} \int_0^T f(\widehat{x}, y(t))dt \le \Delta_1 \tag{6.303}$$

and

$$T^{-1} \int_0^T f(x(t), y(t))dt - T^{-1} \int_0^T f(x(t), \widehat{y})dt \ge -\Delta_2. \tag{6.304}$$

Properties (iii) and (iv), (6.299), and (6.302) imply that for all $t \in [0, T]$,

$$|f(\widehat{x}, y(t)) - f(T^{-1} \int_0^T x(s)ds, y(t))| \le L_1\delta_C, \tag{6.305}$$

$$|f(x(t), \widehat{y}) - f(x(t), T^{-1} \int_0^T y(s)ds)| \le L_2\delta_D. \tag{6.306}$$

In view of (6.300), (6.305), and (6.306),

$$T^{-1}| \int_0^T f(\widehat{x}, y(t))dt - \int_0^T f(x_T, y(t))dt| \le L_1\delta_C, \tag{6.307}$$

$$T^{-1}| \int_0^T f(x(t), \widehat{y})dt - \int_0^T f(x(t), y_T)dt| \le L_2\delta_D. \tag{6.308}$$

By (6.303), (6.304), (6.307), and (6.308),

$$T^{-1}\int_0^T f(x(t),y(t))dt \le \Delta_1 + T^{-1}\int_0^T f(\widehat{x},y(t))dt$$

$$\le \Delta_1 + T^{-1}\int_0^T f(x_T,y(t))dt + L_1\delta_C \tag{6.309}$$

and

$$T^{-1}\int_0^T f(x(t),y(t))dt \ge -\Delta_2 + T^{-1}\int_0^T f(x(t),\widehat{y})dt$$

$$\ge -\Delta_2 - L_2\delta_D + T^{-1}\int_0^T f(x(t),y_T)dt. \tag{6.310}$$

Properties (i) and (ii), (6.300), (6.309), and (6.310) imply that

$$f(x_T,y_T) - T^{-1}\int_0^T f(x(t),y(t))dt$$

$$= f(T^{-1}\int_0^T x(t)dt, y_T) - T^{-1}\int_0^T f(x(t),y(t))dt$$

$$\le T^{-1}\int_0^T f(x(t),y_T)dt - T^{-1}\int_0^T f(x(t),y(t))dt$$

$$\le \Delta_2 + L_2\delta_D,$$

$$f(x_T,y_T) - T^{-1}\int_0^T f(x(t),y(t))dt$$

$$= f(x_T, T^{-1}\int_0^T y(t)dt) - T^{-1}\int_0^T f(x(t),y(t))dt$$

$$\ge T^{-1}\int_0^T f(x_T,y(t))dt - T^{-1}\int_0^T f(x(t),y(t))dt$$

$$\ge -\Delta_1 - L_1\delta_C$$

and

$$|f(x_T,y_T) - T^{-1}\int_0^T f(x(t),y(t))dt|$$

$$\le \max\{\Delta_1 + L_1\delta_C,\ \Delta_2 + L_2\delta_D\}. \tag{6.311}$$

Let

$$z \in C, \ v \in D. \tag{6.312}$$

By (6.295)–(6.298),

$$T^{-1}\int_0^T f(x(t), y(t))dt - T^{-1}\int_0^T f(z, y(t))dt \leq \Delta_1 \tag{6.313}$$

and

$$T^{-1}\int_0^T f(x(t), v)dt - T^{-1}\int_0^T f(x(t), y(t))dt \leq \Delta_2. \tag{6.314}$$

Properties (i) and (ii) and (6.300) imply that

$$T^{-1}\int_0^T f(z, y(t))dt \leq f(z, T^{-1}\int_0^T y(t)dt) = f(z, y_T) \tag{6.315}$$

and

$$T^{-1}\int_0^T f(x(t), v)dt \geq f(T^{-1}\int_0^T x(t)dt, v) = f(x_T, v). \tag{6.316}$$

By (6.313) and (6.315),

$$T^{-1}\int_0^T f(x(t), y(t))dt - f(z, y_T)$$

$$\leq T^{-1}\int_0^T f(x(t), y(t))dt - T^{-1}\int_0^T f(z, y(t))dt \leq \Delta_1.$$

Together with (6.311) this implies that

$$f(z, y_T) \geq T^{-1}\int_0^T f(x(t), y(t))dt - \Delta_1$$

$$\geq f(x_T, y_T) - \Delta_1 - \max\{\Delta_1 + L_1\delta_C, \ \Delta_2 + L_2\delta_D\}.$$

It follows from (6.314) and (6.316),

$$T^{-1}\int_0^T f(x(t), y(t))dt - f(x_T, v)$$

$$\geq T^{-1}\int_0^T f(x(t),y(t))dt - T^{-1}\int_0^T f(x(t),v)dt \geq -\Delta_2.$$

Together with (6.311) this implies that

$$f(x_T,v) \leq T^{-1}\int_0^T f(x(t),y(t))dt + \Delta_2$$

$$\leq f(x_T,y_T) + \Delta_2 + \max\{\Delta_1 + L_1\delta_C,\ \Delta_2 + L_2\delta_D\}.$$

This completes the proof of Theorem 6.12. □

We are interested to make the best choice of the parameter μ. In view of Theorem 6.12, we need to minimize the function

$$\max\{6M_1\mu^{-1}\delta_C + 2\mu(L_1+1)^2,\ 6M_2\mu^{-1}\delta_D + 2\mu(L_2+1)^2\}.$$

It is not difficult to see that the best choice of μ is $c_1\max\{\delta_C,\delta_D\}^{1/2}$ where the constant c_1 depends on L_1, M_1, L_2, M_2. Then the minimal value of the function above is $c_2\max\{\delta_C,\delta_D\}^{1/2}$, where the constant c_2 depends on L_1, M_1, L_2, M_2. The best choice of T is $c_3\max\{\delta_C,\delta_D\}^{1/2}$, where the constant c_3 also depends on L_1, M_1, L_2, M_2.

6.17 A Convergence Result for Games on Unbounded Sets

We continue to use the notation and definitions of Section 14.

Let

$$x_* \in C,\ y_* \in D \tag{6.317}$$

and

$$f(x_*,v) \leq f(x_*,y_*) \leq f(x,y_*) \tag{6.318}$$

for all $x \in C$ and all $v \in D$.

Let $M_0 \geq 1$,

$$D \subset B_Y(0, M_0 - 1), \tag{6.319}$$

$M > 8$,

$$\|x_*\| \leq M/4, \tag{6.320}$$

$\widehat{M} \geq 4M$, for each $y \in V \cap B_Y(0, M_0)$,

$$\{x \in U : f(x, y)$$

$$\leq \sup\{f(x_*, v) : v \in V \cap B_Y(0, M_0 + 1)\} + 4\} \subset B_X(0, \widehat{M}/4), \tag{6.321}$$

$$L_1, L_2 \geq 1 \tag{6.322}$$

and the continuous function $f : U \times V \to R^1$ has the following properties:

(v) for each $v \in V \cap B_Y(0, M_0 + 1)$,

$$|f(u_1, v) - f(u_2, v)| \leq L_1\|u_1 - u_2\|$$

$$\text{for all } u_1, u_2 \in U \cap B_X(0, \widehat{M} + 1);$$

(vi) for each $u \in B_X(0, \widehat{M}) + 1$,

$$|f(u, v_1) - f(u, v_2)| \leq L_2\|v_1 - v_2\|$$

$$\text{for all } v_1, v_2 \in B_Y(0, M_0 + 1);$$

(vii)

$$\{x \in X : d(x, C) \leq 1\} \subset U,$$

$$\{y \in Y : d(y, D) \leq 1\} \subset V.$$

In this section we prove the following result.

Theorem 6.13 *Let* $\mu \in (0, 1]$ *satisfy*

$$\mu(L_1 + 1)^2 \leq 1,$$

$$(\mu^{-1}\delta_C + \delta_{f,1}))(4\widehat{M} + 2\mu(L_1 + 1)) \leq 1, \tag{6.323}$$

$$T > 0,\ x \in W^{1,1}(0, T; X),\ y \in W^{1,1}(0, T; Y),$$

$$x(t) \in U,\ y(t) \in V,\ t \in [0, T],$$

$$d(x(0), C) \leq \delta_C,\ d(y(0), D) \leq \delta_D, \tag{6.324}$$

$$\|x(0)\| \leq M, \tag{6.325}$$

and let for almost every $t \in [0, T]$ *there exist* $\xi(t) \in X$, $\eta(t) \in Y$ *such that*

$$B_X(\xi(t), \delta_{f,1}) \cap \partial_x f(x(t), y(t)) \neq \emptyset, \tag{6.326}$$

$$P_C(x(t) - \mu\xi(t)) \in B_X(x(t) + x'(t), \delta_C), \tag{6.327}$$

$$B_Y(\eta(t), \delta_{f,2}) \cap \partial_y f(x(t), y(t)) \neq \emptyset, \tag{6.328}$$

$$P_D(y(t) + \mu\eta(t)) \in B_Y(yt) + y'(t), \delta_D). \tag{6.329}$$

Let

$$\Delta_1 = 12\widehat{M}\mu^{-1}\delta_C + 4\delta_C(L_1 + 1) + 12\delta_{f,1}\widehat{M}$$

$$+2^{-1}\mu(L_1 + 1)^2 + (2\mu T)^{-1}8\widehat{M}^2 + 4\mu\delta_{f,1}(L_1 + 1), \tag{6.330}$$

$$\Delta_2 = 12M_0\mu^{-1}\delta_D + 4\delta_D(L_2 + 1) + 12\delta_{f,2}M_0$$

$$+2^{-1}\mu(L_2 + 1)^2 + (2\mu T)^{-1}8M_0^2 + 2\mu\delta_{f,2}(L_2 + 1), \tag{6.331}$$

$$x_T = T^{-1}\int_0^T x(t)dt, \ y_T = T^{-1}\int_0^T y(t)dt.$$

Then for almost every $t \in [0, T]$,

$$B_X(x(t), \delta_C) \cap C \neq \emptyset,$$

$$B_Y(y(t), \delta_D) \cap D \neq \emptyset.$$

$$B_X(x_T, \delta_C) \cap C \neq \emptyset,$$

$$B_Y(y_T, \delta_D) \cap D \neq \emptyset,$$

$$x_T \in U, \ y_T \in V,$$

$$|T^{-1}\int_0^T f(x(t), y(t))dt - f(x_*, y_*)| \leq \max\{\Delta_1, \Delta_2\},$$

$$|f(x_T, y_T) - T^{-1}\int_0^T f(x(t), y(t))dt| \leq \max\{\Delta_1, \Delta_2\},$$

and for every $z \in C$ *and every* $v \in D$,

$$f(z, y_T) \geq f(x_T, y_T) - \leq \max\{\Delta_1, \Delta_2\},$$

$$f(x_T, v) \leq f(x_T, y_T) + \max\{\Delta_1, \Delta_2\}.$$

Proof Proposition 6.11 implies that for every $z \in C$, every $v \in D$, and almost every $t \in [0, T]$,

$$(2\mu)^{-1}(d/dt)(\|x(t) - z\|^2)$$

$$+f(x(t), y(t)) - f(z, y(t))$$

$$\leq 4^{-1}\mu\|\xi(t)\|^2 + (\mu^{-1}\delta_C + \delta_{f,1})(2 + 3\|z\|$$

$$+ 3\|x(t)\| + 2\mu\|\xi(t)\|) \tag{6.332}$$

and

$$(2\mu)^{-1}(d/dt)(\|y(t) - v\|^2)$$

$$+f(x(t), v) - f(x(t), y(t))$$

$$\leq 4^{-1}\mu\|\eta(t)\|^2 + (\mu^{-1}\delta_D + \delta_{f,2})(2 + 3\|v\|$$

$$+ 3\|y(t)\| + 2\mu\|\eta(t)\|). \tag{6.333}$$

Set

$$E = \{\tau \in [0, T] : \|x(t) - x_*\| \leq \widehat{M} - M \text{ for all } t \in [0, \tau]\}. \tag{6.334}$$

In view of (6.322), (6.325), and (6.334),

$$E \neq \emptyset.$$

Set

$$\tau_0 = \sup(E).$$

Clearly,

$$\tau_0 \in E. \tag{6.335}$$

Proposition 6.11 implies that for every $t \in [0, T]$,

$$B_X(x(t), \delta_C) \cap C \neq \emptyset,$$

$$B_Y(y(t), \delta_D) \cap D \neq \emptyset. \tag{6.336}$$

In view of (6.319) and (6.334)–(6.336), for all $t \in [0, \tau_0]$,

$$\|y(t)\| \le M_0,$$

$$\|x(t)\| \le \widehat{M}. \tag{6.337}$$

Properties (v) and (vi), (6.326), (6.328), (6.336), and (6.337) imply that for almost every $t \in [0, \tau_0]$,

$$\|\xi(t)\| \le L_1 + 1,$$

$$\|\eta(t)\| \le L_2 + 1. \tag{6.338}$$

We claim that

$$\tau_0 = T.$$

Assume the contrary. Then in view of (6.322) and (6.334)–(6.336),

$$\tau_0 < T, \tag{6.339}$$

$$\|x(\tau_0) - x_*\| = \widehat{M} - M, \tag{6.340}$$

$$\|x(\tau_0)\| > 3\widehat{M}/4. \tag{6.341}$$

By (6.317), (6.320), (6.332), (6.337), and (6.338), for almost every $t \in [0, \tau_0]$,

$$(2\mu)^{-1}(d/dt)(\|x(t) - x_*\|^2)$$

$$+f(x(t), y(t)) - f(x_*, y(t))$$

$$\le 4^{-1}\mu(L_1 + 1)^2$$

$$+ (\mu^{-1}\delta_C + \delta_{f,1})(2 + M + 3\widehat{M} + 2\mu(L_1 + 1)). \tag{6.342}$$

By (6.321) and (6.341),

$$f(x(\tau_0), y(\tau_0)) > \sup\{f(x_*, v) : \ v \in V \cap B_Y(0, M_0 + 1)\} + 4. \tag{6.343}$$

Since the function f is continuous it follows from (6.343) that there exists

$$\tau_1 \in (0, \tau_0)$$

such that for each $t \in [\tau_1, \tau_0]$,

$$f(x(t), y(t)) > \sup\{f(x_*, v) : \ v \in V \cap B_Y(0, M_0 + 1)\} + 4. \tag{6.344}$$

By integrating (6.342) on $[\tau_1, \tau_0]$ we obtain that

$$(2\mu)^{-1}\|x(\tau_0) - x_*\|^2 - (2\mu)^{-1}\|x(\tau_1) - x_*\|^2$$

$$+ \int_{\tau_1}^{\tau_0} f(x(t), y(t))dt - \int_{\tau_1}^{\tau_0} f(x_*, y(t))dt$$

$$\leq (\tau_0 - \tau_1)(4^{-1}\mu(L_1 + 1)^2$$

$$+ (\mu^{-1}\delta_C + \delta_{f,1})(2 + M + 3\widehat{M} + 2\mu(L_1 + 1)). \quad (6.345)$$

It follows from (6.320), (6.322), (6.323), (6.325), (6.337), (6.344), and (6.345) that

$$4(\tau_0 - \tau_1) \leq \int_{\tau_1}^{\tau_0} f(x(t), y(t))dt$$

$$-(\tau_0 - \tau_1)\sup\{f(x_*, v) : v \in V \cap B_Y(0, M_0 + 1)\} \leq 2(\tau_0 - \tau_1),$$

a contradiction. The contradiction we have reached proves that

$$\tau_0 = T. \quad (6.346)$$

By (6.320), (6.322), (6.334)–(6.336), and (6.346),

$$\|x(t) - x_*\| \leq \widehat{M} - M \text{ for all } t \in [0, T],$$

$$\|x(t)\| \leq \widehat{M} - (3/4)M \text{ for all } t \in [0, T]. \quad (6.347)$$

Set

$$\tilde{D} = D, \ \tilde{V} = V,$$

$$\tilde{C} = C \cap B_X(0, \widehat{M} - 1), \ \tilde{U} = U \cap \{x \in X : \|x\| < \widehat{M}\}. \quad (6.348)$$

In view of (6.324), there exists

$$\widehat{x}(0) \in C \text{ such that } \|x(0) - \widehat{x}(0)\| \leq \delta_C. \quad (6.349)$$

It follows from (6.320), (6.347), and (6.349) that

$$\|\widehat{x}(0)\| \leq \|x(0)\| + 1 \leq \widehat{M} - (3/4)M + 1 \leq \widehat{M} - 1. \quad (6.350)$$

By (6.348), (6.350), and the choice of $\widehat{x}(0)$,

$$\widehat{x}(0) \in \tilde{C},$$

$$d(x(0), \tilde{C}) \le \|x(0) - \widehat{x}(0)\| \le \delta_C. \tag{6.351}$$

Lemma 2.2, (6.317), and (6.347) imply that for all $t \in [0, T]$,

$$\|x_* - P_C(x(t))\| \le \|x_* - x(t)\| \le \widehat{M} - M. \tag{6.352}$$

It follows from (6.319), (6.320), (6.323), (6.338), and (6.352) that for all $t \in [0, T]$

$$\|x_* - P_C(x(t) + \mu\xi(t))\| \le \widehat{M} - M + 1,$$

$$\|P_C(x(t) + \mu\xi(t))\| \le \widehat{M} - 3M/4 + 1 \le \widehat{M} - 1. \tag{6.353}$$

In view of (6.348) and (6.353),

$$P_C(x(t) + \mu\xi(t)) \in \tilde{C},$$

$$P_C(x(t) + \mu\xi(t)) = P_{\tilde{C}}(x(t) + \mu\xi(t)). \tag{6.354}$$

By (6.347), (6.348), (6.351), and (6.354) all the assumptions of Theorem 6.12 hold with

$$C = \tilde{C},\ U = \tilde{U},\ D = \tilde{D},\ V = \tilde{V}. \tag{6.355}$$

Together with (6.330) and (6.331) and property (vii) this implies that for all $t \in [0, T]$,

$$B_X(x(t), \delta_C) \cap \tilde{C} \ne \emptyset,$$

$$B_Y(y(t), \delta_D) \cap \tilde{D} \ne \emptyset.$$

$$B_X(x_T, \delta_C) \cap \tilde{C} \ne \emptyset,$$

$$B_Y(y_T, \delta_D) \cap D \ne \emptyset, \tag{6.356}$$

$$x_T \in U,\ y_T \in V, \tag{6.357}$$

$$|T^{-1} \int_0^T f(x(t), y(t))dt - f(x_*, y_*)| \le \max\{\Delta_1, \Delta_2\},$$

$$|f(x_T, y_T) - T^{-1} \int_0^T f(x(t), y(t))dt| \le \max\{\Delta_1, \Delta_2\},$$

for every

$$z \in \tilde{C} = C \cap B_X(0, \widehat{M} - 1)$$

and every $v \in D$,

$$f(z, y_T) \geq f(x_T, y_T) - \max\{\Delta_1, \Delta_2\},$$

$$f(x_T, v) \leq f(x_T, y_T) + \max\{\Delta_1, \Delta_2\}. \tag{6.358}$$

It follows from (6.319) to (6.321), (6.348), and (6.356) to (6.358) that for each $z \in C \setminus \tilde{C}$,

$$f(z, y_T) \geq \sup\{f(x_*, v) : \ v \in V \cap B_X(0, M_0 + 1)\} + 4$$

$$\geq f(x_*, y_T) + 4 \geq f(x_T, y_T) - \max\{\Delta_1, \Delta_2\} + 4.$$

Theorem 6.13 is proved. □

As in the case of Theorem 6.12 we can show that the best choice of μ is $c_1 \max\{\delta_C, \delta_D\}^{1/2}$ where the constant c_1 depends of L_1, M_1, L_2, M_2 and the best choice of T is $c_2 \max\{\delta_C, \delta_D\}^{1/2}$, where the constant c_2 depends of L_1, M_1, L_2, M_2.

Chapter 7
An Optimization Problems with a Composite Objective Function

In this chapter we study an algorithm for minimization of the sum of two functions, the first one being smooth and convex and the second being convex. For this algorithm each iteration consists of two steps. The first step is a calculation of a subgradient of the first function while the second one is a proximal gradient step for the second function. In each of these two steps there is a computational error. In general, these two computational errors are different. We show that our algorithm generates a good approximate solution, if all the computational errors are bounded from above by a small positive constant. Moreover, if we know the computational errors for the two steps of our algorithm, we find out what approximate solution can be obtained and how many iterates one needs for this.

7.1 Preliminaries

Let H be a Hilbert space equipped with an inner product $\langle \cdot, \cdot \rangle$ which induces a complete norm $\| \cdot \|$, $\mathrm{Id} : X \to X$ be the identity operator such that

$$\mathrm{Id}(x) = x, \ x \in X$$

and let a set $D \subset H$ be nonempty.

An operator $T : D \to H$ is called firmly nonexpansive if

$$\|Tx - Ty\|^2 + \|(\mathrm{Id} - T)x - (\mathrm{Id} - T)y\|^2 \le \|x - y\|^2$$

for all $x, y \in D$.

Let $f, h : H \to (-\infty, \infty]$. The infimal convolution of f and g is the function

$$f \Box g : H \to [-\infty, \infty]$$

A. J. Zaslavski, *Convex Optimization with Computational Errors*, Springer Optimization and Its Applications 155, https://doi.org/10.1007/978-3-030-37822-6_7

defined by

$$f\Box g(x) = \inf\{f(y) + g(x-y) : \; y \in H\}$$

[13].

Let $f : H \to (-\infty, \infty]$ and $\gamma > 0$. The Moreau envelope of f of parameter γ [13] is the function

$${}^{\gamma}f := f\Box((2\gamma)^{-1}\|\cdot\|^2).$$

Clearly, for all $x \in H$,

$${}^{\gamma}f(x) = \inf\{f(y) + (2\gamma)^{-1}\|x-y\|^2 : \; y \in H\}. \tag{7.1}$$

Let $f : H \to (-\infty, \infty]$ be a convex and lower semicontinuous function which is not identically infinity. Then $\mathrm{Prox}_f x$ is the unique point in H which satisfies

$$\begin{aligned} {}^{1}f(x) &= \inf\{f(y) + 2^{-1}\|x-y\|^2 : \; y \in H\} \\ &= f(\mathrm{Prox}_f x) + 2^{-1}\|x - \mathrm{Prox}_f x\|^2. \end{aligned} \tag{7.2}$$

Let $h : H \to (-\infty, \infty]$ be a convex and lower semicontinuous function which is not identically infinity and let $x \in H$. In view of (7.2),

$$\begin{aligned} \mathrm{Prox}_{th}(x) &= \mathrm{argmin}\{th(y) + 2^{-1}\|x-y\|^2 : \; y \in H\} \\ &= \mathrm{argmin}\{h(y) + (2t)^{-1}\|x-y\|^2 : \; y \in H\}. \end{aligned} \tag{7.3}$$

In view of (7.3),

$$0 \in \partial h(\mathrm{Prox}_{th}(x)) + t^{-1}(\mathrm{Prox}_{th}(x) - x)$$

and

$$t^{-1}(x - \mathrm{Prox}_{th}(x)) \in \partial h(\mathrm{Prox}_{th}(x)). \tag{7.4}$$

In the sequel we use the following results.

Proposition 7.1 ([13]) *Let $f : H \to (-\infty, \infty]$ be a convex and lower semicontinuous function which is not identically infinity. Then Prox_f and $\mathrm{Id} - \mathrm{Prox}_f$ are firmly nonexpansive mappings.*

Proposition 7.2 ([13]) *Let $f : H \to (-\infty, \infty]$ be a convex and lower semicontinuous function which is not identically infinity and $x, p \in H$. Then*

$$p = Prox_f(x)$$

if and only if

$$\langle y - p, x - p \rangle + f(p) \le f(y)$$

for all $y \in H$.

Proposition 7.3 *Let* $h : H \to (-\infty, \infty]$ *be a convex and lower semicontinuous function which is not identically infinity,* $u, w \in H$, $t > 0$ *and*

$$v = Prox_{th}(u). \tag{7.5}$$

Then

$$h(w) \ge h(v) + (2t)^{-1}(\|u - v\|^2 + \|w - v\|^2 - \|u - w\|^2).$$

Proof Lemma 2.1, Proposition 7.2, and (7.5) imply that

$$th(w) \ge th(v) + \langle w - v, u - v \rangle$$

$$= th(v) + 2^{-1}[\|u - v\|^2 + \|w - v\|^2 - \|u - w\|^2].$$

Proposition 7.3 is proved. □

7.2 The Algorithm and Main Results

Let X be a Hilbert space equipped with an inner product $\langle \cdot, \cdot \rangle$ which induces a complete norm $\| \cdot \|$, $f : X \to R^1$ be a convex Fréchet differentiable function, $g : X \to (-\infty, \infty]$ be a convex and lower semicontinuous function which is not identically infinity and

$$F(x) = f(x) + g(x), \ x \in X. \tag{7.6}$$

We consider the minimization problem

$$F(x) \to \min, \ x \in X$$

studied in [69] for in the finite-dimensional case.

We denote by $f'(x)$ the Fréchet derivative of the function f at the point $x \in X$.

Let g be bounded from below, $\theta \in X$, $M > 1$,

$$\|\theta\| < M - 1, \ g(\theta) < \infty, \tag{7.7}$$

$c_* > 0$,

$$g(x) \geq -c_* \text{ for all } x \in X, \tag{7.8}$$

$$\mathrm{argmin}(F) \cap B_X(0, M) \neq \emptyset, \tag{7.9}$$

$L_1 \geq 1$ be such that

$$|f(z_1) - f(z_2)| \leq L_1 \|z_1 - z_2\|$$

$$\text{for all } z_1, z_2 \in B_X(0, 3M + 2). \tag{7.10}$$

Let

$$M_1 = 6M + 2L_1 + 4 + 2c_* + 2|g(\theta)| + M, \tag{7.11}$$

$L \geq 1$ be such that

$$\|f'(z_1) - f'(z_2)\| \leq L\|z_1 - z_2\|$$

$$\text{for all } z_1, z_2 \in B_X(0, M_1 + 2). \tag{7.12}$$

Let $\delta_f, \delta_G \in (0, 1]$,

$$0 < S_- < S_+ \leq L^{-1}, \tag{7.13}$$

$$S_- \leq S_t \leq S_+ \; t = 0, 1, \ldots. \tag{7.14}$$

Let us describe our algorithm.

Initialization: select an arbitrary $x_0 \in X$.
Iterative Step: given a current iteration vector $x_k \in X$ calculate $\xi_k \in X$ such that

$$\|\xi_k - f'(x_k)\| \leq \delta_f$$

and $x_{k+1} \in X$ such that

$$g(x_{k+1}) + (2S_k)^{-1}\|x_{k+1} - (x_k - S_k\xi_k)\|^2$$

$$\leq g(x) + (2S_k)^{-1}\|x - (x_k - S_k\xi_k)\|^2 + \delta_G \text{ for all } x \in X.$$

Our iterative step consists of two calculations. The second one is a proximal method applied to the function g. It should be mentioned that proximal methods are important tools in analysis and optimization [4, 5, 7, 8, 14, 19, 23–26, 28, 31, 33, 34, 40–45, 47, 48].

In this chapter we prove the following two results.

Theorem 7.4 *Let*

$$\delta_G \le 4^{-1},$$

$$\{x_t\}_{t=0}^{\infty} \subset X, \ \{\xi_t\}_{t=0}^{\infty} \subset X,$$

$$\|x_0\| \le M, \tag{7.15}$$

for each integer $t \ge 0$,

$$\|\xi_t - f'(x_t)\| \le \delta_f \tag{7.16}$$

and

$$g(x_{t+1}) + (2S_t)^{-1}\|x_{t+1} - (x_t - S_t\xi_t)\|^2$$

$$\le g(w) + (2S_t)^{-1}\|w - (x_t - S_t\xi_t)\|^2 + \delta_G \text{ for all } w \in X. \tag{7.17}$$

Let

$$\epsilon_0 = 2\delta_G + 4\delta_f M_1 + S_-^{-1}\delta_G^{1/2}(4M_1 + 1),$$

$$n_0 = \lfloor 4\epsilon_0^{-1} S_-^{-1} M^2 + 1 \rfloor.$$

Then there exists an integer $q \in [0, n_0 + 1]$ *such that*

$$F(x_q) \le \inf(F) + \epsilon_0,$$

$$\|x_t\| \le 3M, \ t = 0, \ldots, q.$$

It is easy to see that

$$\epsilon_0 = c_1 \max\{\delta_G^{1/2}, \delta_f\}$$

and

$$n_0 = \lfloor c_2 \max\{\delta_G^{1/2}, \delta_f\}^{-1} \rfloor,$$

where c_1, c_2 are positive constants which depend on L, M_1 and S_-.

Theorem 7.5 *Let*

$$\delta_f < M_1/8,$$

$$\delta_G < \min\{4^{-1}, \ ((16M_1 + 4)^{-1}S_-)^2\}, \tag{7.18}$$

$$\{x \in X : F(x) \le \inf(F) + 4\} \subset B_X(0, M), \tag{7.19}$$

$\{x_t\}_{t=0}^{\infty} \subset X$, $\{\xi_t\}_{t=0}^{\infty} \subset X$,

$$\|x_0\| \le M,$$

for each integer $t \ge 0$,

$$\|\xi_t - f'(x_t)\| \le \delta_f \tag{7.20}$$

and

$$g(x_{t+1}) + (2S_t)^{-1}\|x_{t+1} - (x_t - S_t\xi_t)\|^2$$

$$\le g(w) + (2S_t)^{-1}\|w - (x_t - S_t\xi_t)\|^2 + \delta_G \text{ for all } w \in X. \tag{7.21}$$

Then for all integers $t \ge 0$

$$\|x_t\| \le 3M$$

and for all pairs of natural numbers m, T,

$$\min\{F(x_t) : t = m+1, \dots, m+T\} - \inf(F),$$

$$f(\sum_{t=m+1}^{m+T} T^{-1}x_t) - \inf(F)$$

$$\le 4T^{-1}M^2 + (2S_-)^{-1}(4M_1 + 1)\delta_G^{1/2} + \delta_G + 2\delta_f M_1.$$

It is easy to see that

$$T = \lfloor c_1 \max\{\delta_G^{1/2}, \delta_f\}^{-1} \rfloor,$$

where c_1 is a positive constant which depends on M, M_1 and S_-, is the best choice of T.

7.3 Auxiliary Results

Lemma 7.6 *Let* $S > 0$, $u_0, u_1, \xi_0, \widehat{u}_1 \in X$,

$$g(u_1) + (2S)^{-1}\|u_1 - (u_0 - S\xi_0)\|^2$$

$$\leq g(x) + (2S)^{-1}\|x - (u_0 - S\xi_0)\|^2 + \delta_G \textit{ for all } x \in X$$

and

$$g(\widehat{u}_1) + (2S)^{-1}\|\widehat{u}_1 - (u_0 - S\xi_0)\|^2$$

$$\leq g(x) + (2S)^{-1}\|x - (u_0 - S\xi_0)\|^2 \textit{ for all } x \in X.$$

Then

$$\|u_1 - \widehat{u}_1\| \leq 2(S\delta_G)^{1/2}.$$

Proof Set

$$h_1(w) = g(w) + (2S)^{-1}\|w - (u_0 - S\xi_0)\|^2, \ w \in X.$$

Clearly, h_1 is a convex function. Since the function g is convex it is not difficult to see that

$$h_1(\widehat{u}_1) \leq h_1(2^{-1}u_1 + 2^{-1}\widehat{u}_1)$$

$$= g(2^{-1}u_1 + 2^{-1}\widehat{u}_1) + (2S)^{-1}\|2^{-1}u_1 + 2^{-1}\widehat{u}_1 - u_0 + S\xi_0\|^2$$

$$\leq 2^{-1}g(u_1) + 2^{-1}g(\widehat{u}_1)$$

$$+(2S)^{-1}\|2^{-1}(u_1 - u_0 + S\xi_0) + 2^{-1}(\widehat{u}_1 - u_0 + S\xi_0)\|^2$$

$$= 2^{-1}g(u_1) + 2^{-1}g(\widehat{u}_1)$$

$$+(2S)^{-1}[2\|2^{-1}(u_1 - u_0 + S\xi_0)\|^2 + 2\|2^{-1}(\widehat{u}_1 - u_0 + S\xi_0)\|^2$$

$$-\|2^{-1}(u_1 - \widehat{u}_1)\|^2]$$

$$= 2^{-1}(g(u_1) + (2S)^{-1}\|u_1 - u_0 + S\xi_0)\|^2)$$

$$+2^{-1}(g(\widehat{u}_1) + (2S)^{-1}\|\widehat{u}_1 - u_0 + S\xi_0)\|^2)$$

$$-(2S)^{-1}\|2^{-1}(u_1 - \widehat{u}_1)\|^2$$

$$= 2^{-1}h_1(u_1) + 2^{-1}h_1(\widehat{u}_1) - (2S)^{-1}\|2^{-1}(u_1 - \widehat{u}_1)\|^2$$

$$\leq h_1(\widehat{u}_1) + \delta_G/2 - (2S)^{-1}\|2^{-1}(u_1 - \widehat{u}_1)\|^2$$

and

$$\|u_1 - \widehat{u}_1\| \le 2(S\delta_G)^{1/2}.$$

Lemma 7.6 is proved. □

Lemma 7.7 *Let $S \in (0, 1]$, $x, y, \xi, p \in X$, $\delta_G \le 4^{-1}$,*

$$\|\xi - f'(y)\| \le \delta_f \tag{7.22}$$

$$\|x\|, \|y\| \le 3M, \tag{7.23}$$

$$g(p) + (2S)^{-1}\|p - (x - S\xi)\|^2$$

$$\le g(z) + (2S)^{-1}\|z - (x - S\xi)\|^2 + \delta_G \tag{7.24}$$

for all $z \in X$. Then

$$\|p\| \le M_1$$

and for every $z \in X$ satisfying $\|z\| \le 3M$,

$$F(z) - F(p)$$

$$\ge ((2S)^{-1} - 2^{-1}L)\|p - z\|^2$$

$$+(2S)^{-1}(\|x - p\|^2 - \|x - z\|^2)$$

$$+\langle p - z, \xi - f'(z)\rangle - \delta_G - (2S)^{1/2}\delta_G^{1/2}(4M_1 + 1),$$

$$F(z) - F(p)$$

$$\ge (2S)^{-1}(\|x - p\|^2 + \|z - p\|^2 - \|x - z\|^2)$$

$$+f(z) - f(y) - 2^{-1}L\|p - y\|^2$$

$$+\langle f'(y), y - z\rangle - 2M_1\delta_f - \delta_G - (2S)^{1/2}\delta_G^{1/2}(4M_1 + 1),$$

$$F(z) - F(p)$$

$$\ge (2S)^{-1}(\|x - p\|^2 + \|z - p\|^2 - \|x - z\|^2)$$

$$-2^{-1}L\|p - y\|^2 - 2M_1\delta_f - \delta_G - (2S)^{1/2}\delta_G^{1/2}(4M_1 + 1).$$

Proof Set

$$p_0 = \mathrm{Prox}_{Sg}(x - S\xi). \tag{7.25}$$

In view of (7.25),

$$g(p_0) + (2S)^{-1}\|p_0 - (x - S\xi)\|^2$$
$$\le g(z) + (2S)^{-1}\|z - (x - S\xi)\|^2 \tag{7.26}$$

for all $z \in X$.

Let

$$z \in B_X(0, 3M).$$

Applying Proposition 7.3 with $v = p_0$, $u = x - S\xi$, $z = w$, $t = S$ we obtain that

$$g(z) - g(p_0)$$
$$\ge (2S)^{-1}(\|x - S\xi - p_0\|^2 + \|z - p_0\|^2 - \|x - S\xi - z\|^2). \tag{7.27}$$

In view of (7.27),

$$g(z) + (2S)^{-1}\|x - S\xi - z\|^2$$
$$-(g(p_0) + (2S)^{-1}\|x - S\xi - p_0\|^2)$$
$$\ge (2S)^{-1}\|z - p_0\|^2. \tag{7.28}$$

Lemma 7.6, (7.24), and (7.26) imply that

$$\|p_0 - p\| \le 2(S\delta_G)^{1/2}. \tag{7.29}$$

By (7.10), (7.22), and (7.23),

$$\|\xi\| \le \|f'(y)\| + 1 \le L_1 + 1. \tag{7.30}$$

It follows from (7.24), (7.28), and (7.29) that

$$g(z) + (2S)^{-1}\|x - S\xi - z\|^2$$
$$-(g(p) + (2S)^{-1}\|x - S\xi - p\|^2)$$
$$\ge g(z) + (2S)^{-1}\|x - S\xi - z\|^2$$
$$-(g(p_0) + (2S)^{-1}\|x - S\xi - p_0\|^2) - \delta_G$$
$$\ge (2S)^{-1}\|z - p_0\|^2 - \delta_G$$

$$\geq (2S)^{-1}\|z-p\|^2 - \delta_G - \|p-p_0\|(\|z-p_0\| + \|z\| + p\|)$$

$$\geq (2S)^{-1}\|z-p\|^2 - \delta_G - 2(S\delta_G)^{1/2}(2\|z\| + 2\|p_0\| + 1). \tag{7.31}$$

In view of (7.8) and (7.24),

$$(2S)^{-1}\|p-(x-S\xi)\|^2$$

$$\leq c_* + g(\theta) + (2S)^{-1}\|\theta - x + S\xi)\|^2 + 1. \tag{7.32}$$

In view of (7.32),

$$\|p-(x-S\xi)\|^2 \leq 2(c_* + |g(\theta)| + 1) + \|\theta - x + S\xi)\|^2. \tag{7.33}$$

It follows from (7.7), (7.11), (7.23), and (7.30) that

$$\|p\| \leq \|x\| + \|\xi\|$$

$$+2(c_* + |g(\theta)| + 1) + \|\theta\| + \|x\| + \|\xi\|$$

$$\leq 3 + M + L_1 + 1 + 2(c_* + |g(\theta)| + 1) + \|\theta\| + 3M + L_1 + 1$$

$$= 6M + 2L_1 + 4 + 2(c_* + |g(\theta)|) + \|\theta\| \leq M_1. \tag{7.34}$$

By (7.11), (7.23), (7.31), and (7.34),

$$g(z) + (2S)^{-1}\|x - S\xi - z\|^2$$

$$-(g(p) + (2S)^{-1}\|x - S\xi - p\|^2)$$

$$\geq (2S)^{-1}\|z-p\|^2 - \delta_G - 2(S\delta_G)^{1/2}(4M_1 + 1). \tag{7.35}$$

In view of (7.35),

$$g(z) - g(p)$$

$$\geq (2S)^{-1}(\|x - S\xi - p\|^2 - \|x - S\xi - z\|^2 + \|z-p\|^2)$$

$$-\delta_G - (2S\delta_G)^{1/2}(4M_1 + 1)$$

$$= (2S)^{-1}(\|x-p\|^2 - \|x-z\|^2 + \|z-p\|^2)$$

$$+ \langle p - z, \xi\rangle - \delta_G - (2S\delta_G)^{1/2}(4M_1 + 1). \tag{7.36}$$

Proposition 4.3 and (7.12) imply that

$$f(z) - f(p) \geq -2^{-1}L\|p - z\|^2 - \langle f'(z), p - z\rangle. \tag{7.37}$$

It follows from (7.6), (7.36), and (7.37) that

$$F(z) - F(p)$$

$$\geq ((2S)^{-1} - 2^{-1}L)\|p - z\|^2 + (2S)^{-1}(\|x - p\|^2 - \|x - z\|^2)$$

$$+ \langle \xi - f'(z), p - z\rangle - \delta_G - (2S\delta_G)^{1/2}(4M_1 + 1). \tag{7.38}$$

In view of (7.38), the first inequality in the statement of the lemma is proved.

Let us prove the second inequality in the statement of the lemma. Proposition 4.3 and (7.12) imply that

$$\langle f'(y), p - y\rangle \geq f(p) - f(y) - 2^{-1}L\|p - y\|^2. \tag{7.39}$$

Combining (7.22), (7.34), (7.36), and (7.39) we obtain that

$$g(z) - g(p)$$

$$\geq (2S)^{-1}(\|x - p\|^2 - \|x - z\|^2 + \|z - p\|^2)$$

$$+f(p) - f(y) - 2^{-1}L\|p - y\|^2$$

$$+\langle f'(y), y - z\rangle + \langle \xi - f'(y), p - z\rangle$$

$$-\delta_G - (2S\delta_G)^{1/2}(4M_1 + 1)$$

$$\geq (2S)^{-1}(\|x - p\|^2 - \|x - z\|^2 + \|z - p\|^2)$$

$$+f(p) - f(y) - 2^{-1}L\|p - y\|^2 + \langle f'(y), y - z\rangle$$

$$- 2M_1\delta_f - \delta_G - (2S\delta_G)^{1/2}(4M_1 + 1). \tag{7.40}$$

We add $f(z) - f(p)$ to (7.40) and obtain the second inequality in the statement of the lemma. Together with the convexity of f this implies the third inequality in the statement of the lemma. Lemma 7.7 is proved. □

Lemma 7.7 implies the following result.

Lemma 7.8 *Let* $S \in (0, 1]$, $x, \xi, p \in X$, $\delta_G \leq 4^{-1}$,

$$S \leq L^{-1},$$

$$\|\xi - f'(x)\| \le \delta_f,$$

$$\|x\| \le M + 3,$$

$$g(p) + (2S)^{-1}\|p - (x - S\xi)\|^2$$

$$\le g(z) + (2S)^{-1}\|z - (x - S\xi)\|^2 + \delta_G$$

for all $z \in X$. Then

$$\|p\| \le M_1$$

and for every $z \in X$ satisfying $\|z\| \le 3M$,

$$F(z) - F(p)$$

$$(2S)^{-1}(\|z - p\|^2 - \|x - z\|^2)$$

$$+((2S)^{-1} - 2^{-1}L)\|x - p\|^2$$

$$-2M_1\delta_f - \delta_G - (2S)^{-1}\delta_G^{1/2}(4M_1 + 1)$$

$$\ge (2S)^{-1}(\|z - p\|^2 - \|x - z\|^2)$$

$$-2M_1\delta_f - \delta_G - (2S)^{-1}\delta_G^{1/2}(4M_1 + 1).$$

7.4 Proof of Theorem 7.4

In view of (7.9), there exists

$$z \in \operatorname{argmin}(F) \cap B_X(0, M). \tag{7.41}$$

By (7.15) and (7.41),

$$\|x_0 - z\| \le 2M. \tag{7.42}$$

If

$$F(x_0) \le \inf(F) + \epsilon_0,$$

then the assertion of the theorem holds. Assume that

$$F(x_0) > \inf(F) + \epsilon_0. \tag{7.43}$$

If

$$F(x_1) \le \inf(F) + \epsilon_0,$$

then by Lemma 7.8,

$$\|x_1\| \le M_1$$

and the assertion of the theorem holds.

Let

$$F(x_1) > \inf(F) + \epsilon_0. \tag{7.44}$$

Assume that $T \ge 0$ is an integer and that for all integers $t = 0, \dots, T$,

$$F(x_{t+1}) - F(z) > \epsilon_0. \tag{7.45}$$

(Note that in view of (7.44), out assumption holds for $T = 0$.)

We show that for all $t = 0, \dots, T$,

$$\|x_t - z\| \le 2M \tag{7.46}$$

and

$$F(z) - F(x_{t+1})$$

$$\ge (2S_t)^{-1}(\|z - x_{t+1}\|^2 - \|z - x_t\|^2)$$

$$- 2M_1\delta_f - \delta_G - (2S_t)^{-1}\delta_G^{1/2}(4M_1 + 1). \tag{7.47}$$

Assume that $t \in \{0, \dots, T\}$ and (7.46) holds. By (7.41) and Lemma 7.8 applied with

$$p = x_{t+1},\ x = x_t,\ \xi = \xi_t,$$

(7.47) is true. It follows from (7.13), (7.45), and (7.47) that

$$\epsilon_0 < (2S_t)^{-1}(\|z - x_t\|^2 - \|z - x_{t+1}\|^2)$$

$$+(2S_t)^{-1}\delta_G^{1/2}(4M_1 + 1) + 2M_1\delta_f + \delta_G$$

$$\le (2S_t)^{-1}(\|z - x_t\|^2 - \|z - x_{t+1}\|^2)$$

$$+ (2S_-)^{-1}\delta_G^{1/2}(4M_1 + 1) + 2M_1\delta_f + \delta_G. \tag{7.48}$$

Relations (7.13), (7.18), and (7.48) imply that

$$\|z - x_{t+1}\| \le \|z - x_t\|.$$

Thus by induction we have shown that (7.46) holds for all $t = 0, \ldots, T + 1$, (7.47) holds for all $t = 0, \ldots, T$ and that

$$\|z - x_{t+1}\| \le \|z - x_t\| \tag{7.49}$$

is true for all $t = 0, \ldots, T$.

By (7.42) and (7.45)–(7.49),

$$(T + 1)\epsilon_0 < (T + 1)(\min\{F(x_t) : t = 1, \ldots, T + 1\} - F(z))$$

$$\le \sum_{t=0}^{T}(F(x_{t+1} - F(z))$$

$$\le \sum_{t=0}^{T}(2S_-)^{-1}(\|z - x_t\|^2 - \|z - x_{t+1}\|^2)$$

$$+(T + 1)[(2S_-)^{-1}\delta_G^{1/2}(4M_1 + 1) + 2M_1\delta_f + \delta_G]$$

$$\le (2S_-)^{-1}4M^2 + (T + 1)[(2S_-)^{-1}\delta_G^{1/2}(4M_1 + 1) + 2M_1\delta_f + \delta_G].$$

Together with the choice of ϵ_0, n_0 this implies that

$$(T + 1)\epsilon_0/2 \le (2S_-)^{-1}4M^2$$

and

$$T \le 4S_-^{-1}M^2\epsilon_0^{-1} < n_0 + 1.$$

Thus we have shown that if T is an integer and (7.45) holds for all integers $t = 0, \ldots, T$, then $T < n_0$ and in view of (7.41) and (7.46)

$$\|x_t\| \le 3M, \ t = 0, \ldots, T + 1.$$

This implies that there exists an integer $q \in [0, n_0 + 1]$ such that

$$F(x_q) \le \inf(F) + \epsilon_0,$$

$$\|x_t\| \le 3M, \ t = 0, \ldots, q.$$

Theorem 7.4 is proved. □

7.5 Proof of Theorem 7.5

In view of (7.9) and (7.19), there exists

$$z \in \operatorname{argmin}(F) \subset B_X(0, M). \tag{7.50}$$

It follows from (7.50) and the assumptions of the theorem that

$$\|x_0 - z\| \leq 2M. \tag{7.51}$$

Assume that $t \geq 0$ is an integer and that

$$\|x_t - z\| \leq 2M. \tag{7.52}$$

By (7.20), (7.50), (7.52), and Lemma 7.8 applied with

$$p = x_{t+1},\ x = x_t,\ \xi = \xi_t,$$

we have

$$F(z) - F(x_{t+1})$$

$$\geq (2S_t)^{-1}(\|z - x_{t+1}\|^2 - \|z - x_t\|^2)$$

$$- 2M_1\delta_f - \delta_G - (2S_t)^{-1}\delta_G^{1/2}(4M_1 + 1). \tag{7.53}$$

There are two cases:

$$\|z - x_{t+1}\| \leq \|z - x_t\|; \tag{7.54}$$

$$\|z - x_{t+1}\| > \|z - x_t\|. \tag{7.55}$$

Assume that (7.54) holds. Then by (7.52) and (7.54),

$$\|z - x_{t+1}\| \leq 2M. \tag{7.56}$$

Assume that (7.55) holds. It follows from (7.18), (7.50), (7.53), and (7.55) that

$$F(x_{t+1}) \leq \inf(F)$$

$$+2M_1\delta_f + \delta_G + (2S_t)^{-1}\delta_G^{1/2}(4M_1 + 1)$$

$$\leq \inf(F) + 1. \tag{7.57}$$

Together with (7.19) and (7.57) this implies that

$$\|x_{t+1}\| \le M,$$

$$\|x_{t+1} - z\| \le 2M$$

and that (7.56) holds in the both cases. Thus (7.52) and (7.53) hold for all integers $t \ge 0$. In view of (7.50) and (7.52),

$$\|x_t\| \le 3M$$

for all integers $t \ge 0$.

Let $m \ge 1$ and $T > 0$ be integers. It follows from (7.50) and (7.53) that

$$\sum_{t=m+1}^{m+T} (F(x_t - \inf(F))$$

$$\le \sum_{t=m}^{m+T-1} (\|z - x_t\|^2 - \|z - x_{t+1}\|^2)$$

$$+T[(2S_-)^{-1}\delta_G^{1/2}(4M_1 + 1) + 2M_1\delta_f + \delta_G].$$

Together with (7.51) this implies that

$$\min\{F(x_t) : t = m+1, \dots, m+T\} - \inf(F),$$

$$f(\sum_{t=m+1}^{m+T} T^{-1}x_t) - \inf(F)$$

$$\le 4T^{-1}M^2 + (2S_-)^{-1}(4M_1 + 1)\delta_G^{1/2} + \delta_G + 2\delta_f M_1.$$

Theorem 7.5 is proved. □

Chapter 8
A Zero-Sum Game with Two Players

In this chapter we study an algorithm for finding a saddle point of a zero-sum game with two players. For this algorithm each iteration consists of two steps. The first step is a calculation of a subgradient while the second one is a proximal gradient step. In each of these two steps there is a computational error. In general, these two computational errors are different. We show that our algorithm generates a good approximate solution, if all the computational errors are bounded from above by a small positive constant. Moreover, if we know the computational errors for the two steps of our algorithm, we find out what approximate solution can be obtained and how many iterates one needs for this.

8.1 The Algorithm and the Main Result

Let $(X, \langle \cdot, \cdot \rangle)$, $(Y, \langle \cdot, \cdot \rangle)$ be Hilbert spaces equipped with the complete norms $\| \cdot \|$ which are induced by their inner products. Denote by Id the identity operator in X. Suppose that $f : X \to R^1$ is a convex Fréchet differentiable function such that for each $r > 0$ its Fréchet derivative $f'(\cdot)$ is Lipschitz on $B_X(0, r)$, $g : Y \to R^1 \cup \{\infty\}$ is convex lower semicontinuous function which is not identically ∞, $A : Y \to X$ is a linear continuous operator, $A^* : X \to Y$ its dual and that

$$\|A\| = \sup\{\|Ay\| : \ y \in Y, \ \|y\| \leq 1\}. \tag{8.1}$$

For each $x \in X$ and each $y \in Y$ define

$$F(x, y) = f(x) + \langle x, Ay \rangle - g(y). \tag{8.2}$$

A. J. Zaslavski, *Convex Optimization with Computational Errors*, Springer Optimization and Its Applications 155, https://doi.org/10.1007/978-3-030-37822-6_8

Suppose that there exists $(u^*, v^*) \in X \times Y$ such that

$$F(u^*, v) \leq F(u^*, v^*) \leq F(u, v^*) \tag{8.3}$$

for all $(u, v) \in X \times Y$.

Fix a constant $\tau > 0$ such that

$$\tau \|A\| \leq 1. \tag{8.4}$$

Let us describe our algorithm.

Initialization: select arbitrary $(u_0, v_0) \in X \times Y$.
Iterative Step: for $t = 1, 2, \ldots$, given current iteration vectors $u_{t-1} \in X$ and $v_{t-1} \in Y$ calculate

$$p_t = u_{t-1} - \tau(Av_{t-1} + f'(u_{t-1})), \tag{8.5}$$

$$v_t = \text{Prox}_g(v_{t-1} + A^* p_t)$$

$$= \text{argmin}\{g(z) + 2^{-1}\|z - v_{t-1} - A^* p_t\|^2 : z \in Y\} \tag{8.6}$$

(see (7.2)),

$$u_t = u_{t-1} - \tau(Av_t + f'(u_{t-1})).$$

This algorithm was considered in [37]. One of its steps (see (8.6)) is a proximal method applied to the function g. It should be mentioned that proximal methods is an important tool in analysis and optimization [49–52, 57, 59, 60, 63, 74, 75, 77, 78, 82, 84, 86–88].

Set

$$G = \text{Id} - \tau A^* A. \tag{8.7}$$

Let $\delta_f, \delta_G \in (0, 1]$.

In this chapter we study the algorithm described above, taking into account the computational errors.

Initialization: select arbitrary $(u_0, v_0) \in X \times Y$.
Iterative Step: for $t = 1, 2, \ldots$, given current iteration vectors $u_{t-1} \in X$ and $v_{t-1} \in Y$ calculate $\xi_{t-1} \in X$ such that

$$\|\xi_{t-1} - f'(u_{t-1})\| \leq \delta_f, \tag{8.8}$$

$$p_t = u_{t-1} - \tau(Av_{t-1} + \xi_{t-1}), \tag{8.9}$$

$v_t \in Y$ such that

$$g(v_t) + 2^{-1}\|v_t - v_{t-1} - A^* p_t\|^2$$

$$\leq g(z) + 2^{-1}\|z - v_{t-1} - A^* p_t\|^2 + \delta_G \text{ for all } z \in Y, \quad (8.10)$$

and

$$u_t = u_{t-1} - \tau(Av_t + \xi_{t-1}). \quad (8.11)$$

Define

$$\mathrm{dom}(g) = \{y \in Y : \ g(y) < \infty\}. \quad (8.12)$$

Assume that the set $\mathrm{dom}(g)$ is bounded and let $M_0 > 0$ be such that

$$\mathrm{dom}(g) \subset B_Y(0, M_0). \quad (8.13)$$

Let $M_1 > 0$ be such that

$$\{x \in X : \ f(x) \leq |f(0)| + 4 + \|x\|\|A\|M_0\}$$

$$\subset B_X(0, M_1) \quad (8.14)$$

and let $L \geq 1$ be such that

$$|f(z_1) - f(z_2)| \leq L\|z_1 - z_2\|$$

$$\text{for all } z_1, z_2 \in B_X(0, M_1 + 1) \quad (8.15)$$

and

$$|f'(z_1) - f'(z_2)| \leq L\|z_1 - z_2\|$$

$$\text{for all } z_1, z_2 \in X. \quad (8.16)$$

In view of (8.3) and (8.13),

$$v^* \in \mathrm{dom}(g) \subset B_Y(0, M_0),$$

$$f(u^*) - g(v^*) + \langle u^*, Av^* \rangle = F(u^*, v^*)$$

$$\leq F(0, v^*) = f(0) - g(v^*). \quad (8.17)$$

By (8.2) and (8.3),

$$f(0) \geq f(u^*) + \langle u^*, A^* v^* \rangle$$

$$\geq f(u^*) - \|u^*\| \|A\| M_0,$$

$$f(u^*) \leq f(0) + M_0 \|A\| \|u^*\|. \tag{8.18}$$

It follows from (8.14) and (8.18) that

$$\|u^*\| \leq M_1. \tag{8.19}$$

In this chapter we prove the following result.

Theorem 8.1 *Let*

$$\delta_{f,1} \leq \min\{\tau M_1^{-1}, \ (\|A\| M_0 + L + 1)^{-1}\}, \tag{8.20}$$

$$\tau \leq L^{-1}, \tag{8.21}$$

Assume that $\{u_t\}_{t=0}^{\infty} \subset X$, $\{v_t\}_{t=0}^{\infty} \subset Y$, $\{\xi_t\}_{t=0}^{\infty} \subset X$, $\{p_t\}_{t=0}^{\infty} \subset X$ *satisfy* (8.8)–(8.11) *for all natural numbers t and that*

$$g(v_0) < \infty, \ \|u_0\| \leq M_1. \tag{8.22}$$

Let

$$\Delta = 2M_1 \delta_f \tau^{-1} + \delta_G^{1/2}(12M_0 + 3 + 4\|A\|(M_1 + M_0 + 2)) \tag{8.23}$$

and for each natural number T let

$$\widehat{u}_T = T^{-1} \sum_{t=1}^{T} u_t, \ \widehat{v}_T = T^{-1} \sum_{t=1}^{T} v_t. \tag{8.24}$$

Then for all integers $t \geq 0$,

$$\|v_t\| \leq M_0, \ \|u_t\| \leq M_1,$$

for all integers $t \geq 1$,

$$\|p_t\| \leq M_1 + M_0 + 2$$

and for all integers $T \geq 1$,

$$\|\widehat{u}_T\| \le M_1,$$

$$\widehat{v}_T \in dom(g),\ \|\widehat{v}_T\| \le M_0,$$

and for each $u \in X$ *and each* $v \in dom(g)$,

$$F(\widehat{u}_T, v) \le F(\widehat{u}_T, \widehat{v}_T) + \Delta + (2\tau T)^{-1}4M_1^2 + (2T)^{-1}4M_0^2\|G\|$$

and

$$F(\widehat{u}_T, \widehat{v}_T) \le F(u, \widehat{v}_T) + \Delta + (2\tau T)^{-1}4M_1^2 + (2T)^{-1}4M_0^2\|G\|.$$

It is clear that the best choice of T is at the same order as

$$(\max\{\delta_f,\ \delta_G^{1/2}\}^{1/2})^{-1}.$$

8.2 Auxiliary Results

Lemma 8.2 *Let a linear continuous operator* $G_0 : Y \to Y$ *satisfy* $G_0^* = G_0$ *and*

$$\langle z, G_0 z\rangle \ge 0 \text{ for all } z \in Y.$$

Then for each $u, v, w \in Y$,

$$2\langle w - v, G_0(u - v)\rangle$$

$$= \langle w - v, G_0(w - v)\rangle - \langle w - u, G_0(w - u)\rangle + \langle u - v, G_0(u - v)\rangle.$$

Proof We have

$$\langle w - v, G_0(w - v)\rangle - \langle w - u, G_0(w - u)\rangle + \langle u - v, G_0(u - v)\rangle$$

$$= \langle w, G_0 w\rangle + \langle v, G_0 v\rangle - 2\langle w, G_0 v\rangle$$

$$-(\langle w, G_0 w\rangle + \langle u, G_0 u\rangle - 2\langle w, G_0 u\rangle)$$

$$+(\langle v, G_0 v\rangle + \langle u, G_0 u\rangle - 2\langle v, G_0 u\rangle)$$

$$2[-\langle w, G_0 v\rangle + \langle v, G_0 v\rangle + \langle w, G_0 u\rangle - \langle v, G_0 u\rangle]$$

$$= 2[\langle G_0 u, w - v\rangle + \langle G_0 v, v - w\rangle] = 2\langle w - v, G_0(u - v)\rangle.$$

Lemma 8.2 is proved. □

Lemma 8.3 *Assume that* $\{u_t\}_{t=0}^{\infty} \subset X$, $\{v_t\}_{t=0}^{\infty} \subset Y$, $\{\xi_t\}_{t=0}^{\infty} \subset X$, $\{p_t\}_{t=0}^{\infty} \subset X$ *satisfy* (8.8)–(8.11) *for all natural numbers* t. *Let* $t \geq 0$ *be an integer and* $\widehat{v}_t \in Y$ *satisfy*

$$g(\widehat{v}_t) + 2^{-1}\|\widehat{v}_t - v_{t-1} - A^* p_t\|^2$$

$$\leq g(z) + 2^{-1}\|z - v_{t-1} - A^* p_t\|^2 \tag{8.25}$$

for all $z \in Y$. *Then*

$$\|v_t - \widehat{v}_t\| \leq 2\delta_G^{1/2}.$$

Proof For all $y \in Y$ set

$$h(y) = g(y) + 2^{-1}\|y - v_{t-1} - A^* p_t\|^2. \tag{8.26}$$

In view of (8.25) and (8.26),

$$\widehat{v}_t \in \mathrm{argmin}(h).$$

By (8.12), (8.25), and (8.26),

$$h(v_t) \leq h(\widehat{v}_t) + \delta_G. \tag{8.27}$$

Since the function h is convex it follows from (8.25) to (8.27) that

$$h(\widehat{v}_t) \leq h(2^{-1}v_t + 2^{-1}\widehat{v}_t)$$

$$= g(2^{-1}v_t + 2^{-1}\widehat{v}_t)$$

$$+2^{-1}\|2^{-1}v_t + 2^{-1}\widehat{v}_t - v_{t-1} - A^* p_t\|^2$$

$$\leq g(2^{-1}v_t + 2^{-1}\widehat{v}_t)$$

$$+2^{-1}\|2^{-1}(v_t - v_{t-1} - A^* p_t) + 2^{-1}(\widehat{v}_t - v_{t-1} - A^* p_t)\|^2$$

$$\leq 2^{-1}g(v_t) + 2^{-1}g(\widehat{v}_t)$$

$$+2^{-1}[2\|2^{-1}(v_t - v_{t-1} - A^* p_t)\|^2$$

$$+2\|2^{-1}(\widehat{v}_t - v_{t-1} - A^* p_t)\|^2 - \|2^{-1}(v_t - \widehat{v}_t)\|^2]$$

$$= 2^{-1}h(v_t) + 2^{-1}h(\widehat{v}_t) - 2^{-1}\|2^{-1}(v_t - \widehat{v}_t)\|^2$$

$$\leq 2^{-1}h(\widehat{v}_t) + 2^{-1}\delta_G + 2^{-1}h(\widehat{v}_t) - 2^{-1}\|2^{-1}(v_t - \widehat{v}_t)\|^2,$$

$$\|v_t - \widehat{v_t}\|^2 \le 4\delta_G$$

and

$$\|v - \widehat{v_t}\| \le 2\delta_G^{1/2}.$$

Lemma 8.3 is proved. □

Lemma 8.4 *Assume that* $\{u_t\}_{t=0}^{\infty} \subset X$, $\{v_t\}_{t=0}^{\infty} \subset Y$, $\{\xi_t\}_{t=0}^{\infty} \subset X$, $\{p_t\}_{t=0}^{\infty} \subset X$ *satisfy* (8.8)–(8.11) *for all natural numbers t,*

$$\|u_0\| \le M_1,$$

$$\|v_0\| \le M_0,\ g(v_0) < \infty. \tag{8.28}$$

Then the following assertions hold.

1.

$$\|v_t\| \le M_0 \textit{ for all integers } t \ge 0$$

and

$$v_t \in dom(g) \textit{ for all integers } t \ge 0.$$

2. Let $t \ge 0$ *be an integer. Then for all* $u \in X$,

$$F(u_t, v_t) - F(u, v_t)$$

$$\le (2\tau)^{-1}(\|u - u_{t-1}\|^2 - \|u - u_t\|^2)$$

$$-2^{-1}(\tau^{-1} - L)\|u_t - u_{t-1}\|^2 + \tau^{-1}\delta_f\|u_t - u\|.$$

3. Let $t \ge 0$ *be an integer. Then for all* $v \in dom(g)$,

$$F(u_t, v) - F(u_t, v_t)$$

$$\le \delta_G^{1/2}(12M_0 + 3 + 4\|A\|\|p_t\|)$$

$$+2^{-1}[\langle v - v_{t-1}, G(v - v_{t-1})\rangle - \langle v - v_t, G(v - v_t)\rangle$$

$$-\langle v_{t-1} - v_t, G(v_{t-1} - v_t)\rangle].$$

Proof In view of (8.10), (8.13), and (8.28),

$$g(v_t) < \infty, \ t = 0, 1, \ldots.$$

Clearly, $\|v_t\| \le M_0$ for all integers $t \ge 0$. Let us prove assertion 2. For every $u \in X$ set

$$h(u) = F(u, v_t) = f(u) + \langle u, Av_t \rangle - g(v_t). \tag{8.29}$$

By (8.29), the function h is Fréchet differentiable, for all $u \in X$,

$$h'(u) = f'(u) + Av_t \tag{8.30}$$

and for all $u_1, u_2 \in X$,

$$\|h'(u_1) - h'(u_2)\| = \|f'(u_1) - f'(u_2)\|. \tag{8.31}$$

Proposition 4.3, (8.16), and (8.31) imply that

$$h(u_t) \le h(u_{t-1}) + \langle h'(u_{t-1}), u_t - u_{t-1} \rangle$$

$$+ 2^{-1} L \|u_t - u_{t-1}\|^2. \tag{8.32}$$

Let $u \in X$. Since the function h is convex we have

$$h(u) \ge h(u_{t-1}) + \langle h'(u_{t-1}), u - u_{t-1} \rangle. \tag{8.33}$$

It follows from (8.32) and (8.33) that

$$h(u_t) \le h(u) - \langle h'(u_{t-1}), u - u_{t-1} \rangle$$

$$+ \langle h'(u_{t-1}), u_t - u_{t-1} \rangle + 2^{-1} L \|u_t - u_{t-1}\|^2$$

$$= h(u) + \langle h'(u_{t-1}), u_t - u \rangle + 2^{-1} L \|u_t - u_{t-1}\|^2. \tag{8.34}$$

By (8.29), (8.30), and (8.34),

$$F(u_t, v_t) - F(u, v_t)$$

$$\le \langle f'(u_{t-1}) + Av_t, u_t - u \rangle + 2^{-1} L \|u_t - u_{t-1}\|^2. \tag{8.35}$$

In view of (8.11),

$$\|\tau^{-1}(u_{t-1} - u_t) - (f'(u_{t-1}) + Av_t)\|$$

$$= \|\tau^{-1}(\xi_{t-1} - f'(u_{t-1}))\| \le \tau^{-1} \delta_f. \tag{8.36}$$

It follows from (8.35) and (8.36) that

$$F(u_t, v_t) - F(u, v_t)$$
$$\leq \langle \tau^{-1}(u_{t-1} - u_t), u_t - u\rangle + \tau^{-1}\delta_f \|u_t - u\|$$
$$+2^{-1}L\|u_t - u_{t-1}\|^2.$$

Lemma 2.1 and the relation above imply that

$$F(u_t, v_t) - F(u, v_t)$$
$$(2\tau)^{-1}(\|u - u_{t-1}\|^2 - \|u - u_t\|^2 - \|u_t - u_{t-1}\|^2)$$
$$+\tau^{-1}\delta_f \|u_t - u\| + 2^{-1}L\|u_t - u_{t-1}\|^2.$$

Thus assertion 2 holds.

Let us prove assertion 3. In view of (8.10),

$$g(v_t) + 2^{-1}\|v_t - v_{t-1} - A^* p_t\|^2$$
$$\leq g(z) + 2^{-1}\|z - v_{t-1} - A^* p_t\|^2 + \delta_G \tag{8.37}$$

for all $z \in Y$.

Let

$$\widehat{v} = \text{argmin}\{g(z) + 2^{-1}\|z - v_{t-1} - A^* p_t\|^2 : \ z \in Y\}. \tag{8.38}$$

Lemma 8.3, (8.37), and (8.38) imply that

$$\|v_t - \widehat{v}\| \leq 2\delta_G^{1/2}. \tag{8.39}$$

In view of (8.3), for every $v \in Y$,

$$-F(p_t, v) = -f(p_t) - \langle p_t, Av\rangle + g(v). \tag{8.40}$$

By (8.40), for every $v \in Y$,

$$g(v) + 2^{-1}\|v - v_{t-1} - A^* p_t\|^2$$
$$= g(v) + 2^{-1}\|v - v_{t-1}\|^2 - \langle v - v_{t-1}, A^* p_t\rangle + \|A^* p_t\|^2$$
$$= g(v) - \langle p_t, Av\rangle + \langle Av_{t-1}, p_t\rangle$$

$$+2^{-1}\|v - v_{t-1}\|^2 + \|A^* p_t\|^2$$

$$= -F(p_t, v) + 2^{-1}\|v - v_{t-1}\|^2$$

$$+ f(p_t) + \langle Av_{t-1}, p_t\rangle + \|A^* p_t\|^2. \tag{8.41}$$

It follows from (8.38) and (8.41) that

$$\widehat{v_t} = \operatorname{argmin}\{-F(p_t, v) + 2^{-1}\|v - v_{t-1}\|^2 : v \in Y\}. \tag{8.42}$$

Set

$$h(v) = -F(p_t, v), \ v \in Y. \tag{8.43}$$

Proposition 7.2 and (8.43) imply that for every $y \in Y$,

$$F(p_t, \widehat{v}) - F(p_t, v) = h(v) - h(\widehat{v_t})$$

$$\geq \langle v - \widehat{v}, v_{t-1} - \widehat{v}\rangle. \tag{8.44}$$

Let $v \in \mathrm{dom}(g)$. In view of (8.2),

$$F(u_t, v) - F(u_t, v_t) + F(p_t, v_t) - F(p_t, v)$$

$$= f(u_t) + \langle u_t, Av\rangle - g(v)$$

$$-(f(u_t) + \langle u_t, Av_t\rangle - g(v_t))$$

$$+f(p_t) + \langle p_t, Av_t\rangle - g(v_t)$$

$$-(f(p_t) + \langle p_t, Av\rangle - g(v))$$

$$= \langle u_t - p_t, Av - Av_t\rangle.$$

By the relation above, (8.9)–(8.11) and (8.44),

$$F(u_t, v) - F(u_t, v_t)$$

$$= F(p_t, v) - F(p_t, v_t) + \langle u_t - p_t, A(v - v_t)\rangle$$

$$= F(p_t, v) - F(p_t, v_t) + \tau\langle A(v_{t-1} - v_t), A(v - v_t)\rangle$$

$$= F(p_t, v) - F(p_t, v_t) + \tau\langle v_{t-1} - v_t, A^* A(v - v_t)\rangle. \tag{8.45}$$

In view of (8.2) and (8.38),

$$F(p_t, \widehat{v}) - F(p_t, v_t)$$

$$= f(p_t) + \langle p_t, A\widehat{v}\rangle - g(\widehat{v})$$

$$-(f(p_t) + \langle p_t, Av_t\rangle - g(v_t))$$

$$= \langle p_t, A(\widehat{v} - v_t)\rangle + g(v_t) - g(\widehat{v}). \tag{8.46}$$

It is clear that

$$g(v_t) - g(\widehat{v})$$

$$= g(v_t) + 2^{-1}\|v_t - v_{t-1} - A^* p_t\|^2$$

$$-(g(\widehat{v}) + 2^{-1}\|\widehat{v} - v_{t-1} - A^* p_t\|^2)$$

$$-(2^{-1}\|v_t - v_{t-1} - A^* p_t\|^2 - 2^{-1}\|\widehat{v} - v_{t-1} - A^* p_t\|^2). \tag{8.47}$$

It follows from (8.37), (8.39), and (8.47) that

$$|g(v_t) - g(\widehat{v})|$$

$$\leq |g(v_t) + 2^{-1}\|v_t - v_{t-1} - A^* p_t\|^2$$

$$-(g(\widehat{v}) + 2^{-1}\|\widehat{v} - v_{t-1} - A^* p_t\|^2|$$

$$+2^{-1}|\|v_t - v_{t-1} - A^* p_t\|^2 - \|\widehat{v} - v_{t-1} - A^* p_t\|^2|$$

$$\leq \delta_G + 2^{-1}\|v_t - \widehat{v}\|(\|v_t\| + 2\|v_{t-1}\| + 2\|A\|\|p_t\| + \|\widehat{v}\|)$$

$$\leq \delta_G + \delta_G^{1/2}(2 + 2\|v_t\| + 2\|v_{t-1}\| + 2\|A\|\|p_t\|). \tag{8.48}$$

Relations (8.39), (8.46), and (8.48) imply that

$$|F(p_t, \widehat{v}) - F(p_t, v_t)|$$

$$\leq 2\|p_t\|\|A\|\delta_G^{1/2} + \delta_G$$

$$+ \delta_G^{1/2}(2 + 2\|v_t\| + 2\|v_{t-1}\| + 2\|A\|\|p_t\|). \tag{8.49}$$

By (8.44) and (8.45),

$$
\begin{aligned}
&F(u_t, v) - F(u_t, v_t) \\
&= F(p_t, v) - F(p_t, v_t) + \tau\langle v_{t-1} - v_t, A^*A(v - v_t)\rangle \\
&= F(p_t, v) - F(p_t, \widehat{v}) + F(p_t, \widehat{v}) - F(p_t, v_t) \\
&+\tau\langle v_{t-1} - v_t, A^*A(v - v_t)\rangle \\
&\le \langle \widehat{v} - v_{t-1}, v - \widehat{v}\rangle \\
&+\delta_G^{1/2}(2 + 2\|v_t\| + 2\|v_{t-1}\| + 2\|A\|\|p_t\|) \\
&+\delta_G + 2\delta_G^{1/2}\|p_t\|\|A\| \\
&+\tau\langle v_{t-1} - v_t, A^*A(v - v_t)\rangle \\
&= \delta_G + 2\delta_G^{1/2}\|p_t\|\|A\| \\
&+\delta_G^{1/2}(2 + 2\|v_t\| + 2\|v_{t-1}\| + 2\|A\|\|p_t\|) \\
&+\tau\langle v_{t-1} - v_t, A^*A(v - v_t)\rangle \\
&+\langle v_t - v_{t-1}, v - v_t\rangle \\
&+ [\langle v_{t-1} - v_t, v - v_t\rangle + \langle \widehat{v} - v_{t-1}, v - \widehat{v}\rangle]. \qquad (8.50)
\end{aligned}
$$

By (8.13), (8.39), assertion 1, and the inclusion $v \in \mathrm{dom}(g)$,

$$
\begin{aligned}
&|\langle \widehat{v} - v_{t-1}, v - \widehat{v}\rangle - \langle \widehat{v}_t - v_{t-1}, v - v_t\rangle| \\
&\le |\langle \widehat{v} - v_{t-1}, v - \widehat{v}\rangle - \langle \widehat{v} - v_{t-1}, v - v_t\rangle| \\
&+|\langle \widehat{v} - v_{t-1}, v - v_t\rangle - \langle v_t - v_{t-1}, v - v_t\rangle| \\
&\le \|\widehat{v} - v_{t-1}\|\|v - \widehat{v}\| \\
&+\|v - v_t\|\|\widehat{v} - v_t\| \\
&\le 2\delta_G^{1/2}(2M_0 + 2M_0 + \|v\|). \qquad (8.51)
\end{aligned}
$$

Assertion 1, (8.50), and (8.51) imply that

$$
\begin{aligned}
&F(u_t, v) - F(u_t, v_t) \\
&\le \delta_G^{1/2}(4M_0 + 3 + 4\|A\|\|p_t\|) + 8M_0\delta_G^{1/2}
\end{aligned}
$$

$$+\tau\langle v_{t-1} - v_t, A^*A(v - v_t)\rangle$$

$$+\langle v_t - v_{t-1}, v - v_t\rangle$$

$$= \delta_G^{1/2}(12M_0 + 3 + 4\|A\|\|p_t\|)$$

$$+ \langle v_t - v_{t-1}, (\mathrm{Id} - \tau A^*A)(v - v_t)\rangle. \tag{8.52}$$

Recall (see (8.7)) that

$$G = \mathrm{Id} - \tau A^*A. \tag{8.53}$$

In view of (8.4) and (8.53),

$$G^* = G$$

and for each $z \in Y$,

$$\langle z, Gz\rangle = \langle z, z - \tau A^*Az\rangle$$

$$\|z\|^2 - \tau\|Az\|^2 \geq \|z\|^2(1 - \tau\|A\|) \geq 0. \tag{8.54}$$

By (8.52)–(8.54) and Lemma 8.2,

$$F(u_t, v) - F(u_t, v_t)$$

$$\leq \delta_G^{1/2}(12M_0 + 3 + 4\|A\|\|p_t\|)$$

$$+2^{-1}(\langle v - v_{t-1}, G(v - v_{t-1})\rangle$$

$$-\langle v - v_t, G(v - v_t)\rangle$$

$$-\langle v_{t-1} - v_t, G(v_{t-1} - v_t)\rangle).$$

Thus assertion 3 holds. This completes the proof of Lemma 8.4. □

8.3 Proof of Theorem 8.1

Assertion 1 of Lemma 8.4 implies that for all integers $t \geq 0$,

$$v_t \in \mathrm{dom}(g),\ \|v_t\| \leq M_0. \tag{8.55}$$

Assertion 2 of Lemma 8.4 and (8.21) imply that for all integers $t \geq 0$ and all $u \in X$,

$$F(u_t, v_t) - F(u, v_t)$$

$$\leq (2\tau)^{-1}(\|u - u_{t-1}\|^2 - \|u - u_t\|^2) + L^{-1}\delta_f\|u_t - u\|. \tag{8.56}$$

By Assertion 3 of Lemma 8.4, for all integers $t \geq 0$ and all $v \in \mathrm{dom}(g)$,

$$F(u_t, v) - F(u_t, v_t)$$

$$\leq \delta_G^{1/2}(12M_0 + 3 + 4\|A\|\|p_t\|)$$

$$+ 2^{-1}[\langle v - v_{t-1}, G(v - v_{t-1})\rangle - \langle v - v_t, G(v - v_t)\rangle]. \tag{8.57}$$

We show that for all integers $t \geq 0$,

$$\|u_t\| \leq M_1,$$

$$\|p_{t+1}\| \leq M_1 + M_0 + 2. \tag{8.58}$$

Assume that $t \geq 1$ is an integer and that

$$\|u_{t-1}\| \leq M_1. \tag{8.59}$$

(Note that in view of (8.22), inequality (8.59) holds for $t = 1$.) By (8.56),

$$F(u_t, v_t) \leq F(0, v_t)$$

$$+ (2\tau)^{-1}(\|u_{t-1}\|^2 - \|u_t\|^2) + \tau^{-1}\delta_f\|u_t\|. \tag{8.60}$$

There are two cases:

$$\|u_t\| \leq \|u_{t-1}\|; \tag{8.61}$$

$$\|u_t\| > \|u_{t-1}\|. \tag{8.62}$$

Assume that (8.61) holds. In view of (8.59) and (8.61),

$$\|u_t\| \leq \|u_{t-1}\| \leq M_1.$$

Assume that (8.62) holds. By (8.2), (8.60), and (8.62),

$$f(u_t) + \langle u_t, Av_t\rangle \leq f(0) + \tau^{-1}\delta_f\|u_t\|. \tag{8.63}$$

Relations (8.55) and (8.63) imply that

$$f(u_t) \leq f(0) + \|A\|M_0\|u_t\| + \delta_f\tau^{-1}\|u_t\|. \tag{8.64}$$

It follows from (8.8), (8.11), (8.15), (8.55), and (8.59) that

$$\|u_t\| \le \|u_{t-1}\| + \tau(\|A\|\|v_t\| + \|\xi_{t-1}\|)$$

$$\le M_1 + \tau(\|A\|M_0 + L + 1). \tag{8.65}$$

By (8.20) and (8.65),

$$\delta_f \tau^{-1}\|u_t\| \le \delta_f \tau^{-1} M_1 + \delta_f(\|A\|M_0 + L + 1) \le 2. \tag{8.66}$$

It follows from (8.14), (8.64), and (8.66) that

$$f(u_t) \le f(0) + \|A\|M_0\|u_t\| + 2. \tag{8.67}$$

In view of (8.67),

$$\|u_t\| \le M_1.$$

Thus by induction we showed that

$$\|u_t\| \le M_1 \text{ for all integers } t \ge 0. \tag{8.68}$$

Relations (8.8), (8.15), and (8.68) imply that for all integers $t \ge 0$,

$$\|\xi_t\| \le L + 1. \tag{8.69}$$

By (8.4), (8.9), (8.21), (8.55), (8.68), and (8.69),

$$\|p_t\| = \|u_{t-1} - \tau(Av_{t-1} + \xi_{t-1})\|$$

$$\le \|u_{t-1}\| + \tau(\|A\|\|v_{t-1}\| + \|\xi_{t-1}\|)$$

$$\le M_1 + \tau(\|A\|M_0 + L + 1) \le M_1 + M_0 + 2.$$

Thus (8.58) holds.

In view of (8.56), (8.58), and (8.68), for all integers $t \ge 0$, all $u \in B_X(0, M_1)$, and all $v \in \mathrm{dom}(g)$,

$$F(u_t, v_t) - F(u, v_t)$$

$$\le (2\tau)^{-1}(\|u - u_{t-1}\|^2 - \|u - u_t\|^2) + 2\tau^{-1}\delta_f M_1 \tag{8.70}$$

and

$$F(u_t, v) - F(u_t, v_t)$$

$$\leq 2^{-1}[\langle v - v_{t-1}, G(v - v_{t-1})\rangle - \langle v - v_t, G(v - v_t)\rangle]$$

$$+ \delta_G^{1/2}(12M_0 + 3 + 4\|A\|(M_1 + M_0 + 2)). \quad (8.71)$$

It follows from (8.7) and (8.70) that for all integers $t \geq 0$, all $u \in B_X(0, M_1)$, and all $v \in \mathrm{dom}(g)$,

$$F(u_t, v) - F(u, v_t)$$

$$\leq (2\tau)^{-1}(\|u - u_{t-1}\|^2 - \|u - u_t\|^2) + 2\tau^{-1}\delta_f M_1$$

$$+2^{-1}[\langle v - v_{t-1}, G(v - v_{t-1})\rangle - \langle v - v_t, G(v - v_t)\rangle]$$

$$+ 4\delta_G^{1/2}(12M_0 + 3 + 4\|A\|(M_1 + M_0 + 2)). \quad (8.72)$$

Let T be a natural number. By (8.24), (8.62), and (8.72), for all $u \in B_X(0, M_1)$ and all $v \in \mathrm{dom}(g)$,

$$F(\widehat{u}_T, v) - F(u, \widehat{v}_T)$$

$$= F(\sum_{t=1}^{T} T^{-1}u_t, v) - F(u, \sum_{t=1}^{T} T^{-1}v_t)$$

$$\leq T^{-1}\sum_{t=1}^{T} F(u_t, v) - T^{-1}\sum_{t=1}^{T} F(u, v_t)$$

$$\leq 2M_1\tau^{-1}\delta_f + \delta_G^{1/2}(12M_0 + 3 + 4\|A\|(M_1 + M_0 + 2))$$

$$+(2T\tau)^{-1}\sum_{t=1}^{T}(\|u - u_{t-1}\|^2 - \|u - u_t\|^2)$$

$$+(2T)^{-1}\sum_{t=1}^{T}(\langle v - v_{t-1}, G(v - v_{t-1})\rangle - \langle v - v_t, G(v - v_t)\rangle)$$

$$\leq 2M_1\tau^{-1}\delta_f + \delta_G^{1/2}(12M_0 + 3 + 4\|A\|(M_1 + M_0 + 2))$$

$$+(2T\tau)^{-1}4M_1^2 + (2T)^{-1}4M_0^2\|G\|$$

$$= \Delta + (2T\tau)^{-1}4M_1^2 + (2T)^{-1}4M_0^2\|G\|. \quad (8.73)$$

Set

$$\Delta_T = \Delta + (2T\tau)^{-1}4M_1^2 + (2T\tau)^{-1}4M_0^2\|G\|. \quad (8.74)$$

In view of (8.73) and (8.74), with $v = v^*$ and $u = u^*$,

$$F(\widehat{u}_T, v^*) - F(u^*, \widehat{v}_T) \le \Delta_T. \tag{8.75}$$

Relations (8.3) and (8.75) imply that

$$0 \le F(\widehat{u}_T, v^*) - F(u^*, v^*) \le \Delta_T, \tag{8.76}$$

$$0 \le F(u^*, v^*) - F(u^*, v_T) \le \Delta_T. \tag{8.77}$$

By (8.73) and (8.74), for each $u \in B_X(0, M_1)$ and each $v \in \mathrm{dom}(g)$,

$$F(\widehat{u}_T, \widehat{v}_T) - F(u, \widehat{v}_T) \le \Delta_T, \tag{8.78}$$

$$F(\widehat{u}_T, v) - F(\widehat{u}_T, \widehat{v}_T) \le \Delta_T. \tag{8.79}$$

Assume that

$$u \in X \setminus B_X(0, M_1). \tag{8.80}$$

In view of (8.2) and (8.78),

$$F(\widehat{u}_T, \widehat{v}_T) \le \Delta_T + F(0, \widehat{v}_T) = \Delta_T + f(0) + g(\widehat{v}_T). \tag{8.81}$$

It follows from (8.2), (8.14), (8.24), (8.55), and (8.80) that

$$\begin{aligned} &F(u, \widehat{v}_T) - F(0, \widehat{v}_T) \\ &= f(u) + \langle u, A\widehat{v}_T \rangle - f(0) \\ &\ge f(u) - \|u\| \|A\| M_0 - |f(0)| \ge 4. \end{aligned} \tag{8.82}$$

By (8.81) and (8.82),

$$\begin{aligned} F(\widehat{u}_T, \widehat{v}_T) &\le \Delta_T + F(0, \widehat{v}_T) \\ &\le \Delta_T + F(u, \widehat{v}_T) - 4. \end{aligned} \tag{8.83}$$

In view of (8.83) and (8.88),

$$F(\widehat{u}_T, \widehat{v}_T) \le \Delta_T + F(u, \widehat{v}_T)$$

for all $u \in X$. Theorem 8.1 is proved. □

Chapter 9
PDA-Based Method for Convex Optimization

In this chapter we use predicted decrease approximation (PDA) for constrained convex optimization. For PDA-based method each iteration consists of two steps. In each of these two steps there is a computational error. In general, these two computational errors are different. We show that our algorithm generates a good approximate solution, if all the computational errors are bounded from above by a small positive constant. Moreover, if we know the computational errors for the two steps of our algorithm, we find out what approximate solution can be obtained and how many iterates one needs for this.

9.1 Preliminaries and the Main Result

Let X be a Hilbert space equipped with an inner product $\langle \cdot, \cdot \rangle$ which induces a complete norm $\| \cdot \|$, $F : X \to R^1$ be a convex Fréchet differentiable function, $G : X \to R^1 \cup \{\infty\}$ be a convex lower semicontinuous function which is not identically infinity,

$$\mathrm{dom}(G) = \{x \in X : \ G(x) < \infty\},$$

$$H(x) = F(x) + G(x), \ x \in X. \tag{9.1}$$

We denote by $F'(x)$ the Fréchet derivative of F at $x \in X$.

We suppose that $\mathrm{dom}(G)$ is a bounded set and that

$$\inf(H) = \inf\{H(x) : \ x \in X\}$$

is a finite number. Let

A. J. Zaslavski, *Convex Optimization with Computational Errors*, Springer Optimization and Its Applications 155, https://doi.org/10.1007/978-3-030-37822-6_9

$$D := \text{diam}(\text{dom}(G))$$

$$= \sup\{\|x - y\| : x, y \in \text{dom}(G)\}, \tag{9.2}$$

$x_* \in X$ satisfy

$$H(x_*) = \inf(H). \tag{9.3}$$

Given a function $f : X \to R^1 \cup \{\infty\}$, its convex conjugate is the function

$$f^*(y) = \sup\{\langle x, y\rangle - f(x) : x \in X\}, \ y \in X. \tag{9.4}$$

Clearly, for all $x, y \in X$,

$$f(x) + f^*(u) \geq \langle x, u\rangle \tag{9.5}$$

(Fenchel inequality).

For every $y \in X$ define

$$S(y) = \sup\{\langle F'(y), y - p\rangle + G(y) - G(p) : p \in X\}. \tag{9.6}$$

By (9.4) and (9.6), for $y \in X$,

$$S(y) = G(y) + \langle F'(y), y\rangle$$

$$+ \sup\{-\langle F'(y), p\rangle - G(p) : p \in X\}$$

$$G(y) + \langle F'(y), y\rangle + G^*(-F'(y)). \tag{9.7}$$

It follows from Fenchel inequality (9.5) and (9.7) that for every $y \in \text{dom}(G)$,

$$S(y) \geq 0. \tag{9.8}$$

Let $y \in \text{dom}(G)$. In view (9.6) and (9.8), the following four properties are equivalent:

$$S(y) = 0; \tag{9.9}$$

$$G(p) \geq G(y) + \langle -F'(y), p - y\rangle \text{ for all } p \in X; \tag{9.10}$$

$$-F'(y) \in \partial G(y); \tag{9.11}$$

$$y \text{ is a minimizer of } H. \tag{9.12}$$

For each $x, y \in X$ set

$$[x, y] = \{\alpha x + (1 - \alpha)y : \alpha \in [0, 1]\}.$$

Lemma 9.1 ([16]) *For every $y \in X$,*

$$H(y) - \inf(H) \leq S(y).$$

Proof Let $y \in Y$. If $S(y) = \infty$, then the lemma holds. Assume that

$$S(y) < \infty. \tag{9.13}$$

In view of (9.7) and (9.13),

$$\sup\{-\langle F'(y), p\rangle - G(p) : p \in X\} < \infty \tag{9.14}$$

and

$$G(y) < \infty. \tag{9.15}$$

By (9.14), there exists $p_\epsilon \in X$ such that

$$\langle -F'(y), p_\epsilon\rangle - G(p_\epsilon) \geq -\epsilon - \langle F'(y), p\rangle - G(p) \text{ for all } p \in X. \tag{9.16}$$

It follows from (9.6) and (9.16) that

$$S(y) \geq \langle F'(y), y - p_\epsilon\rangle + G(y) - G(p_\epsilon)$$

$$= \langle F'(y), y\rangle + G(y) - [\langle F'(y), p_\epsilon\rangle + G(p_\epsilon)]$$

$$\geq \langle F'(y), y\rangle + G(y) - \langle F'(y), x_*\rangle + G(x_*) - \epsilon$$

$$= \langle F'(y), y - x_*\rangle + G(y) - G(x_*) - \epsilon.$$

Since ϵ is any positive number in view of (9.3),

$$S(y) \geq \langle F'(y), y - x_*\rangle + G(y) - G(x_*)$$

$$\geq F(y) - F(x_*) + G(y) - G(x_*) = H(y) - \inf(H).$$

Lemma 9.1 is proved. □

Suppose that $L > 0$ and

$$\|F'(x) - F'(y)\| \leq L\|x - y\| \text{ for all } x, y \in X. \tag{9.17}$$

For every $\gamma \geq 1$ and every $\bar{y} \in \mathrm{dom}(G)$, we say that

$$u(\bar{y}) \in \mathrm{dom}(G)$$

is a γ^{-1}-predicted decrease approximation (PDA) vector of H at $\bar{y}$ [16] if

$$\gamma^{-1} S(\bar{y}) \leq \langle F'(\bar{y}), \bar{y} - u(\bar{y}) \rangle + G(\bar{y}) - G(u(\bar{y})). \tag{9.18}$$

Note that for any $\tilde{\gamma} \geq \gamma \geq 1$ any γ^{-1}-PDA vector is also a $\tilde{\gamma}^{-1}$-PDA vector.

Let $\gamma \geq 1$, $\bar{y} \in \mathrm{dom}(G)$, $\epsilon > 0$. We say that

$$u(\bar{y}) \in \mathrm{dom}(G)$$

is a (γ^{-1}, ϵ)-predicted decrease approximation (PDA) vector of H at $\bar{y}$ if

$$\gamma^{-1} S(\bar{y}) \leq \langle F'(\bar{y}), \bar{y} - u(\bar{y}) \rangle + G(\bar{y}) - G(u(\bar{y})) + \epsilon. \tag{9.19}$$

For $x, y \in X$ define

$$Q(y, x) = F(x) + \langle F'(x), y - x \rangle + 2^{-1} L x - y\|^2. \tag{9.20}$$

Let $\gamma \geq 1$, $\delta_1, \delta_2 \in (0, 1]$.

Let us describe our algorithm.

Initialization: select an arbitrary $y_0 \in \mathrm{dom}(G)$.
Iterative Step: given a current iteration vector $y_k \in \mathrm{dom}(G)$ do the following:

(i) calculate $u(y_k)$ which is (γ^{-1}, δ_1)-predicted decrease approximation (PDA) vector of H at y_k such that

$$u(y_k) \in \mathrm{dom}(G), \tag{9.21}$$

$$\gamma^{-1} S(y_k) - \delta_1 \leq \langle F'(y_k), y_k - u(y_k) \rangle + G(y_k) - G(u(y_k)); \tag{9.22}$$

(ii) choose a bounded set X_k such that

$$[y_k, u(y_k)] \subset X_k \tag{9.23}$$

and calculate

$$y_{k+1} \in X_k \tag{9.24}$$

for which at least one of the following inequalities holds:

$$Q(y_{k+1}, y_k) + G(y_{k+1}) \leq Q(y, y_k) + G(y) + \delta_2 \text{ for all } y \in X_k \tag{9.25}$$

(local modal update);

$$H(y_{k+1}) \leq H(y) + \delta_2 \text{ for all } y \in X_k \tag{9.26}$$

(global exact model update).

Note that an exact version of this algorithm (without computational errors δ_1, δ_2) was studied in [16].

Theorem 9.2 *Assume that* $\{y_k\}_{k=0}^{\infty}, \{u(y_k)\}_{k=0}^{\infty} \subset X$, $X_k \subset X$, $k = 0, 1, \ldots$, *(9.21)–(9.24) hold for all* $k = 0, 1, \ldots$ *and that for all nonnegative integers* $k \geq 0$, *at least one of the relations* (9.25) *and* (9.26) *hold. Then for every integer* $k \geq 0$,

$$H(y_{k+1}) - \inf(H)$$

$$\leq 2\gamma(k + 2\gamma)^{-1}((2\gamma - 2)(k + 1)^{-1}(H(y_0) - \inf(H)) + LD^2\gamma)$$

$$+(\delta_1 + \delta_2)(2\gamma + k - 1).$$

Theorem 9.2 is proved in Sect. 9.3. Its prototype without computational errors was obtained in [16].

9.2 Auxiliary Results

Lemma 9.3 *Assume that* $\{y_k\}_{k=0}^{\infty}, \{u(y_k)\}_{k=0}^{\infty} \subset X$, $X_k \subset X$, $k = 0, 1, \ldots$, *(9.21)–(9.24) hold for all* $k = 0, 1, \ldots$ *and that for all nonnegative integers k, at least one of the relations* (9.25) *and* (9.26) *hold.*

Let $\{t_k\}_{k=0}^{\infty} \subset (0, 1]$. *Then for every integer* $k \geq 0$,

$$H(y_{k+1}) \leq H(y_k) + 2^{-1}t_k^2 LD^2 - \gamma^{-1}t_k S(y_k) + \delta_1 + \delta_2.$$

Proof Let $k \geq 0$ be an integer and set

$$u_k = u(y_k). \tag{9.27}$$

By (9.21)–(9.23) and (9.27),

$$[y_k, u_k] \subset X_k, \tag{9.28}$$

$$\gamma^{-1}S(y_k) - \delta_1 \leq \langle F'(y_k), y_k - u_k\rangle + G(y_k) - G(u_k). \tag{9.29}$$

Since the function G is convex it follows from (9.2), (9.20), (9.23), (9.28), and (9.29) that

$$\begin{aligned}
&\inf\{Q(y, y_k) + G(y) : \ y \in X_k\} \\
&\leq \inf\{Q(y, y_k) + G(y) : \ y \in [y_k, u_k]\} \\
&\leq \inf\{Q(tu_k + (1-t)y_k, y_k) + G(tu_k + (1-t)y_k) : \ t \in [0, 1]\} \\
&\leq Q(t_k u_k + (1-t_k)y_k, y_k) + G(t_k u_k + (1-t_k)y_k) \\
&= F(y_k) + t_k\langle F'(y_k), u_k - y_k\rangle + 2^{-1}Lt_k^2\|y_k - u_k\|^2 \\
&+G(t_k u_k + (1-t)y_k) \\
&\leq F(y_k) + G(y_k) \\
&+t_k(\langle F'(y_k), u_k - y_k\rangle + G(u_k) - G(y_k)) \\
&+2^{-1}t_k^2 L\|y_k - u_k\|^2 \\
&\leq F(y_k) + G(y_k) - t_k\gamma^{-1}S(y_k) + \delta_1 + 2^{-1}Lt_k^2 D^2. \qquad (9.30)
\end{aligned}$$

Proposition 4.3, (9.17), and (9.20) imply that

$$F(y) \leq Q(y, y_k) \text{ for all } y \in X_k \tag{9.31}$$

and

$$F(y) + G(y) \leq Q(y, y_k) + G(y), \ \ y \in X_k. \tag{9.32}$$

In the case of the local update it follows from (9.25) and (9.32) that

$$\begin{aligned}
&F(y_{k+1}) + G(y_{k+1}) \\
&\leq Q(y_{k+1}, y_k) + G(y_{k+1}) \\
&\leq \inf\{Q(y, y_k) + G(y) : \ y \in X_k\} + \delta_2.
\end{aligned}$$

In the case of the exact model update it follows from (9.26) and (9.32) that

$$\begin{aligned}
&F(y_{k+1}) + G(y_{k+1}) \\
&\leq \inf\{F(y) + G(y) : \ y \in X_k\} + \delta_2 \\
&\leq \inf\{Q(y, y_k) + G(y) : \ y \in X_k\} + \delta_2.
\end{aligned}$$

Therefore in both cases in view of (9.30),

$$F(y_{k+1}) + G(y_{k+1})$$

$$\leq \inf\{Q(y, y_k) + G(y) : \ y \in X_k\} + \delta_2.$$

$$\leq F(y_k) + G(y_k) + 2^{-1}t_k^2 LD^2 - \gamma^{-1}t_k S(y_k) + \delta_1 + \delta_2.$$

Lemma 9.3 is proved. □

Lemma 9.4 *Let $c > 0$, $\gamma \geq 1$, $\delta \in (0, 2]$, $\{a_k\}_{k=0}^{\infty}$, $\{b_k\}_{k=0}^{\infty} \subset R^1$,*

$$0 \leq a_k \leq b_k, \ k = 0, 1, \ldots,$$

$$t_k = 2\gamma(2\gamma + k)^{-1}, \ k = 0, 1, \ldots. \tag{9.33}$$

Assume that for all integers $k \geq 0$,

$$a_{k+1} \leq a_k - t_k b_k \gamma^{-1} + 2^{-1} c t_k^2 + \delta. \tag{9.34}$$

Then for all integers $k \geq 0$,

$$a_{k+1} + (\sum_{i=0}^{k}(b_i - a_i)(i + 2\gamma - 1))(\sum_{i=0}^{k}(i + 2\gamma - 1))^{-1}$$

$$\leq 2\gamma(k + 2\gamma)^{-1}((2\gamma - 2)(k + 1)^{-1}a_0 + c\gamma) + \delta(2\gamma + k - 1).$$

Proof In view of (9.33) and (9.34), for any integer $i \geq 0$,

$$b_i \leq \gamma t_i^{-1}(a_i - a_{i+1} + 2^{-1}t_i^2 c + \delta)^2$$

and

$$b_i - a_i \leq (\gamma t_i^{-1} - 1)a_i$$

$$-\gamma t_i^{-1}a_{i+1} + 2^{-1}c\gamma t_i + \delta\gamma t_i^{-1}$$

$$= 2^{-1}a_i(2\gamma + i - 2) - 2^{-1}a_{i+1}(2\gamma + i)$$

$$+(2\gamma + i)^{-1}c\gamma^2 + 2^{-1}\delta(2\gamma + i).$$

Multiplying the relation above by $i + 2\gamma - 1 \geq 0$ we obtain that

$$(b_i - a_i)(i + 2\gamma - 1)$$

$$\leq 2^{-1}(i + 2\gamma - 2)(i + 2\gamma - 1)a_i - 2^{-1}(i + 2\gamma)(i + 2\gamma - 1)a_{i+1}$$

$$+c\gamma^2(i+2\gamma-1)(i+2\gamma)^{-1}+2^{-1}\delta(2\gamma+i)(i+2\gamma-1). \tag{9.35}$$

Let $k \geq 0$ be integer. Summing up (9.35) for $i = 0, \ldots, k$ we obtain that

$$\sum_{i=0}^{k}(b_i - a_i)(i+2\gamma-1)$$

$$\leq 2^{-1}(2\gamma-2)(2\gamma-1)a_0 - 2^{-1}(k+2\gamma)(k+2\gamma-1)a_{k+1}$$

$$+c(k+1)\gamma^2 + 2^{-1}\delta(k+1)(2\gamma+k)(k+2\gamma-1). \tag{9.36}$$

Dividing both sides of (9.36) by

$$2^{-1}(k+1)(k+2\gamma)$$

we obtain that for any integer $k \geq 0$,

$$(\sum_{i=0}^{k}(b_i - a_i)(i+2\gamma-1))(2^{-1}(k+1)(k+2\gamma))^{-1}$$

$$+(k+2\gamma-1)(k+1)^{-1}a_{k+1}$$

$$\leq 2\gamma(k+2\gamma)^{-1}((2\gamma-2)(2\gamma-1)(2\gamma)^{-1}(k+1)^{-1}a_0 + c\gamma)$$

$$+\delta(2\gamma+k-1). \tag{9.37}$$

Clearly,

$$\sum_{i=0}^{k}(i+2\gamma-1) = 2^{-1}(k+1)(k+4\gamma-2)$$

$$\geq 2^{-1}(k+1)(k+2\gamma). \tag{9.38}$$

By (9.37) and (9.38),

$$\sum_{i=0}^{k}(b_i - a_i)(i+2\gamma-1)(\sum_{i=0}^{k}(i+2\gamma-1))^{-1} + a_{k+1}$$

$$\leq \sum_{i=0}^{k}(b_i - a_i)(i+2\gamma-1)(2^{-1}(k+1)(k+2\gamma))^{-1}$$

$$+(k+2\gamma-1)(k+1)^{-1}a_{k+1}$$

$$\leq 2\gamma(k+2\gamma)^{-1}((2\gamma-2)(k+1)^{-1}a_0+c\gamma)+\delta(2\gamma+k-1).$$

Lemma 9.4 is proved. □

9.3 Proof of Theorem 9.2 and Examples

Theorem 9.2 follows from Lemmas 9.1, 9.3, and 9.4 applied with

$$\delta=\delta_1+\delta_2,$$

$$a_k=H(y_k)-\inf(H),\ b_k=S(y_k),\ k=0,1,\dots,$$

$$t_k=2\gamma(2\gamma+k)^{-1},\ k=0,1,\dots,$$

$$c=LD^2.$$

Clearly, the best choice of k is at the same order as $(\delta_1+\delta_2)^{-1/2}$ and in this case the right-hand side of the final equation of Theorem 9.2 is $c_1(\delta_1+\delta_2)^{1/2}$, where c_1 is a positive constant.

We consider two special cases of the PDA-based method.

Let $\gamma=1$ and for a given integer $k\geq 0$ define

$$y_k\in X,\ u_k\in \mathrm{dom}(G)$$

such that

$$\langle F'(y_k),u_k\rangle+G(u_k))\leq \langle F'(y_k),u\rangle+G(u))+\delta_1 \tag{9.39}$$

for all $u\in X$. By (9.6), for every integer $k\geq 0$,

$$S(y_k)=\sup\{\langle F'(y_k),y_k-u\rangle+G(y_k)-G(u):\ u\in X\}$$

$$=\langle F'(y_k),y_k\rangle+G(y_k)$$

$$+\sup\{-\langle F'(y_k),u\rangle-G(u):\ u\in X\}$$

$$\leq \langle F'(y_k),y_k\rangle+G(y_k)$$

$$-\langle F'(y_k),u_k\rangle-G(u_k)+\delta_1$$

$$=\langle F'(y_k),y_k-u_k\rangle+G(y_k)-G(u_k)+\delta_1$$

and (9.22) holds with $u(y_k) = u_k$. Then we chose $t_k \in [0, 1]$ such that

$$H(y_k + t_k(u_k - y_k)) \le H(y_k + t(u_k - y_k)) + \delta_2$$

for all $t \in [0, 1]$ and set

$$y_{k+1} = y_k + t_k(u_k - y_k).$$

Clearly, in this case we have a generalized conditional algorithm.

Consider now another special case of the PDA-based method with $X_k = X$, $k = 0, 1, \dots$

For any integer $k \ge 0$ find $y_{k+1} \in X$ such that

$$F(y_k) + \langle F'(y_k), y_{k+1} - y_k \rangle + 2^{-1}L\|y_k - y_{k+1}\|^2 + G(y_{k+1})$$

$$\le F(y_k) + \langle F'(y_k), y - y_k \rangle + 2^{-1}L\|y_k - y\|^2 + G(y) + \delta_2$$

for all $y \in X$ which is equivalent to the relation

$$L^{-1}G(y_{k+1}) + 2^{-1}\|y_{k+1} - (y_k - L^{-1}F'(y_k))\|^2$$

$$\le L^{-1}G(y) + 2^{-1}\|y - (y_k - L^{-1}F'(y_k))\|^2 + L^{-1}\delta_2$$

for all $y \in X$.

Chapter 10
Minimization of Quasiconvex Functions

In this chapter we study minimization of a quasiconvex function. Our algorithm has two steps. In each of these two steps there is a computational error. In general, these two computational errors are different. We show that our algorithm generates a good approximate solution, if all the computational errors are bounded from above by a small positive constant. Moreover, if we know the computational errors for the two steps of our algorithm, we find out what approximate solution can be obtained and how many iterates one needs for this.

10.1 Preliminaries

Let X be a Hilbert space equipped with an inner product $\langle \cdot, \cdot \rangle$ which induces a complete norm $\| \cdot \|$ and let $f : X \to R^1$. We consider the problem

$$f(x) \to \min, \ x \in X$$

using the algorithm considered in [53].

Recall that

$$\inf(f) = \inf\{f(x) : \ x \in X\},$$

$$\operatorname{argmin}(f) = \{x \in X : \ f(x) = \inf(f)\}.$$

Let $F \subset X$. Denote by $\mathrm{int}(F)$, $\mathrm{cl}(F)$, and $\mathrm{bd}(F)$ its interior, closure, and boundary respectively. For every $x \in X$,

$$N(F, x) = \{q \in X : \ \langle q, y - x \rangle \le 0 \text{ for all } y \in F\}. \tag{10.1}$$

A. J. Zaslavski, *Convex Optimization with Computational Errors*, Springer Optimization and Its Applications 155, https://doi.org/10.1007/978-3-030-37822-6_10

Set

$$S_X(0,1) = \{z \in X : \|z\| = 1\}. \quad (10.2)$$

Suppose that for each $\alpha \in R^1$ the set

$$\{x \in X : \ f(x) \le \alpha\}$$

is convex. In other words, the function f is quasiconvex. Assume that

$$\inf(f) > -\infty. \quad (10.3)$$

For every $\epsilon > 0$ and every $x \in X$ set

$$G_\epsilon(x) = \{z \in X : \ f(z) < f(x) - \epsilon\}. \quad (10.4)$$

Let

$$G(x) = G_0(x). \quad (10.5)$$

For every $x \in X$ set

$$N(G(x), x) = \{q \in X : \ \langle q, y - x\rangle \le 0$$

$$\text{for all } y \in X \text{ satisfying } f(y) < f(x)\}. \quad (10.6)$$

Let

$$\beta > 0, \ L \ge 1, \ x_* \in \text{argmin}(f). \quad (10.7)$$

Assume that for every $z \in Y$,

$$|f(z) - f(x_*)| \le L\|z - x_*\|^\beta. \quad (10.8)$$

10.2 An Auxiliary Result

Proposition 10.1 *Let $\epsilon \ge 0$, $x \in X$, and*

$$f(x) - \inf(f) > \epsilon. \quad (10.9)$$

Then

$$f(x) - \inf(f) \le L\langle q, x - x_*\rangle^\beta + \epsilon$$

$$\textit{for all } q \in S_X(0,1) \cap N(G_\epsilon(x), x).$$

Proof Since the function f is continuous and quasiconvex the set $G_\epsilon(x)$ is nonempty, open, and convex. Set

$$r = \inf\{\|z - x_*\| : z \in \mathrm{bd}(G_\epsilon(x))\}. \tag{10.10}$$

There exist sequences

$$\{z_k\}_{k=1}^\infty \subset \mathrm{bd}(G_\epsilon(x)) \tag{10.11}$$

and $\{\delta_k\}_{k=1}^\infty \subset (0, \infty)$ such that

$$\lim_{k\to\infty} \delta_k = 0,$$

$$\|z_k - x_*\| \le r + \delta_k, \ k = 1, 2, \ldots. \tag{10.12}$$

By (10.8), (10.11), and (10.12), for all integers $k \ge 0$,

$$\begin{aligned} &f(x) - \inf(f) - \epsilon \\ &\le f(z_k) - \inf(f) \\ &\le L\|z_k - x_*\|^\beta \le L(r + \delta_k)^\beta. \end{aligned} \tag{10.13}$$

It follows from (10.11) and (10.13) that

$$f(x) - \inf(f) \le Lr^\beta + \epsilon. \tag{10.14}$$

By (10.7), (10.9), and (10.10),

$$x_* + rz \in \mathrm{cl}(G_\epsilon(x)) \text{ for all } z \in S_X(0,1). \tag{10.15}$$

In view of (10.1), (10.4), and (10.15), for every

$$q \in S_X(0,1) \cap N(G_\epsilon(x), x)$$

we have

$$\langle q, x_* + rz - x\rangle \le 0.$$

Applying the inequality above with $z = q$ we obtain that

$$0 \geq \langle q, x_* + rq - x \rangle$$

and

$$\langle q, x - x_* \rangle \geq r \text{ for all } q \in S_X(0, 1) \cap N(G_\epsilon(x), x). \tag{10.16}$$

Let

$$q \in S_X(0, 1) \cap N(G_\epsilon(x), x).$$

By (10.14) and (10.16),

$$f(x) - \inf(f) \leq Lr^\beta + \epsilon \leq L\langle q, x - x_* \rangle^\beta + \epsilon.$$

Proposition 10.1 is proved. □

10.3 The Main Result

Let $\delta_1, \delta_2 > 0$, $\gamma \in (0, 2)$.

Let us describe our algorithm.

Initialization: select an arbitrary $x_0 \in X$.
Iterative Step: Let $x_k \in X$ be given a current iteration vector. If $G_{\delta_1}(x_k) = \emptyset$, then

$$\inf(f) \geq f(x_k) - \delta_1$$

and x_k is considered as an approximate solution of our problem. If

$$G_{\delta_1}(x_k) \neq \emptyset,$$

then find $q_k \in X$ such that

$$B_X(q_k, \delta_2) \cap S_X(0, 1) \cap N(G_{\delta_1}(x_k), x_k) \neq \emptyset$$

and set

$$\lambda_k = \gamma[(f(x_k) - \inf(f) - \delta_1)L^{-1}]^{1/\beta},$$

$$x_{k+1} = x_k - \lambda_k q_k.$$

Let $M > 0$ be such that

$$\|x_*\| \leq M. \tag{10.17}$$

In this chapter we prove the following result.

Theorem 10.2 *Let*

$$T = \lfloor (4\gamma)^{-1}(2-\gamma)\delta_2^{-2} \rfloor + 1, \tag{10.18}$$

$\{x_t\}_{t=0}^{T}$, $\{q_t\}_{t=0}^{T-1} \subset X$, $\{\lambda_t\}_{t=0}^{T} \subset (0, \infty)$,

$$\|x_0\| \leq M, \tag{10.19}$$

$$G_{\delta_1}(x_t) \neq \emptyset,\ t = 0, \ldots, T-1 \tag{10.20}$$

and for $t = 0, \ldots, T-1$,

$$B_X(q_t, \delta_2) \cap S_X(0, 1) \cap N(G_{\delta_1}(x_t), x_t) \neq \emptyset, \tag{10.21}$$

$$\lambda_t = \gamma[(f(x_t) - \inf(f) - \delta_1)L^{-1}]^{1/\beta}, \tag{10.22}$$

$$x_{t+1} = x_t - \lambda_t q_t. \tag{10.23}$$

Then there exists an integer $t \in [0, T]$ *such that*

$$f(x_t) \leq (8\delta_2 M)^{\beta}(2-\gamma)^{-\beta} + \inf(f) + \delta_1.$$

Proof Assume that the theorem is not true. Then for all $t = 0, \ldots, T$,

$$f(x_t) > (8\delta_2 M)^{\beta}(2-\gamma)^{-\beta} + \inf(f) + \delta_1. \tag{10.24}$$

In view of (10.24) for all $t = 0, \ldots, T$,

$$((2-\gamma)(f(x_t) - \inf(f) - \delta_1))^{1/\beta} > 8\delta_2 M. \tag{10.25}$$

By (10.17) and (10.19),

$$\|x_0 - x_*\| \leq 2M. \tag{10.26}$$

Assume that $S \in \{0, \ldots, T\}$, $S < T$ and that

$$\|x_t - x_*\| \leq 2M,\ t = 0, \ldots, S. \tag{10.27}$$

(Note that in view of (10.26), (10.27) holds for $S = 0$.) Let $t \in [0, S]$ be an integer. By (10.21) and (10.23),

$$\|x_{t+1} - x_*\|^2 = \|x_t - \lambda_t q_t - x_*\|^2$$
$$= \|x_t - x_*\|^2 - 2\lambda_t \langle q_t, x_t - x_* \rangle + \lambda_t^2. \tag{10.28}$$

In view of (10.21), there exists

$$\widehat{q_t} \in B_X(q_t, \delta_2) \cap S_X(0, 1) \cap N(G_{\delta_1}(x_t), x_t). \tag{10.29}$$

Proposition 10.1, (10.4), (10.20), (10.22), (10.25), (10.28), (10.29), and the inequality

$$f(x_t) - \inf(f) > \delta_1$$

imply that

$$\|x_{t+1} - x_*\|^2$$
$$\leq \|x_t - x_*\|^2 - 2\lambda_t \langle \widehat{q_t}, x_t - x_* \rangle + \lambda_t^2$$
$$+2\|\widehat{q_t} - q_t\|\lambda_t\|x_t - x_*\|$$
$$\leq \|x_t - x_*\|^2 - 2\lambda_t \langle \widehat{q_t}, x_t - x_* \rangle + \lambda_t^2$$
$$+2\delta_2\lambda_t\|x_t - x^*\|$$
$$\leq \|x_t - x_*\|^2 - 2\gamma((f(x_t) - \inf(f) - \delta_1)L^{-1})^{1/\beta}$$
$$\times(f(x_t) - \inf(f) - \delta_1)^{1/\beta}$$
$$+\gamma^2((f(x_t) - \inf(f) - \delta_1)L^{-1})^{2/\beta}$$
$$+ 2\delta_2\gamma((f(x_t) - \inf(f) - \delta_1)L^{-1})^{1/\beta}\|x_t - x^*\|. \tag{10.30}$$

By (10.24), (10.27), and (10.30),

$$\|x_{t+1} - x_*\|^2$$
$$\leq \|x_t - x_*\|^2 - \gamma(f(x_t) - \inf(f) - \delta_1)^{2/\beta}(2L^{-1/\beta} - \gamma L^{-2/\beta})$$
$$+2\delta_2\gamma((f(x_t) - \inf(f) - \delta_1)L^{-1})^{1/\beta}\|x_t - x^*\|$$
$$\leq \|x_t - x_*\|^2$$

$$-\gamma(f(x_t) - \inf(f) - \delta_1)^{1/\beta} L^{-1/\beta}$$

$$\times(2 - \gamma L^{-1/\beta})((f(x_t) - \inf(f) - \delta_1)^{1/\beta} - 4\delta_2 M)$$

$$\leq \|x_t - x_*\|^2 - (f(x_t) - \inf(f) - \delta_1)^{1/\beta}\gamma(4\delta_2 M)$$

$$\leq \|x_t - x_*\|^2 - 4\delta_2 M\gamma(2-\gamma)^{-1} 8\delta_2 M. \tag{10.31}$$

In view of (10.27) and (10.31),

$$\|x_{t+1} - x_*\| \leq \|x_t - x_*\| \leq 2M.$$

By induction we have shown that

$$\|x_t - x_*\| \leq 2M, \ t = 0, \ldots, T$$

and that for all $t = 0, \ldots, T-1$,

$$\|x_{t+1} - x_*\|^2 \leq \|x_t - x_*\|^2 - (4\delta_2 M)^2\gamma(2-\gamma)^{-1}. \tag{10.32}$$

It follows from (10.26) and (10.32) that

$$4M^2 \geq \|x_0 - x_*\|^2 - \|x_T - x_*\|^2$$

$$= \sum_{t=0}^{T-1}(\|x_t - x_*\|^2 - \|x_{t+1} - x_*\|^2)$$

$$\geq T(\delta_2 M)^2\gamma(2-\gamma)^{-1}$$

and

$$T \leq 4\gamma^{-1}(2-\gamma)4^{-2}\delta^{-2}.$$

This contradicts (10.18). The contradiction we have reached completes the proof of Theorem 10.2. □

Chapter 11
Minimization of Sharp Weakly Convex Functions

In this chapter we study the subgradient projection algorithm for minimization of sharp weakly convex functions, under the presence of computational errors. The problem is described by an objective function and a set of feasible points. For this algorithm each iteration consists of two steps. The first step is a calculation of a subgradient of the objective function while in the second one we calculate a projection on the feasible set. In each of these two steps there is a computational error. In general, these two computational errors are different. We show that our algorithm generates a good approximate solution, if all the computational errors are bounded from above by a small positive constant. Moreover, if we know the computational errors for the two steps of our algorithm, we find out what approximate solution can be obtained and how many iterates one needs for this.

11.1 Preliminaries

The problem and the algorithm studied in this chapter were considered in [35] in a finite-dimensional setting and without computational error.

Let X be a Hilbert space equipped with an inner product $\langle \cdot, \cdot \rangle$ which induces a complete norm $\| \cdot \|$. Let $C \subset X$ be a nonempty closed convex set,

$$\rho \geq 0, \mu > 0$$

$$g : X \to R^1$$

be a locally Lipschitz function such that the function

$$g(x) + 2^{-1}\rho\|x\|^2, \ x \in X \tag{11.1}$$

A. J. Zaslavski, *Convex Optimization with Computational Errors*, Springer Optimization and Its Applications 155, https://doi.org/10.1007/978-3-030-37822-6_11

is convex. Following [35] the function g is called ρ-weakly convex.

Set

$$C_{min} = \text{argmin}(g, C)$$

$$= \{x \in C : g(x) = \inf(g, C)\}. \tag{11.2}$$

Suppose that

$$C_{min} \neq \emptyset. \tag{11.3}$$

Let

$$M_1 > 2 + 2\mu\rho^{-1},$$

$$C_{min} \subset B_X(0, M_1 - 2 - \mu\rho^{-1}). \tag{11.4}$$

Let $L \geq 2$ be such that

$$|g(z_1) - g(z_2)| \leq L\|z_1 - z_2\|$$

$$\text{for all } z_1, z_2 \in B_X(0, M_1 + 4). \tag{11.5}$$

For every $x \in X$, denote by $\partial g(x)$ the set of all $v \in X$ such that

$$g(y) \geq g(x) + \langle v, y - x\rangle + o(\|y - x\|) \text{ as } y \to 0. \tag{11.6}$$

We can show (see Proposition 11.1) that for every $x \in X$,

$$\partial(g + 2^{-1}\rho\|\cdot\|^2)(x) = \partial g(x) + \rho x, \tag{11.7}$$

where the left-hand side of (11.7) is the subdifferential of the convex function.

Suppose that the following sharpness property holds [35]:
for all $z \in C$

$$g(z) - \inf(g, C) \geq \mu d(z, C_{min}). \tag{11.8}$$

A point $\bar{x} \in C$ is called stationary if

$$g(x) - g(\bar{x}) \geq o(\|x - \bar{x}\|) \text{ as } x \to \bar{x} \text{ in } C. \tag{11.9}$$

11.2 The Subdifferential of Weakly Convex Functions

Proposition 11.1 *Let $x \in X$. Then $v \in \partial g(x)$ if and only if*

$$v + \rho x \in \partial(g + 2^{-1}\rho\|\cdot\|^2)(x).$$

Moreover, if $v \in \partial g(x)$, then for every $y \in X$,

$$g(y) \geq g(x) + \langle v, y - x\rangle - 2^{-1}\rho\|y - x\|^2.$$

Proof Let

$$v \in \partial g(x) \tag{11.10}$$

and $h \in X \setminus \{0\}$. In view of (11.6), for all $t > 0$,

$$g(x + th) \geq g(x) + \langle v, th\rangle + o(\|th\|) \text{ as } t \to 0^+.$$

This implies that

$$t^{-1}(g(x + th) - g(x)) \geq \langle v, h\rangle + o(\|th\|)t^{-1}$$

and

$$\liminf_{t\to 0+} t^{-1}(g(x + th) - g(x)) \geq \langle v, h\rangle. \tag{11.11}$$

Thus we have shown that the following property holds:

(i) if $v \in \partial g(x)$, then (11.11) holds for every $h \in X$.

Note that the function

$$g(x) + 2^{-1}\rho\|x\|^2, \ x \in X$$

is convex. For every $t > 0$ and every $h \in X$, we have

$$t^{-1}(g(x + th) + 2^{-1}\rho\|x + th\|^2 - g(x) - 2^{-1}\rho\|x\|^2)$$

$$= t^{-1}(g(x + th) - g(x)) + t^{-1}(2^{-1}\rho\langle 2th, x\rangle + 2^{-1}\rho t^2\|h\|^2). \tag{11.12}$$

By (11.12),

$$\lim_{t\to 0^+} t^{-1}(g(x + th) + 2^{-1}\rho\|x + th\|^2 - g(x) - 2^{-1}\rho\|x\|^2)$$

$$= \lim_{t \to 0^+} (t^{-1}(g(x + th) - g(x)) + \rho\langle h, x\rangle). \tag{11.13}$$

It follows from (11.13) that

$$\lim_{t \to 0^+} (t^{-1}(g(x + th) - g(x))$$

exists and

$$\lim_{t \to 0^+} (t^{-1}(g(x + th) - g(x))$$

$$= \lim_{t \to 0^+} t^{-1}((g + 2^{-1}\rho\|\cdot\|^2)(x + th) - (g + 2^{-1}\rho\|\cdot\|^2)(x)) - \rho\langle h, x\rangle. \tag{11.14}$$

Property (i) implies that if

$$v \in \partial g(x),$$

then (11.11) holds and in view of (11.14),

$$v + \rho x \in \partial(g + 2^{-1}\rho\|\cdot\|^2)(x). \tag{11.15}$$

Let $v \in X$ and (11.15) hold. By (11.15), for every $y \in X$,

$$g(y) + 2^{-1}\rho\|y\|^2 \geq g(x) + 2^{-1}\rho\|x\|^2 + \langle v + \rho x, y - x\rangle$$

and

$$g(y) \geq g(x) + \langle v, y - x\rangle$$

$$+2^{-1}\rho\|x\|^2 - 2^{-1}\|y\|^2 + \rho\langle x, y - x\rangle$$

$$= g(x) + \langle v, y - x\rangle - 2^{-1}\rho\|y\|^2 - 2^{-1}\|x\|^2 + \rho\langle x, y\rangle$$

$$= g(x) + \langle v, y - x\rangle - 2^{-1}\rho\|y - x\|^2. \tag{11.16}$$

Relation (11.16) implies that $v \in \partial g(x)$. Thus (11.15) implies the inclusion above. Proposition 11.1 is proved. □

11.3 An Auxiliary Result

Lemma 11.2 *Let*

$$\delta \in (0, 1],\ \delta_0 \in (0, 1],\ \delta_1 = \mu/16, \tag{11.17}$$

$$x, \xi \in X, \ \|x\| \leq M_1 + 3, \tag{11.18}$$

$$B(x, \delta) \cap C \neq \emptyset, \tag{11.19}$$

$$\xi \in \partial f(x), \tag{11.20}$$

$$\|\xi\| \leq \delta_1. \tag{11.21}$$

Then at least one of the following inequalities holds:

$$d(x, C_{min}) \leq 16\delta; \tag{11.22}$$

$$d(x, C_{min}) \leq 16L\delta\mu^{-1}; \tag{11.23}$$

$$d(x, C_{min}) \geq 2^{-1}3\mu\rho^{-1}.$$

Proof There exists

$$\{x_i\}_{i=1}^{\infty} \subset C_{min} \tag{11.24}$$

such that

$$\lim_{i \to \infty} \|x_i - x\| = d(x, C_{min}). \tag{11.25}$$

In view of (11.19), there exists $\widehat{x} \in C$ such that

$$\|x - \widehat{x}\| \leq \delta. \tag{11.26}$$

By (11.5), (11.8), (11.17), (11.18), (11.24), and (11.26), for $i = 1, 2, \ldots$

$$L\delta + g(x) - g(x_i)$$

$$\geq g(\widehat{x}) - g(x) + g(x) - g(x_i)$$

$$g(\widehat{x}) - g(x_i)$$

$$\geq \mu d(\widehat{x}, C_{min}). \tag{11.27}$$

Proposition 11.1 and (11.20) imply that

$$\xi + \rho x \in \partial(g + 2^{-1}\rho\|\cdot\|^2)(x). \tag{11.28}$$

In view of (11.21) and (11.28), for every $y \in X$,

$$g(y) - g(x)$$
$$\geq \langle \xi, y - x \rangle - 2^{-1}\rho\|y\|^2 + 2^{-1}\rho\|x\|^2 - \rho\|x\|^2 + \rho\langle x, y \rangle$$
$$\geq -\delta_1\|y - x\| - 2^{-1}\rho\|x - y\|^2. \tag{11.29}$$

By (11.29), for $i = 1, 2, \ldots,$

$$g(x) - g(x_i) \leq \delta_1\|x_i - x\| + 2^{-1}\rho\|x_i - x\|^2. \tag{11.30}$$

It follows from (11.27) and (11.30) that for all integers $i \geq 1$,

$$\mu d(\widehat{x}, C_{min}) \leq L\delta + g(x) - g(x_i)$$
$$\leq L\delta + \delta_1\|x_i - x\| + 2^{-1}\rho\|x_i - x\|^2. \tag{11.31}$$

By (11.24)–(11.26) and (11.31),

$$-\mu\delta + \mu d(x, C_{min}) \leq \mu d(\widehat{x}, C_{min})$$
$$\leq L\delta + \delta_1 d(x, C_{min}) + 2^{-1}\rho d(x, C_{min})^2. \tag{11.32}$$

In view of (11.32),

$$\mu d(x, C_{min})$$
$$\leq \mu\delta + L\delta + \delta_1 d(x, C_{min}) + 2^{-1}\rho d(x, C_{min})^2. \tag{11.33}$$

Assume that (11.22) and (11.23) do not hold. Together with (11.17) and (11.33) this implies that

$$\mu d(x, C_{min}) \leq (\mu/16)d(x, C_{min})$$
$$+(\mu/16)d(x, C_{min}) + (\mu/16)d(x, C_{min}) + 2^{-1}\rho d(x, C_{min})^2$$
$$\leq (\mu/4)d(x, C_{min}) + 2^{-1}\rho d(x, C_{min})^2. \tag{11.34}$$

By (11.34),

$$\mu \leq \mu/4 + 2^{-1}\rho d(x, C_{min})$$

and

$$\mu \leq (2/3)\rho d(x, C_{min}).$$

Lemma 11.2 is proved. □

11.4 The First Main Result

Let

$$\gamma \in (0, 1),\ \delta, \delta_0 \in (0, 1],\ \delta_0 \le \mu/16.$$

Let us describe our algorithm.

Initialization: select an arbitrary $x_0 \in X$ such that

$$d(x_0, C_{min}) < \gamma\mu\rho^{-1},\ d(x_0, C) \le \delta.$$

Iterative Step: given a current iteration vector $x_t \in X$, if

$$g(x_t) \le \inf(g, C)$$

$$+ \max\{(16\delta(L+1)^3\gamma(1-\gamma)^{-1}\rho^{-1})^{1/2},$$

$$16\delta L,\ 16L^2\delta\mu^{-1},\ 4\delta L(\mu + L)\mu^{-1}(1-\gamma)^{-1}\},$$

then x_t is considered as an approximate solution; otherwise calculate $\xi_t \in X$ such that

$$B(\xi_t, \delta_0) \cap \partial f(x_t) \neq \emptyset$$

(note that by Lemma 11.2, $\xi_t \neq 0$) and calculate $x_{t+1} \in X$ such that

$$\|x_{t+1} - P_C(x_t - \|\xi_t\|^{-2}(g(x_t) - \inf(g, C))\xi_t)\| \le \delta.$$

Proposition 11.3 *Let*

$$\gamma \in (0, 1),\ \delta, \delta_0 \in (0, 1],$$

$$\delta_0 < \min\{\mu(1-\gamma)/4,\ \mu/32\}, \tag{11.35}$$

$T > 0$ *be an integer*, $\{x_t\}_{t=0}^{T} \subset X$, $\{\xi_t\}_{t=0}^{T-1} \subset X$,

$$d(x_0, C) \le \delta, \tag{11.36}$$

$$d(x_0, C_{min}) < \gamma\mu\rho^{-1}, \tag{11.37}$$

for all $t = 0, \dots, T-1$,

$$g(x_t) > \inf(g, C) + (16\delta(L+1)^3\gamma(1-\gamma)^{-1}\rho^{-1})^{1/2}, \tag{11.38}$$

$$g(x_t) > \inf(g, C)$$

$$+ \max\{16\delta L,\ 16L^2\delta\mu^{-1},\ \delta(\mu + L)L,\ 4\delta L(\mu + L)\mu^{-1}(1 - \gamma)^{-1}\}, \tag{11.39}$$

$$B_X(\xi_t, \delta_0) \cap \partial f(x_t) \neq \emptyset \tag{11.40}$$

and

$$\|x_{t+1} - P_C(x_t - \|\xi_t\|^{-2}(g(x_t) - \inf(g, C))\xi_t)\| \leq \delta. \tag{11.41}$$

Then

$$\|\xi_t\| \geq \mu/32,\ t = 0, \ldots, T,$$

$$T \leq 2M_1\delta^{-1}.$$

Proof By (11.4) and (11.37),

$$\|x_0\| \leq M_1. \tag{11.42}$$

Assume that $t \in [0, T)$ is an integer, x_t is well defined, and

$$d(x_t, C_{min}) < \gamma\mu\rho^{-1}. \tag{11.43}$$

(Note that in view of (11.37), inequality (11.43) holds for $t = 0$.) By (11.4) and (11.43),

$$\|x_t\| \leq M_1. \tag{11.44}$$

It follows from (11.4), (11.5), (11.43), and (11.44),

$$|g(x_t) - \inf(g, C)| < Ld(x_t, C_{min}). \tag{11.45}$$

In view of (11.45),

$$g(x_t) < \inf(g, C) + Ld(x_t, C_{min}). \tag{11.46}$$

Relations (11.39) and (11.46) imply that

$$d(x_t, C_{min}) > 16\delta,\ 16\delta L\mu^{-1},\ 4\delta(\mu + L)\mu^{-1}(1 - \gamma)^{-1}. \tag{11.47}$$

By (11.40), there exists

$$\eta_t \in \partial f(x_t) \tag{11.48}$$

such that

$$\|\eta_t - \xi_t\| \leq \delta_0. \tag{11.49}$$

Lemma 11.2, (11.17), (11.36), (11.41), (11.43), (11.44), (11.47), and (11.48) imply that

$$\|\eta_t\| > \mu/16. \tag{11.50}$$

It follows from (11.35), (11.49), and (11.50) that

$$\|\xi_t\| \geq \mu/32. \tag{11.51}$$

Clearly, x_{t+1} is well defined by (11.41). (Thus we showed by induction that x_i, $i = 0, \dots, T$, ξ_i, $i = 0, \dots, T-1$ are well defined.) There exists a sequence

$$\{z_i\}_{i=1}^{\infty} \subset C_{min} \tag{11.52}$$

such that

$$\lim_{i \to \infty} \|x_t - z_i\| = d(x_t, C_{min}). \tag{11.53}$$

Lemma 2.2, (11.41), and (11.52) imply that for all natural numbers i,

$$\|x_{t+1} - z_i\|$$

$$\leq \delta + \|P_C(x_t - \|\xi_t\|^{-2}(g(x_t) - \inf(g, C))\xi_t) - z_i\|$$

$$\leq \delta + \|x_t - \|\xi_t\|^{-2}(g(x_t) - \inf(g, C))\xi_t - z_i\|. \tag{11.54}$$

Proposition 11.1, (11.6), (11.48), (11.49), (11.51), and (11.52) imply that

$$\|x_t - z_i - \|\xi_t\|^{-2}(g(x_t) - \inf(g, C))\xi_t)\|^2$$

$$= \|x_t - z_i\|^2$$

$$+2\|\xi_t\|^{-2}(g(x_t) - \inf(g, C))\langle \xi_t, z_i - x_t \rangle$$

$$+\|\xi_t\|^{-2}(g(x_t) - \inf(g, C))^2$$

$$\leq \|x_t - z_i\|^2$$

$$+(g(x_t) - \inf(g, C))^2\|\xi_t\|^{-2}$$

$$+2(g(x_t) - \inf(g, C))\|\xi_t\|^{-2}\langle \xi_t - \eta_t, z_i - x_t\rangle$$

$$+2(g(x_t) - \inf(g, C))\|\xi_t\|^{-2}\langle \eta_t, z_i - x_t\rangle$$

$$\leq \|x_t - z_i\|^2$$

$$+(g(x_t) - \inf(g, C))^2\|\xi_t\|^{-2}$$

$$+2(g(x_t) - \inf(g, C))\|\xi_t\|^{-2}\delta_0\|z_i - x_t\|$$

$$+2(g(x_t) - \inf(g, C) - \delta)\|\xi_t\|^{-2}(g(z_i) - g(x_t) + 2^{-1}\rho\|z_i - x_t\|^2)$$

$$= \|x_t - z_i\|^2$$

$$+(g(x_t) - g(z_i))\|\xi_t\|^{-2}(\rho\|x_t - z_i\|^2 - (g(x_t) - g(z_i))$$

$$+2\delta_0\|z_i - x_t\|\|\xi_t\|^{-2}(g(x_t) - \inf(g, C))$$

$$= \|x_t - z_i\|^2$$

$$+(g(x_t) - g(z_i))\|\xi_t\|^{-2}(\rho\|x_t - z_i\|^2 - (g(x_t) - g(z_i))$$

$$+ 2\delta_0\|z_i - x_t\|). \tag{11.55}$$

By (11.41) and (11.46), there exists

$$\widehat{x} \in C \tag{11.56}$$

such that

$$\|\widehat{x} - x_t\| \leq \delta. \tag{11.57}$$

By (11.5), (11.44), and (11.57), for all integers $i = 1, 2, \ldots,$

$$|(g(x_t) - g(z_i)) - (g(\widehat{x}) - g(z_i))\| \leq L\delta. \tag{11.58}$$

Let $i \geq 0$ be an integer. It follows from (11.8), (11.52), (11.56), and (11.57) that

$$g(\widehat{x}) - g(z_i) \geq \mu d(\widehat{x}, C_{min}) \geq \mu d(x_t, C_{min}) - \delta. \tag{11.59}$$

In view of (11.58) and (11.59),

$$g(x_t) - g(z_i) \geq -L\delta - \mu\delta + \mu d(x_t, C_{min}). \tag{11.60}$$

By (11.55) and (11.60),

$$\|x_t - z_i - \|\xi_t\|^{-2}(g(x_t) - \inf(g, C))\xi_t)\|^2$$

$$\le \|x_t - z_i\|^2$$

$$+\|\xi_t\|^{-2}(g(x_t) - g(z_i))(\rho\|x_t - z_i\|^2$$

$$- \mu d(x_t, C_{min}) + L\delta + \mu\delta + 2\delta_0\|z_i - x_t\|). \tag{11.61}$$

It follows from (11.47) and (11.62) that for all sufficiently large natural numbers i,

$$16\delta \le d(x_t, C_{min}) \le \|z_i - x_t\| \le 2d(x_t, C_{min}). \tag{11.62}$$

By (11.35), (11.43), (11.52), (11.53), and (11.61),

$$d(x_t - \|\xi_t\|^{-2}(g(x_t) - \inf(g, C))\xi_t, C_{min})^2$$

$$\le \limsup_{i\to\infty} \|x_t - z_i - \|\xi_t\|^{-2}(g(x_t) - g(z_i))\xi_t\|^2$$

$$\le d(x_t, C_{min})^2 + \|\xi_t\|^{-2}(g(x_t) - \inf(g, C))$$

$$\times(\rho d(x_t, C_{min})^2 - \mu d(x_t, C_{min})$$

$$+L\delta + \mu\delta + 2\delta_0 d(x_t, C_{min}))$$

$$= d(x_t, C_{min})^2 + \|\xi_t\|^{-2}(g(x_t) - \inf(g, C))$$

$$\times(\rho d(x_t, C_{min})^2 - (\mu - 2\delta_0)d(x_t, C_{min}) + \delta(\mu + L))$$

$$= d(x_t, C_{min})^2$$

$$+\|\xi_t\|^{-2}\rho(g(x_t) - \inf(g, C))d(x_t, C_{min})(d(x_t, C_{min})$$

$$-(\mu - 2\delta_0)\rho^{-1} + d(x_t, C_{min})^{-1}\delta(\mu + L))\rho^{-1}$$

$$\le d(x_t, C_{min})^2$$

$$+\|\xi_t\|^{-2}\rho(g(x_t) - \inf(g, C))d(x_t, C_{min})(\gamma\mu\rho^{-1}$$

$$-\mu\rho^{-1} + 2\delta_0\rho^{-1} + \delta(\mu + L)d(x_t, C_{min})^{-1})\rho^{-1})$$

$$\le d(x_t, C_{min})^2$$

$$-\|\xi_t\|^{-2}\rho(g(x_t)-\inf(g,C))(d(x_t,C_{min})((1-\gamma)\mu\rho^{-1}$$
$$-2\delta_0\rho^{-1})-\delta(\mu+L)\rho^{-1})$$
$$\leq d(x_t,C_{min})^2-\|\xi_t\|^{-2}\rho(g(x_t)-\inf(g,C))$$
$$\times(d(x_t,C_{min})(1-\gamma)\mu\rho^{-1}/2-\delta(\mu+L)\rho^{-1}). \quad (11.63)$$

By (11.47) and (11.63),

$$d(x_t-\|\xi_t\|^{-2}(g(x_t)-\inf(g,C))\xi_t,C_{min})^2$$
$$\leq d(x_t,C_{min})^2-\|\xi_t\|^{-2}\rho(g(x_t)$$
$$-\inf(g,C))d(x_t,C_{min})(1-\gamma)\mu\rho^{-1}/4. \quad (11.64)$$

In view of (11.39) and (11.64),

$$d(x_t-\|\xi_t\|^{-2}(g(x_t)-\inf(g,C))\xi_t,C_{min})^2\leq d(x_t,C_{min})^2. \quad (11.65)$$

It follows from (11.43), (11.64), and (11.65) that

$$d(x_t,C_{min})-d(x_t-\|\xi_t\|^{-2}(g(x_t)-\inf(g,C))\xi_t,C_{min})$$
$$=(d(x_t,C_{min})^2-d(x_t-\|\xi_t\|^{-2}(g(x_t)-\inf(g,C))\xi_t,C_{min})^2)$$
$$\times(d(x_t,C_{min})+d(x_t-\|\xi_t\|^{-2}(g(x_t)-\inf(g,C))\xi_t,C_{min}))^{-1}$$
$$\geq d(x_t,C_{min})^2-d(x_t-\|\xi_t\|^{-2}(g(x_t)$$
$$-\inf(g,C))\xi_t,C_{min})^2(2\gamma\mu/\rho)^{-1}$$
$$\geq d(x_t,C_{min})\|\xi_t\|^{-2}(g(x_t)-\inf(g,C))8^{-1}\gamma^{-1}(1-\gamma)\rho. \quad (11.66)$$

Lemma 2.2, (11.41), and (11.66) imply that

$$d(x_{t+1},C_{min})$$
$$\leq\delta+d(P_C(x_t-\|\xi_t\|^{-2}(g(x_t)-\inf(g,C))\xi_t,C_{min})$$
$$\leq\delta+d(x_t-\|\xi_t\|^{-2}(g(x_t)-\inf(g,C))\xi_t,C_{min})$$
$$\leq\delta+d(x_t,C_{min})$$
$$-d(x_t,C_{min})\|\xi_t\|^{-2}(g(x_t)-\inf(g,C))8^{-1}\gamma^{-1}(1-\gamma)\rho. \quad (11.67)$$

By (11.5), (11.44), (11.48), and (11.49),

$$\|\xi_t\| \leq L + 1. \tag{11.68}$$

In view of (11.4), (11.5), and (11.44),

$$g(x_t) - \inf(g, C) \leq Ld(x_t, C_{min}). \tag{11.69}$$

It follows from (11.38), (11.68), and (11.69) that

$$d(x_t, C_{min})\|\xi_t\|^{-2}(g(x_t) - \inf(g, C))8^{-1}\gamma^{-1}(1 - \gamma)\rho$$

$$\geq (g(x_t) - \inf(g, C))L^{-1}8^{-1}\gamma^{-1}(1 - \gamma)\rho(L + 1)^{-2} \geq 2\delta. \tag{11.70}$$

By (11.67) and (11.70),

$$d(x_{t+1}, C_{min}) \leq d(x_t, C_{min}) - \delta. \tag{11.71}$$

We assume that (11.43) holds and obtained (11.71). This implies that for all integers $t = 0, \ldots, T - 1$,

$$d(x_{t+1}, C_{min}) \leq d(x_t, C_{min}) - \delta$$

and

$$d(x_t, C_{min}) \leq \gamma\mu\rho^{-1}$$

for all integers $t = 0, \ldots, T$. Together with (11.4) and (11.42) this implies that

$$2M_1 \geq d(x_0, C_{min})$$

$$\geq d(x_0, C_{min}) - d(x_T, C_{min})$$

$$= \sum_{t=0}^{T-1}(d(x_t, C_{min}) - d(x_{t+1}, C_{min})) \geq T\delta$$

and

$$T \leq 2M_1\delta^{-1}.$$

This completes the proof of Proposition 11.3. □

Proposition 11.3 implies the following result.

Theorem 11.4 *Let*

$$\gamma \in (0, 1),\ \delta, \delta_0 \in (0, 1],$$

$$\delta_0 < \min\{\mu(1-\gamma)/4,\ \mu/32\},$$

$$T = \lfloor 2M_1\delta^{-1} \rfloor,$$

$\{x_t\}_{t=0}^{T} \subset X$, $\{\xi_t\}_{t=0}^{T-1} \subset X$,

$$d(x_0, C) \le \delta,$$

$$d(x_0, C_{min}) < \gamma\mu\rho^{-1},$$

for all $t = 0, \dots, T-1$,

$$B(\xi_t, \delta_0) \cap \partial f(x_t) \neq \emptyset,$$

if

$$\xi_t \neq 0,$$

then

$$\|x_{t+1} - P_C(x_t - \|\xi_t\|^{-2}(g(x_t) - \inf(g, C))\xi_t)\| \le \delta,$$

otherwise

$$x_{t+1} = x_t.$$

Then there exists $t \in \{0, \dots, T\}$, *such that*

$$g(x_t) \le \inf(g, C) + \max\{(16\delta(L+1)^3\gamma(1-\gamma)^{-1}\rho^{-1})^{1/2},$$

$$16\delta L,\ 16L^2\delta\mu^{-1},\ \delta(\mu+L)L,\ 4\delta L(\mu+L)\mu^{-1}(1-\gamma)^{-1}\}.$$

11.5 An Algorithm with Constant Step Sizes

Let

$$\gamma \in (0, 1),\ \delta, \delta_0 \in (0, 1],\ \alpha > 0.$$

Let us describe our algorithm.

Initialization: select an arbitrary $x_0 \in X$ such that

$$d(x_0, C_{min}) < \gamma \mu \rho^{-1}, \; d(x_0, C) \le \delta.$$

Iterative Step: given a current iteration vector $x_t \in X$, calculate $\xi_t \in X$ such that

$$B(\xi_t, \delta_0) \cap \partial f(x_t) \neq \emptyset,$$

if

$$\xi_t \neq 0,$$

then calculate $x_{t+1} \in X$ such that

$$\|x_{t+1} - P_C(x_t - \alpha \|\xi_t\|^{-1} \xi_t)\| \le \delta;$$

otherwise set

$$x_{k+1} = x_k.$$

11.6 An Auxiliary Result

Lemma 11.5 *Let* $\alpha > 0$, $\delta, \delta_0 \in (0, 1]$, $\gamma \in (0, 1)$ *satisfy*

$$\alpha \le 2^{-1} \mu (L + 1)^{-1},$$

$$\delta_0 < \mu/32, \; \delta_0 \le 64^{-1} \alpha \rho, \; \delta < \mu \rho^{-1},$$

$$\delta \le \min\{64^{-2} \alpha \mu (L + \mu)^{-1}, \; 2^{-1} \alpha^2 L^{-2} \mu^{-1} \rho\}, \tag{11.72}$$

$x \in X$ *satisfy*

$$d(x, C) < \delta, \; d(x, C_{min}) < \gamma \mu \rho^{-1}, \tag{11.73}$$

$$d(x, C_{min}) > \max\{16\delta, 16 L \delta \mu^{-1}, \; 6 \alpha \mu^{-1} L\}. \tag{11.74}$$

Then for every $\eta \in \partial f(x)$,

$$\|\eta\| \ge \mu/16 \tag{11.75}$$

and the following assertion holds.

Let $\xi \in X$, $x^+ \in X$,

$$B_X(\xi, \delta_0) \cap \partial f(x) \neq \emptyset, \tag{11.76}$$

$$\|x^+ - P_C(x - \alpha\|\xi\|^{-1}\xi)\| \leq \delta. \tag{11.77}$$

Then

$$\|\xi\| \geq \mu/32,$$

$$d(x^+, C_{min})^2 \leq d(x, C_{min})^2 - 2^{-1}\alpha^2.$$

Proof By (11.4) and (11.73),

$$\|x\| \leq M_1. \tag{11.78}$$

Lemma 11.2, (11.73), (11.74), and (11.78) imply that for every $\eta \in \partial f(x)$ (11.76) holds.

Let us prove the assertion. In view of (11.76), there exists

$$\eta \in \partial f(x) \tag{11.79}$$

such that

$$\|\xi - \eta\| \leq \delta_0. \tag{11.80}$$

It follows from (11.5), (11.72), (11.75), (11.76), and (11.78)–(11.80) that

$$\|\xi\| \geq \mu/32, \tag{11.81}$$

$$\|\xi\| \leq L + 1.$$

Let

$$\{x_i\}_{i=1}^{\infty} \subset C_{min} \tag{11.82}$$

and

$$\|x - x_i\| \to d(x, C_{min}) \text{ as } i \to \infty. \tag{11.83}$$

In view of (11.73), we may assume that for all integers $i = 1, 2, \ldots,$

$$\|x_i - x\| < \gamma\mu\rho^{-1}. \tag{11.84}$$

Let $i \geq 1$ be an integer. Lemma 2.2, (11.77), and (11.82) imply that

$$
\begin{aligned}
&\|x^+ - x_i\| \\
&\leq \delta + \|P_C(x - \alpha\|\xi\|^{-1}\xi) - x_i\| \\
&\leq \delta + \|x - x_i - \alpha\|\xi\|^{-1}\xi\|.
\end{aligned} \tag{11.85}
$$

Proposition 11.1, (11.6), (11.72), (11.79)–(11.81), and (11.84) imply that

$$
\begin{aligned}
&\|x - x_i - \alpha\|\xi\|^{-1}\xi\|^2 \\
&= \|x - x_i\|^2 + 2\alpha\|\xi\|^{-1}\langle \xi, x_i - x\rangle + \alpha^2 \\
&\leq \|x - x_i\|^2 + 2\alpha\|\xi\|^{-1}\|\eta - \xi\|\|x_i - x\| + 2\alpha\|\xi\|^{-1}\langle \eta, x_i - x\rangle + \alpha^2 \\
&\leq \|x - x_i\|^2 + 64\alpha\mu^{-1}\delta_0\mu\rho^{-1} + \alpha^2 \\
&+ 2\alpha\|\xi\|^{-1}(g(x_i) - g(x) + 2^{-1}\rho\|x - x_i\|^2).
\end{aligned} \tag{11.86}
$$

In view of (11.73), there exists

$$
\widehat{x} \in C \cap B_X(x, \delta). \tag{11.87}
$$

It follows from (11.5), (11.78), and (11.87) that

$$
\|g(x) - g(\widehat{x})| \leq L\delta. \tag{11.88}
$$

By (11.5), (11.8), (11.78)–(11.81), and (11.86)–(11.88),

$$
\begin{aligned}
&\|x - x_i - \alpha\|\xi\|^{-1}\xi\|^2 \\
&\|x - x_i\|^2(1 + \alpha\rho\|\xi\|^{-1}) + \alpha^2 + 64\alpha\delta_0\rho^{-1} \\
&+2\alpha\|\xi\|^{-1}(g(x_i) - g(\widehat{x}) + g(\widehat{x}) - g(x)) \\
&\leq \|x - x_i\|^2(1 + \alpha\rho\|\xi\|^{-1}) \\
&+\alpha^2 + 64\alpha\delta_0\rho^{-1} + 2\alpha\|\xi\|^{-1}(L\delta - \mu d(\widehat{x}, C_{min})) \\
&\leq \|x - x_i\|^2(1 + \alpha\rho\|\xi\|^{-1}) \\
&+\alpha^2 + 64\alpha\delta_0\rho^{-1} + 2\alpha\|\xi\|^{-1}(L\delta + \delta\mu - \mu d(x, C_{min})) \\
&\leq \|x - x_i\|^2(1 + \alpha\rho\|\xi\|^{-1}) \\
&-2\alpha\mu d(x, C_{min})\|\xi\|^{-1} \\
&+\alpha^2 + 64\alpha\delta_0\rho^{-1} + 2\alpha\|\xi\|^{-1}(L + \mu)\delta.
\end{aligned}
$$

It follows from the relation above, (11.72), (11.73), and (11.81)–(11.83) that

$$d(x - \alpha\|\xi\|^{-1}\xi, C_{min})^2$$

$$\leq d(x, C_{min})^2(1 + \alpha\rho\|\xi\|^{-1})$$

$$-2\alpha\mu d(x, C_{min})\|\xi\|^{-1}$$

$$+\alpha^2 + 64\alpha\delta_0\rho^{-1} + 2\alpha\|\xi\|^{-1}(L + \mu)\delta. \tag{11.89}$$

By (11.89),

$$d(x, C_{min})^2 - d(x - \alpha\|\xi\|^{-1}\xi, C_{min})^2$$

$$\geq d(x, C_{min})(2\alpha\mu\|\xi\|^{-1} - \alpha\rho\|\xi\|^{-1}d(x, C_{min}))$$

$$-\alpha^2 - 64\alpha\delta_0\rho^{-1} - 64\alpha\mu^{-1}(L + \mu)\delta$$

$$\geq d(x, C_{min})\alpha\mu\|\xi\|^{-1} - \alpha^2 - 64\alpha\delta_0\rho^{-1} - 64\alpha\mu^{-1}(L + \mu)$$

$$\geq d(x, C_{min})\alpha\mu(L + 1)^{-1} - 3\alpha^2$$

$$\geq d(x, C_{min})\alpha\mu(2L + 2)^{-1}. \tag{11.90}$$

By (11.90),

$$d(x - \alpha\|\xi\|^{-1}\xi, C_{min}) \leq d(x, C_{min}). \tag{11.91}$$

Lemma 2.2 and (11.77) imply that for every $z \in C_{min}$,

$$d(x^+, C_{min}) \leq \|x^+ - z\|$$

$$\leq \delta + \|P_C(x - \alpha\|\xi\|^{-1}\xi) - z\|$$

$$\leq \delta + \|x - \alpha\|\xi\|^{-1}\xi - z\|.$$

This implies that

$$d(x^+, C_{min}) \leq \delta + d(x - \alpha\|\xi\|^{-1}\xi, C_{min}). \tag{11.92}$$

By (11.72), (11.73), (11.91), and (11.92),

$$d(x^+, C_{min}) \leq \delta + d(x, C_{min})$$

$$\leq \delta + \gamma\mu\rho^{-1} \leq 2\mu\rho^{-1}. \tag{11.93}$$

It follows from (11.73) and (11.91)–(11.93) that

$$d(x^+, C_{min})^2 - d(x - \alpha\|\xi\|^{-1}\xi, C_{min})^2$$
$$= (d(x^+, C_{min}) - d(x - \alpha\|\xi\|^{-1}\xi, C_{min}))$$
$$\times(d(x^+, C_{min}) + d(x - \alpha\|\xi\|^{-1}\xi, C_{min}))$$
$$\leq 3\delta\mu\rho^{-1}. \tag{11.94}$$

Lemma 2.2, (11.72), (11.74), (11.90), and (11.94) imply that

$$d(x^+, C_{min})^2 \leq d(x - \alpha\|\xi\|^{-1}\xi, C_{min})^2 + 3\delta\mu\rho^{-1}$$
$$\leq d(x, C_{min})^2 - d(x, C_{min})\alpha\mu(2L+2)^{-1} + 3\delta\mu\rho^{-1}$$
$$\leq d(x, C_{min})^2 - 6\alpha^2\mu^{-1}L\mu(2L+2)^{-1} + 3\mu\rho^{-1}(2^{-1}\alpha^2L^{-2}\mu^{-1}\rho)$$
$$\leq d(x, C_{min})^2 - \alpha^2 + 2\alpha^2L^{-2}$$
$$\leq d(x, C_{min})^2 - \alpha^2/2.$$

Lemma 11.5 is proved. □

11.7 The Second Main Result

Theorem 11.6 *Let*

$$\gamma \in (0, 1),\ \delta, \delta_0 \in (0, 1],\ \alpha > 0,$$
$$\alpha \leq 2^{-1}\mu(L+1)^{-1},$$
$$\delta_0 < \mu/32,\ \delta_0 \leq 64^{-1}\rho\alpha,$$
$$\delta < \mu/\rho,\ \delta \leq \min\{2^{-1}\alpha^2L^{-2}\mu^{-1}\rho,\ 64^{-1}\mu(L+\mu)^{-1}\alpha\},$$

$x_0 \in X$ *satisfy*

$$d(x_0, C) \leq \delta,\ d(x_0, C_{min}) < \gamma\mu\rho^{-1}, \tag{11.95}$$
$$T = 1 + \lfloor 2(\mu/\rho)^2\alpha^{-2}\rfloor,$$

$\{x_t\}_{t=0}^{T} \subset X$, $\{\xi_t\}_{t=0}^{T-1} \subset X$, *for all* $t = 0, \ldots, T-1$,

$$B(\xi_t, \delta_0) \cap \partial f(x_t) \neq \emptyset,$$

if

$$\xi_t \neq 0,$$

then

$$\|x_{t+1} - P_C(x_t - \|\xi_t\|^{-1}\alpha\xi_t)\| \leq \delta,$$

otherwise

$$x_{t+1} = x_t.$$

Then there exists $t \in \{0, \ldots, T\}$, *such that*

$$\|\xi_s\| \neq 0$$

for all nonnegative integers $s < t$,

$$g(x_t) \leq \inf(g, C) + \max\{16\delta,\ 16L\mu^{-1}\delta,\ 6\alpha\mu^{-1}L\}.$$

Proof Assume that the theorem does not hold. Then for all integers $t = 0, \ldots, T$,

$$d(x_t, C_{min}) > \max\{16\delta,\ 16\delta\mu^{-1}L,\ 6\alpha\mu^{-1}L\} + \inf(g, C).$$

Applying by induction Lemma 11.5 we obtain that for all integers $t = 0, \ldots, T-1$

$$\|\xi_t\| \geq \mu/32,$$

$$d(x_{t+1}, C_{min})^2 \leq d(x_t, C_{min})^2 - 2^{-1}\alpha^2.$$

By the relation above and (11.95),

$$(\mu/\rho)^2 \geq d(x_0, C_{min})^2$$

$$\geq d(x_0, C_{min})^2 - d(x_T, C_{min})^2$$

$$\sum_{t=0}^{T-1}(d(x_t, C_{min})^2 - d(x_{t+1}, C_{min})^2)$$

$$\geq 2^{-1}T\alpha^2$$

and

$$T \leq 2(\mu/\rho)^2\alpha^{-2}.$$

This contradicts the choice of T. The contradiction we have reached completes the proof of Theorem 11.6. □

11.8 Convex Problems

Let $g : X \to R^1$ be a continuous convex function, $\mu > 0$, $M > 4$, $\gamma \in (0, 1)$ for every $z \in C$,

$$g(z) - \inf(g, C) \geq \mu d(z, \text{argmin}(g, C)) \tag{11.96}$$

and

$$\text{argmin}(g, C) \subset B_X(0, M - 4). \tag{11.97}$$

Fix $\alpha > 0$. Let $L \geq 1$ be such that

$$|f(z_1) - f(z_2)| \leq L\|z_1 - z_2\|$$

$$\text{for all } z_1, z_2 \in B_X(0, M + 1) \tag{11.98}$$

and

$$\delta, \delta_0 \in (0, 1].$$

We prove the following result.

Theorem 11.7 *Let*

$$\alpha \leq 2^{-1}(L + 1)^{-1},$$

$$\alpha \leq 4^{-1}\mu(L + 1)^{-2},$$

$$\delta_0 \leq 4^{-1}M^{-1}\alpha^2(L + 1)^2,$$

$$\delta \leq (2L)^{-1}\alpha(L + 1)^2, \tag{11.99}$$

$\{x_t\}_{t=0}^{\infty} \subset X$, $\{\xi_t\}_{t=0}^{\infty} \subset X$,

$$\|x_0\| \leq M, \; d(x_0, C) \leq \delta, \tag{11.100}$$

and for all integer $t \geq 0$,

$$B(\xi_t, \delta_0) \cap \partial f(x_t) \neq \emptyset,$$

$$\|x_{t+1} - P_C(x_t - \alpha\xi_t)\| \leq \delta. \tag{11.101}$$

Then there exists a nonnegative integer $\tau \leq \lfloor 2M^2\alpha^{-2}(L+1)^{-2} \rfloor$ *such that for all integers* $t \in [0, \tau] \setminus \{\tau\}$,

$$d(x_t, C_{min}) \leq d(x_t, C_{min})$$

and for all integers $t \geq \tau$,

$$d(x_t, C_{min}) \leq \alpha(L+1)((4L+4)\mu^{-1} + 3).$$

11.9 An Auxiliary Result

Lemma 11.8 *Let*

$$\alpha < (2L+2)^{-1},$$

$$4\delta_0 M < \alpha^2(L+1)^2,$$

$$\delta \leq 2^{-1}\alpha(L+1), \tag{11.102}$$

$x \in X$,

$$\|x\| \leq M, \tag{11.103}$$

$$d(x, C) \leq \delta, \tag{11.104}$$

$\xi \in X$,

$$B_X(\xi, \delta_0) \cap \partial f(x) \neq \emptyset, \tag{11.105}$$

$y \in X$,

$$\|y - P_C(x - \alpha\xi)\| \leq \delta. \tag{11.106}$$

Then

$$d(y, C_{min})^2 \leq d(x, C_{min})^2 + 6\alpha^2(L+1)^2 - 2\alpha\mu d(x, C_{min}).$$

Proof Clearly, C_{min} is a closed convex set. In view of (11.105), there exists

$$\eta \in \partial g(x) \tag{11.107}$$

such that

$$\|\xi - \eta\| \leq \delta_0. \tag{11.108}$$

By (11.98), (11.103), and (11.107),

$$\|\eta\| \leq L. \tag{11.109}$$

It follows from (11.108) and (11.109) that

$$\|\xi\| \leq L + 1. \tag{11.110}$$

In view of (11.104), there exists

$$\widehat{x} \in C \tag{11.111}$$

such that

$$\|x - \widehat{x}\| \leq \delta. \tag{11.112}$$

Lemma 2.2 and (11.106) imply that

$$\begin{aligned} &\|y - P_{C_{min}}(x)\| \\ &\leq \|y - P_C(x - \alpha\xi)\| + \|P_C(x - \alpha\xi) - P_{C_{min}}(x)\| \\ &\leq \delta + \|x - \alpha\xi - P_{C_{min}}(x)\|. \end{aligned} \tag{11.113}$$

By (11.96), (11.98), (11.103), (11.104), (11.107), (11.108), and (11.110)–(11.112),

$$\begin{aligned} &\|x - \alpha\xi - P_{C_{min}}(x)\|^2 \\ &= \|x - P_{C_{min}}(x)\|^2 + \alpha^2\|\xi\|^2 + 2\alpha\langle\xi, P_{C_{min}}(x) - x\rangle \\ &\leq \|x - P_{C_{min}}(x)\|^2 + \alpha^2\|\xi\|^2 + 2\alpha\langle\eta, P_{C_{min}}(x) - x\rangle \\ &\quad +2\alpha\|\eta - \xi\|\|P_{C_{min}}(x) - x\| \\ &\leq \|x - P_{C_{min}}(x)\|^2 + \alpha^2(L+1)^2 + 4\alpha\delta_0 M + 2\alpha\langle\eta, P_{C_{min}}(x) - x\rangle \end{aligned}$$

$$\leq d(x, C_{min})^2 + \alpha^2(L+1)^2 + 4\alpha\delta_0 M$$

$$+2\alpha(g(P_{C_{min}}(x)) - g(x))$$

$$\leq d(x, C_{min})^2 + \alpha^2(L+1)^2 + 4\alpha\delta_0 M$$

$$+2\alpha(\inf(g, C) - g(\widehat{x})) + 2\alpha(g(\widehat{x}) - g(x))$$

$$\leq d(x, C_{min})^2 + \alpha^2(L+1)^2 + 4\alpha\delta_0 M + 2\alpha\delta L$$

$$-2\alpha\mu d(\widehat{x}, C_{min})$$

$$\leq d(x, C_{min})^2 + \alpha^2(L+1)^2 + 4\alpha\delta_0 M + 2\alpha\delta L$$

$$-2\alpha\mu d(x, C_{min}) + 2\alpha\delta\mu$$

$$\leq d(x, C_{min})^2 + 4\alpha^2(L+1)^2 - 2\alpha\mu d(x, C_{min}). \tag{11.114}$$

It follows from (11.97), (11.102), (11.103), (11.110), (11.113), and (11.114) that

$$\|y - P_{C_{min}}(x)\|^2$$

$$\leq \|x - \alpha\xi - P_{C_{min}}(x)\|^2 + \delta^2 + 2\delta\|x - \alpha\xi - P_{C_{min}}(x)\|$$

$$\leq d(x, C_{min})^2 + 4\alpha^2(L+1)^2 - 2\alpha d(x, C_{min})\mu + \delta^2 + 2\delta(2M+2)$$

$$\leq d(x, C_{min})^2 + 6\alpha^2(L+1)^2 - 2\alpha d(x, C_{min})\mu.$$

Lemma 11.8 is proved. □

11.10 Proof of Theorem 11.7

Assume that $t \geq 0$ is an integer,

$$\|x_t\| \leq M. \tag{11.115}$$

Applying Lemma 11.8 with

$$x = x_t,\ y = x_{t+1},\ \xi = \xi_t$$

we obtain that

$$d(x_{t+1}, C_{min})^2 \le d(x_t, C_{min})^2 + 6\alpha^2(L+1)^2 - 2\alpha d(x, C_{min})\mu. \tag{11.116}$$

Assume in addition that

$$d(x_t, C_{min}) \ge 4\alpha(L+1)^2\mu^{-1}. \tag{11.117}$$

By (11.116) and (11.117),

$$d(x_{t+1}, C_{min})^2 \le d(x_t, C_{min})^2 - 2\alpha^2(L+1)^2. \tag{11.118}$$

Thus we have shown that the following property holds:

(a) if an integer $t \ge 0$ and (11.115) holds, then (11.116) holds and if (11.115) and (11.117) are valid, then (11.118) is true.

Assume that $\tau \ge 1$ is an integer and that for all integers $t = 0, \dots, \tau - 1$, (11.117) holds. By induction, we apply property (a) and (11.110) and show that (11.118) holds for all $t = 0, \dots, \tau - 1$ and that (11.115) holds for all $t = 0, \dots, \tau$. It follows from (11.97), (11.100), and (11.118) holding for all $t = 0, \dots, \tau - 1$ that

$$4M^2 \ge d(x_0, C_{min})^2$$

$$\ge d(x_0, C_{min})^2 - d(x_\tau, C_{min})^2$$

$$= \sum_{t=0}^{\tau-1} (d(x_t, C_{min})^2 - d(x_{t+1}, C_{min})^2)$$

$$\ge 2\alpha^2(L+1)^2\tau$$

and

$$\tau \le 2M^2\alpha^{-2}(L+1)^{-2}.$$

This implies that there exists an integer

$$\tau \le 2M^2\alpha^{-2}(L+1)^2 \tag{11.119}$$

such that

$$d(x_\tau, C_{min}) < 4\alpha(L+1)^2\mu^{-1}, \tag{11.120}$$

if an integer $t \in [0, \tau] \setminus \{\tau\}$, then (11.117) holds and

$$d(x_{t+1}, C_{min}) \le d(x_t, C_{min}).$$

We show that for all integers $t \geq \tau$,

$$d(x_t, C_{min}) \leq 4\alpha(L+1)^2\mu^{-1} + 3\alpha(L+1).$$

Assume the contrary. Then in view of (11.120) there exists an integer $s > \tau$ such that

$$d(x_s, C_{min}) > 4\alpha(L+1)^2\mu^{-1} + 3\alpha(L+1) \tag{11.121}$$

(11.121) does not hold for all $t = \tau, \dots, s-1$ and in particular,

$$d(x_{s-1}, C_{min}) \leq 4\alpha(L+1)^2\mu^{-1} + 3\alpha(L+1). \tag{11.122}$$

By (11.97), (11.99), and (11.122),

$$\|x_{s-1}\| \leq M. \tag{11.123}$$

There are two cases:

$$d(x_{s-1}, C_{min}) > 4\alpha(L+1)^2\mu^{-1}; \tag{11.124}$$

$$d(x_{s-1}, C_{min}) \leq 4\alpha(L+1)^2\mu^{-1}. \tag{11.125}$$

Assume that (11.124) holds. Together with property (a) and (11.123) this implies that (11.118) holds with $t = s-1$ and that

$$d(x_s, C_{min}) \leq d(x_{s-1}, C_{min}) \leq 4\alpha(L+1)^2\mu^{-1} + 3\alpha(L+1).$$

Assume that (11.125) holds. Property (a), (11.116) with $t = s-1$, (11.213) and (11.125) imply that

$$\begin{aligned} d(x_s, C_{min}) &\leq d(x_{s-1}, C_{min}) + 3\alpha(L+1) \\ &\leq 4\alpha(L+1)^2\mu^{-1} + 3\alpha(L+1). \end{aligned}$$

Thus in both cases

$$d(x_s, C_{min}) \leq 4\alpha(L+1)^2\mu^{-1} + 3\alpha(L+1).$$

This contradicts (11.121). The contradiction we have reached proves that

$$\begin{aligned} d(x_t, C_{min}) &\leq 4\alpha(L+1)^2\mu^{-1} + 3\alpha(L+1) \\ &= \alpha(L+1)((4L+4)\mu^{-1} + 3) \end{aligned}$$

for all integers $t \geq \tau$. This completes the proof of Theorem 11.7. □

Chapter 12
A Projected Subgradient Method for Nonsmooth Problems

In this chapter we study the convergence of the projected subgradient method for a class of constrained optimization problems in a Hilbert space. For this class of problems, an objective function is assumed to be convex but a set of admissible points is not necessarily convex. Our goal is to obtain an ϵ-approximate solution in the presence of computational errors, where ϵ is a given positive number.

12.1 Preliminaries and Main Results

Let $(X, \langle \cdot, \cdot \rangle)$ be a Hilbert space with an inner product $\langle \cdot, \cdot \rangle$ which induces a complete norm $\| \cdot \|$.

For each $x \in X$ and each nonempty set $A \subset X$ put

$$d(x, A) = \inf\{\|x - y\| : y \in A\}.$$

Assume that $f : X \to R^1$ is a convex continuous function which is Lipschitz on all bounded subsets of X.

For each point $x \in X$ and each positive number ϵ let

$$\partial f(x) = \{l \in X : f(y) - f(x) \geq \langle l, y - x \rangle \text{ for all } y \in X\}$$

be the subdifferential of f at x and let

$$\partial_\epsilon f(x) = \{l \in X : f(y) - f(x) \geq \langle l, y - x \rangle - \epsilon \text{ for all } y \in X\}$$

be the ϵ-subdifferential of f at x.

Let C be a closed nonempty subset of the space X.

Assume that

A. J. Zaslavski, *Convex Optimization with Computational Errors*, Springer Optimization and Its Applications 155, https://doi.org/10.1007/978-3-030-37822-6_12

$$\lim_{\|x\|\to\infty} f(x) = \infty. \tag{12.1}$$

It means that for each $M_0 > 0$ there exists $M_1 > 0$ such that if a point $x \in X$ satisfies the inequality $\|x\| \geq M_1$, then $f(x) > M_0$.

Define

$$\inf(f, C) = \inf\{f(z) : z \in C\}.$$

Since the function f is Lipschitz on all bounded subsets of the space X it follows from (12.1) that $\inf(f, C)$ is finite.

Set

$$C_{min} = \{x \in C : f(x) = \inf(f, C)\}.$$

It is well known that if the set C is convex, then the set C_{min} is nonempty. Clearly, the set $C_{min} \neq \emptyset$ if the space X is finite-dimensional.

In this chapter we assume that

$$C_{min} \neq \emptyset. \tag{12.2}$$

It is clear that C_{min} is a closed subset of X.

We suppose that the following assumption holds.

(A1) For every positive number ϵ there exists $\delta > 0$ such that if a point $x \in C$ satisfies the inequality $f(x) \leq \inf(f, C) + \delta$, then $d(x, C_{min}) \leq \epsilon$.

(It is clear that (A1) holds if the space X is finite-dimensional.)

We also suppose $P_C : X \to X$ satisfies for all $x \in C$ and all $y \in X$,

$$\|x - P_C(y)\| \leq \|x - y\| \tag{12.3}$$

and that

$$P_C(x) = x \text{ for all } x \in C. \tag{12.4}$$

(Note that in [85] we assumed that $P_C(X) = X$.)

In this chapter we also use the following assumption.

(A2) For each $M > 0$ and each $r > 0$ there exists $\delta > 0$ such that for each $x \in B_X(0, M)$ satisfying $d(x, C) \geq r$ and each $z \in B_X(0, M) \cap C$, we have

$$\|P_C(x) - z\| \leq \|x - z\| - \delta.$$

In Section 3.13, Chapter 3 of [93] mappings satisfying (A2), (12.3) and (12.4) are called (C)-quasi-contractive and it is shown that in appropriate spaces of mappings satisfying (12.3) and (12.4) most of the mappings are (C)-quasi-contractive.

In the chapter we also use the following assumption.

(A3) The mapping P_C is uniformly continuous on all bounded subsets of X and for every $x \in X$, $P_C^n(x)$ converges in the norm topology to a point of C uniformly on every bounded subset of X.

Clearly, if (A3) holds, then $\lim_{n\to\infty} P_C^n(x)$ is a fixed point of P_C for every $x \in X$. Set

$$P^0 x = x, \ x \in X.$$

For every number $\epsilon \in (0, \infty)$ set

$$\phi(\epsilon) = \sup\{\delta \in (0, 1] : \text{ if } x \in C \text{ satisfies } f(x) \le \inf(f, C) + \delta,$$

$$\text{then } d(x, C_{min}) \le \min\{1, \epsilon\}\}. \tag{12.5}$$

In view of (A1), $\phi(\epsilon)$ is well defined for every positive number ϵ.

In this chapter we will prove the following two results.

Theorem 12.1 *Suppose that at least one of the assumptions (A2) and (A3) holds. Let* $\{\alpha_i\}_{i=0}^{\infty} \subset (0, 1]$ *satisfy*

$$\lim_{i\to\infty} \alpha_i = 0, \ \sum_{i=1}^{\infty} \alpha_i = \infty \tag{12.6}$$

and let $M, \epsilon > 0$*. Then there exist a natural number* n_0 *and* $\delta > 0$ *such that the following assertion holds.*

Assume that an integer $n \ge n_0$,

$$\{x_k\}_{k=0}^{n} \subset X, \ \|x_0\| \le M,$$

$$v_k \in \partial_\delta f(x_k) \setminus \{0\}, \ k = 0, 1, \dots, n-1,$$

$$\{\eta_k\}_{k=0}^{n-1}, \ \{\xi_k\}_{k=0}^{n-1} \subset B_X(0, \delta),$$

and that for $k = 0, \dots, n-1$,

$$x_{k+1} = P_C(x_k - \alpha_k \|v_k\|^{-1} v_k - \alpha_k \xi_k) - \alpha_k \eta_k.$$

Then the inequality $d(x_k, C_{min}) \le \epsilon$ *hods for all integers* k *satisfying* $n_0 \le k \le n$.

Theorem 12.2 *Suppose that at least one of the assumptions (A2) and (A3) holds. Let* $M, \epsilon > 0$*. Then there exists* $\beta_0 \in (0, 1)$ *such that for each* $\beta_1 \in (0, \beta_0)$ *there exist a natural number* n_0 *and* $\delta > 0$ *such that the following assertion holds.*

Assume that an integer $n \ge n_0$,

$$\{x_k\}_{k=0}^{n} \subset X, \ \|x_0\| \le M,$$

$$v_k \in \partial_\delta f(x_k) \setminus \{0\},\ k = 0, 1, \ldots, n-1,$$

$$\{\alpha_k\}_{k=0}^{n-1} \subset [\beta_1, \beta_0],$$

$$\{\eta_k\}_{k=0}^{n-1},\ \{\xi_k\}_{k=0}^{n-1} \subset B_X(0, \delta)$$

and that for $k = 0, \ldots, n-1$,

$$x_{k+1} = P_C(x_k - \alpha_k \|v_k\|^{-1} v_k - \alpha_k \xi_k) - \eta_k.$$

Then the inequality $d(x_k, C_{min}) \le \epsilon$ *holds for all integers* k *satisfying* $n_0 \le k \le n$.

In this chapter we use the following definitions and notation.

Define

$$X_0 = \{x \in X :\ f(x) \le \inf(f, C) + 4\}. \tag{12.7}$$

In view of (12.1), there exists a number $\bar{K} > 0$ such that

$$X_0 \subset B_X(0, \bar{K}). \tag{12.8}$$

Since the function f is Lipschtiz on all bounded subsets of the space X there exists a number $\bar{L} > 1$ such that

$$|f(z_1) - f(z_2)| \le \bar{L}\|z_1 - z_2\|$$

$$\text{for all } z_1, z_2 \in B_X(0, \bar{K} + 4). \tag{12.9}$$

12.2 Auxiliary Results

Proposition 12.3 *Let* $\epsilon \in (0, 1]$. *Then for each* $x \in X$ *satisfying*

$$d(x, C) < \min\{\bar{L}^{-1} 2^{-1} \phi(\epsilon/2),\ \epsilon/2\}, \tag{12.10}$$

$$f(x) \le \inf(f, C) + \min\{2^{-1}\phi(\epsilon/2),\ \epsilon/2\}, \tag{12.11}$$

the inequality $d(x, C_{min}) \le \epsilon$ *holds.*

Proof In view of the definition of ϕ, $\phi(\epsilon/2) \in (0, 1]$ and the following property holds:

(i)

$$\text{if } x \in C \text{ satisfies } f(x) < \inf(f, C) + \phi(\epsilon/2),$$

$$\text{then } d(x, C_{min}) \le \min\{1, \epsilon/2\}. \tag{12.12}$$

Assume that a point $x \in X$ satisfies (12.10) and (12.11). In view of (12.10), there exists a point $y \in C$ which satisfies

$$\|x - y\| < 2^{-1}\bar{L}^{-1}\phi(\epsilon/2) \text{ and } \|x - y\| < \epsilon/2. \tag{12.13}$$

Relations (12.5), (12.7), (12.8), (12.11), and (12.13) imply that

$$x \in B_X(0, \bar{K}), \ \ y \in B_X(0, \bar{K} + 1). \tag{12.14}$$

By (12.13), (12.14), and the definition of $\bar{L}$ (see (12.9)),

$$|f(x) - f(y)| \le \bar{L}\|x - y\| < \phi(\epsilon/2)2^{-1}. \tag{12.15}$$

It follows from the choice of the point y, (12.11), and (12.15) that

$$y \in C$$

and

$$f(y) < f(x) + \phi(\epsilon/2)2^{-1} \le \inf(f, C) + \phi(\epsilon/2).$$

Combined with property (i) this implies that

$$d(y, C_{min}) \le \epsilon/2.$$

Together with (12.13) this implies that

$$d(x, C_{min}) \le \|x - y\| + d(y, C_{min}) \le \epsilon.$$

This completes the proof of Proposition 12.3. □

Lemma 12.4 *Assume that $\epsilon > 0$, $x \in X$, $y \in X$,*

$$f(x) > \inf(f, C) + \epsilon, \ \ f(y) \le \inf(f, C) + \epsilon/4, \tag{12.16}$$

$$v \in \partial_{\epsilon/4} f(x). \tag{12.17}$$

Then $\langle v, y - x\rangle \le -\epsilon/2$.

Proof In view of (12.17),

$$f(u) - f(x) \ge \langle v, u - x\rangle - \epsilon/4 \text{ for all } u \in X. \tag{12.18}$$

By (12.16) and (12.18),

$$-(3/4)\epsilon \geq f(y) - f(x) \geq \langle v, y - x \rangle - \epsilon/4.$$

This completes the proof of Lemma 12.4. □

Lemma 12.5 *Let*

$$\bar{x} \in C_{min}, \tag{12.19}$$

$K_0 > 0$, $\epsilon \in (0, 16\bar{L}]$, $\alpha \in (0, 1]$, *let a number* $\delta \in (0, 1]$ *satisfy*

$$\delta(K_0 + \bar{K} + 1) \leq (8\bar{L})^{-1}\epsilon, \tag{12.20}$$

let a point $x \in X$ *satisfy*

$$\|x\| \leq K_0, \tag{12.21}$$

$$f(x) > \inf(f, C) + \epsilon, \tag{12.22}$$

$$\eta, \ \xi \in B_X(0, \delta), \tag{12.23}$$

$$v \in \partial_{\epsilon/4} f(x) \setminus \{0\}, \tag{12.24}$$

and let

$$y = P_C(x - \alpha\|v\|^{-1}v - \alpha\xi) - \eta. \tag{12.25}$$

Then

$$\|y - \bar{x}\|^2 \leq \|x - \bar{x}\|^2 - \alpha(4\bar{L})^{-1}\epsilon + 2\alpha^2 + \|\eta\|^2 + 2\|\eta\|(K_0 + \bar{K} + 2).$$

Proof In view of (12.7)–(12.9) and (12.19), for every point

$$z \in B_X(\bar{x}, 4^{-1}\epsilon\bar{L}^{-1}),$$

we have

$$\begin{aligned} f(z) &\leq f(\bar{x}) + \bar{L}\|z - \bar{x}\| \\ &\leq f(\bar{x}) + \epsilon/4 = \inf(f, C) + \epsilon/4. \end{aligned} \tag{12.26}$$

Lemma 12.4, (12.22), (12.24), and (12.26) imply that for every point

$$z \in B_X(\bar{x}, 4^{-1}\epsilon\bar{L}^{-1}),$$

we have

$$\langle v, z - x\rangle \le -\epsilon/2.$$

Combined with (8.22) the inequality above implies that

$$\langle \|v\|^{-1}v, z - x\rangle < 0 \text{ for all } z \in B(\bar{x}, (4\bar{L})^{-1}\epsilon). \tag{12.27}$$

Set

$$\tilde{z} = \bar{x} + 4^{-1}\bar{L}^{-1}\epsilon\|v\|^{-1}v. \tag{12.28}$$

It is easy to see that

$$\tilde{z} \in B(\bar{x}, 4^{-1}\bar{L}^{-1}\epsilon). \tag{12.29}$$

Relations (12.27)–(12.29) imply that

$$0 > \langle \|v\|^{-1}v, \tilde{z} - x\rangle = \langle \|v\|^{-1}v, \bar{x} + 4^{-1}\bar{L}^{-1}\epsilon\|v\|^{-1}v - x\rangle. \tag{12.30}$$

By (12.30),

$$\langle \|v\|^{-1}v, \bar{x} - x\rangle < -4^{-1}\bar{L}^{-1}\epsilon. \tag{12.31}$$

Set

$$y_0 = x - \alpha\|v\|^{-1}v - \alpha\xi. \tag{12.32}$$

It follows from (12.7), (12.8), (12.19)–(12.21), (12.23), (12.31), and (12.32) that

$$\|y_0 - \bar{x}\|^2 = \|x - \alpha\|v\|^{-1}v - \alpha\xi - \bar{x}\|^2$$

$$= \|x - \alpha\|v\|^{-1}v - \bar{x}\|^2 + \alpha^2\|\xi\|^2$$

$$-2\alpha\langle \xi, x - \alpha\|v\|^{-1}v - \bar{x}\rangle$$

$$\le \|x - \alpha\|v\|^{-1}v - \bar{x}\|^2$$

$$+\alpha^2\delta^2 + 2\alpha\delta(K_0 + \bar{K} + 1)$$

$$\le \|x - \bar{x}\|^2 - 2\langle x - \bar{x}, \alpha\|v\|^{-1}v\rangle$$

$$+\alpha^2 + \alpha^2\delta^2 + 2\alpha\delta(K_0 + \bar{K} + 1)$$

$$< \|x - \bar{x}\|^2 - 2\alpha(4^{-1}\bar{L}^{-1}\epsilon)$$

$$+\alpha^2(1+\delta^2)+2\alpha\delta(K_0+\bar{K}+1)$$

$$\leq \|x-\bar{x}\|^2-\alpha(4\bar{L})^{-1}\epsilon+2\alpha^2. \tag{12.33}$$

In view of (12.7), (12.8), (12.19), (12.21), and (12.33),

$$\|y_0-\bar{x}\|^2 \leq (K_0+\bar{K})^2+2$$

and

$$\|y_0-\bar{x}\| \leq K_0+\bar{K}+2. \tag{12.34}$$

By (12.3), (12.19), (12.25), (A2), (12.32), (12.33), and (12.34),

$$\|y-\bar{x}\|^2 = \|P_C(y_0)-\eta-\bar{x}\|^2$$

$$\leq \|P_C(y_0)-\bar{x}\|^2+\|\eta\|^2+2\|\eta\|\|P_C(y_0)-\bar{x}\|$$

$$\leq \|y_0-\bar{x}\|^2+\|\eta\|^2+2\|\eta\|\|y_0-\bar{x}\|$$

$$\leq \|x-\bar{x}\|^2-\alpha(4\bar{L})^{-1}\epsilon$$

$$+2\alpha^2+\|\eta\|^2+2\|\eta\|(K_0+\bar{K}+2).$$

This completes the proof of Lemma 12.5. □

Lemma 12.5 implies the following result.

Lemma 12.6 *Let* $K_0>0$, $\epsilon\in(0,16\bar{L}]$, $\alpha\in(0,1]$, *a number* $\delta\in(0,1]$ *satisfy*

$$\delta(K_0+\bar{K}+1)\leq(8\bar{L})^{-1}\epsilon,$$

let $x\in X$ *satisfy*

$$\|x\|\leq K_0,\ \ f(x)>\inf(f,C)+\epsilon,$$

let

$$\eta,\ \xi\in B_X(0,\delta),\ \ v\in\partial_{\epsilon/4}f(x)\setminus\{0\},$$

and let

$$y=P_C(x-\alpha\|v\|^{-1}v-\alpha\xi)-\eta.$$

Then

$$d(y, C_{min})^2 \le d(x, C_{min})^2 - \alpha(4\bar{L})^{-1}\epsilon$$

$$+2\alpha^2 + \|\eta\|^2 + 2\|\eta\|(K_0 + \bar{K} + 2).$$

Lemma 12.6 applied with $\epsilon = 16\bar{L}$ implies the next result.

Lemma 12.7 *Let $K_0 > 0$, $\alpha \in (0, 1]$, a number $\delta \in (0, 1]$ satisfy*

$$\delta(K_0 + \bar{K} + 1) \le 2,$$

let $x \in X$ satisfy

$$\|x\| \le K_0, \; f(x) > \inf(f, C) + 16\bar{L},$$

let

$$\eta, \; \xi \in B_X(0, \delta), \; v \in \partial_4 f(x) \setminus \{0\},$$

and let

$$y = P_C(x - \alpha\|v\|^{-1}v - \alpha\xi) - \eta.$$

Then

$$d(y, C_{min})^2 \le d(x, C_{min})^2 - 2\alpha + 2\|\eta\|(K_0 + \bar{K} + 3).$$

12.3 An Auxiliary Result with Assumption A2

Lemma 12.8 *Assume that (A2) holds, $K_0, \epsilon > 0$ and*

$$\bar{x} \in C_{min}. \tag{12.35}$$

Then there exist a natural number m_0 and $\delta_0 \in (0, 1)$ such that for each integer $n \ge m_0$ and each finite sequence $\{x_i\}_{i=0}^n \subset X$ satisfying

$$\|x_i\| \le K_0, \; i = 0, \dots, n \tag{12.36}$$

and

$$B_X(x_{i+1}, \delta_0) \cap P_C(B_X(x_i, \delta_0)) \ne \emptyset, \; i = 0, \dots, n-1 \tag{12.37}$$

the inequality

$$d(x_i, C) \le \epsilon$$

holds for all integers $i \in [m_0, n]$.

Proof Set

$$\bar{K}_1 = 2K_0 + \bar{K} + 5. \tag{12.38}$$

Assumption (A2) implies that there exists $\gamma_0 \in (0, 1)$ such that the following property holds:

(ii) for each $x \in X$ satisfying

$$\|x\| \le K_0 + 1$$

and

$$d(x, C) \ge \epsilon/8$$

and each $z \in B_X(0, \bar{K}_1) \cap C$, we have

$$\|P_C(x) - z\| \le \|x - z\| - \gamma_0.$$

Choose a natural number

$$m_0 > 2(K_0 + \bar{K} + 1)\gamma_0^{-1} + 2 \tag{12.39}$$

and a positive number

$$\delta_0 < \min\{\epsilon/8, 1, \gamma_0/4\}. \tag{12.40}$$

Assume that an integer $n \ge m_0$ and that a finite sequence $\{x_i\}_{i=0}^n \subset X$ satisfies (12.36) and (12.37) for all integers $i = 0, \dots, n-1$. We show that there exists $j \in \{0, \dots, m_0\}$ such that

$$d(x_j, C) \le \epsilon/4.$$

We may assume without loss of generality that

$$d(x_0, C) > \epsilon/4. \tag{12.41}$$

By (12.37), there exists

$$y_0 \in B_X(x_0, \delta_0) \tag{12.42}$$

such that

$$\|x_1 - P_C(y_0)\| \le \delta_0. \tag{12.43}$$

By (12.40)–(12.42),

$$d(y_0, C) \ge d(x_0, C) - \delta_0 > \epsilon/4 - \delta_0 > \epsilon/8. \tag{12.44}$$

In view of (12.36), (12.40), and (12.42),

$$\|y_0\| \le \|x_0\| + \delta_0 \le K_0 + 1. \tag{12.45}$$

Property (ii), (12.7), (12.8), (12.35), (12.38), (12.42), (12.44), and (12.45) imply that

$$\|P_C(y_0) - \bar{x}\| \le \|y_0 - \bar{x}\| - \gamma_0$$

$$\le \|x_0 - \bar{x}\| + \|x_0 - y_0\| - \gamma_0$$

$$\le \|x_0 - \bar{x}\| + \delta_0 - \gamma_0. \tag{12.46}$$

By (12.40), (12.43), and (12.46),

$$\|x_1 - \bar{x}\| \le \|x_1 - P_C(y_0)\| + \|P_C(y_0) - \bar{x}\|$$

$$\le \delta_0 + \|x_0 - \bar{x}\| - \gamma_0 + \delta_0$$

$$= \|x_0 - \bar{x}\| - \gamma_0 + 2\delta_0 \le \|x_0 - \bar{x}\| - \gamma_0/2$$

and

$$\|x_1 - \bar{x}\| \le \|x_0 - \bar{x}\| - \gamma_0/2. \tag{12.47}$$

We may assume without loss of generality that

$$d(x_1, C) > \epsilon/4. \tag{12.48}$$

Assume that p is a natural number such that for all $i = 0, \ldots, p-1$,

$$\|x_{i+1} - \bar{x}\| \le \|x_i - \bar{x}\| - \gamma_0/2. \tag{12.49}$$

(In view of (12.47), this assumption holds for $p = 1$.) It follows from (12.7), (12.8), (12.35), (12.36), and (12.49) that

$$K_0 + \bar{K} \ge \|x_0 - \bar{x}\|$$

$$\ge \|x_0 - \bar{x}\| - \|x_p - \bar{x}\|$$

$$= \sum_{i=0}^{p-1} (\|x_i - \bar{x}\| - \|x_{i+1} - \bar{x}\|) \geq (\gamma_0/2)p$$

and

$$p \leq 2(K_0 + \bar{K})\gamma_0^{-1}.$$

Together with (12.39) this implies that

$$p < m_0. \tag{12.50}$$

In view of (12.50), we may assume without loss of generality that p is the largest natural number such that (12.49) holds for $i = 0, \ldots, p - 1$. Thus

$$\|x_{p+1} - \bar{x}\| > \|x_p - \bar{x}\| - \gamma_0/2. \tag{12.51}$$

Assume that

$$d(x_p, C) > \epsilon/4. \tag{12.52}$$

By (12.37), there exists

$$y_p \in B_X(x_p, \delta_0) \tag{12.53}$$

such that

$$\|x_{p+1} - P_C(y_p)\| \leq \delta_0. \tag{12.54}$$

Relations (12.36), (12.40), and (12.53) imply that

$$\|y_p\| \leq K_0 + 1. \tag{12.55}$$

By (12.40), (12.52), and (12.53),

$$d(y_p, C) \geq d(x_p, C) - \|x_p - y_p\| > \epsilon/4 - \delta_0 \geq \epsilon/8. \tag{12.56}$$

It follows from (12.7), (12.8), (12.35), (12.53), (12.55), and (12.56) that

$$\|P_C(y_p) - \bar{x}\| \leq \|y_p - \bar{x}\| - \gamma_0$$

$$\leq \|x_p - \bar{x}\| + \|y_p - x_p\| - \gamma_0$$

$$\leq \|x_p - \bar{x}\| - \gamma_0 + \delta_0. \tag{12.57}$$

By (12.40), (12.54), and (12.57),

$$\|x_{p+1} - \bar{x}\| \leq \|x_{p+1} - P_C(y_p)\| + \|P_C(y_p) - \bar{x}\|$$

$$\leq \|x_p - \bar{x}\| - \gamma_0 + 2\delta_0$$

$$\leq \|x_p - \bar{x}\| - \gamma_0/2.$$

This contradicts (12.51). The contradiction we have reached proves that

$$d(x_p, C) \leq \epsilon/4.$$

Thus we proved the existence of an integer $p \geq 0$ such that

$$p < m_0, \tag{12.58}$$

$$d(x_p, C) \leq \epsilon/4. \tag{12.59}$$

We show that

$$d(x_j, C) \leq \epsilon \text{ for all integers } j \in [p, n].$$

Assume that $j \geq p$ is an integer, $j < n$ and that

$$d(x_j, C) \leq \epsilon. \tag{12.60}$$

There are two cases:

$$d(x_j, C) \leq \epsilon/4; \tag{12.61}$$

$$d(x_j, C) > \epsilon/4. \tag{12.62}$$

By (12.37), there exists

$$y \in B_X(x_j, \delta_0) \tag{12.63}$$

such that

$$\|x_{j+1} - P_C(y)\| \leq \delta_0. \tag{12.64}$$

Assume that (12.61) holds. It follows from (12.3), (12.40), (12.61), (12.63), and (12.64) that

$$d(x_{j+1}, C) \leq \|x_{j+1} - P_C(y)\| + d(P_C(y), C)$$

$$\leq \delta_0 + d(y, C)$$

$$\leq \delta_0 + d(x_j, C) + \|y - x_j\|$$

$$\leq 2\delta_0 + d(x_j, C) \leq \epsilon/4 + 2\delta_0 \leq \epsilon. \tag{12.65}$$

Assume that (12.62) holds. By (12.7), (12.8), (12.35), (12.36), (12.40), and (12.63),

$$d(y, C) \leq \|y - \bar{x}\| \leq K_0 + \bar{K} + 1. \tag{12.66}$$

In view of (12.40) and (12.66), there exists $z \in C$ such that

$$\|z - y\| \leq d(y, C) + \delta_0 \leq K_0 + \bar{K} + 2. \tag{12.67}$$

By (12.36), (12.40), (12.63), and (12.67),

$$\|z\| \leq K_0 + \bar{K} + 2 + \|y\| \leq K_0 + \bar{K} + 2$$

$$+ \|x_j\| + \|y - x_j\| \leq 2K_0 + \bar{K} + 3. \tag{12.68}$$

It follows from (12.36), (12.40), and (12.63) that

$$\|y\| \leq \|x_j\| + 1 \leq K_0 + 1. \tag{12.69}$$

By (12.40), (12.62), and (12.63),

$$d(y, C) \geq d(x_j, C) - \|x_j - y\| \geq \epsilon/4 - \delta_0 > \epsilon/8. \tag{12.70}$$

Property (ii), the inclusion $z \in C$, (12.38), and (12.68)–(12.70) imply that

$$\|P_C(y) - z\| \leq \|y - z\| - \gamma_0. \tag{12.71}$$

The inclusion $z \in C$, (12.40), (12.63)–(12.65), (12.67), and (12.71) imply that

$$d(x_{j+1}, C) \leq \|x_{j+1} - z\|$$

$$\leq \|P_C(y) - z\| + \|x_{j+1} - P_C(y)\|$$

$$\leq \|y - z\| - \gamma_0 + \delta_0$$

$$\leq d(y, C) + \delta_0 - \gamma_0 + \delta_0$$

$$\leq d(x_j, C) + \|y - x_j\| + \delta_0 - \gamma_0 + \delta_0$$

$$\leq d(x_j, C) + 2\delta_0 - \gamma_0 + \delta_0 \leq d(x_j, C) - \gamma_0/4 \leq \epsilon.$$

Thus in both cases

$$d(x_{j+1}, C) \le \epsilon.$$

Therefore $d(x_i, C) \le \epsilon$ for all $i = j, \ldots, n$. Lemma 12.8 is proved. □

12.4 An Auxiliary Result with Assumption A3

Lemma 12.9 *Assume that (A3) holds,* $K_0 > 0$, $\epsilon \in (0, 1)$, *and*

$$\bar{x} \in C_{min}. \tag{12.72}$$

Then there exist a natural number m_0 *and* $\delta_0 \in (0, 1)$ *such that for each integer* $n \ge m_0$ *and each finite sequence* $\{x_i\}_{i=0}^n \subset X$ *satisfying*

$$\|x_i\| \le K_0,\ i = 0, \ldots, n \tag{12.73}$$

and

$$B_X(x_{i+1}, \delta_0) \cap P_C(B_X(x_i, \delta_0)) \ne \emptyset,\ i = 0, \ldots, n-1 \tag{12.74}$$

the inequality

$$d(x_i, C) \le \epsilon$$

holds for all integers $i \in [m_0, n]$.

Proof Assumption (A3) implies that for every $x \in X$ there exists

$$\lim_{i\to\infty} P_C^i(x) \in C. \tag{12.75}$$

Set

$$Q(x) = \lim_{i\to\infty} P_C^i(x),\ x \in X. \tag{12.76}$$

(A3) implies that there exists a natural number m_0 such that

$$\|P^{m_0}(y) - Q(y)\| \le \epsilon/4 \text{ for all } x \in B_X(0, K_0 + \bar{K} + 4). \tag{12.77}$$

Set

$$\delta_{m_0} = \epsilon/4. \tag{12.78}$$

In view of (A3), there exists

$$\delta_{m_0-1} \in (0, \delta_{m_0}/4)$$

such that

$$\|P_C(z_1) - P_C(z_2)\| \leq \delta_{m_0}/4$$

for all $z_1, z_2 \in B_X(0, K_0 + 1)$ satisfying

$$\|z_1 - z_2\| \leq 4\delta_{m_0-1}.$$

By induction we construct a finite sequence $\{\delta_i\}_{i=0}^{m_0} \subset (0, \infty)$ such that for each integer $i \in [0, m_0 - 1]$ we have

$$\delta_i < \delta_{i+1}/4 \tag{12.79}$$

and for each $z_1, z_2 \in B_X(0, K_0 + 1)$ satisfying

$$\|z_1 - z_2\| \leq 4\delta_i \tag{12.80}$$

we have

$$\|P_C(z_1) - P_C(z_2)\| \leq \delta_{i+1}/4. \tag{12.81}$$

Assume that $n \geq m_0$ is an integer,

$$\{x_i\}_{i=0}^{n} \subset B_X(0, K_0) \tag{12.82}$$

and that (12.74) holds for each integer $i \in \{0, \ldots, n - 1\}$.

Let $k \in [m_0, n]$ be an integer. In order to complete the proof it is sufficient to show that

$$d(x_k, C) \leq \epsilon.$$

By (12.75)–(12.77) and (12.82),

$$\|P^{m_0}(x_{k-m_0}) - Q(x_{k-m_0})\| \leq \epsilon/4$$

and

$$d(P^{m_0}(x_{k-m_0}), C) \leq \epsilon/4. \tag{12.83}$$

In view of (12.74), there exists

$$y_{k-m_0} \in B_X(x_{k-m_0}, \delta_0) \tag{12.84}$$

such that

$$\|P_C(y_{k-m_0}) - x_{k-m_0+1}\| \le \delta_0. \tag{12.85}$$

By (12.80)–(12.82) and (12.84),

$$\|P_C(y_{k-m_0}) - P_C(x_{k-m_0})\| \le \delta_1/4. \tag{12.86}$$

It follows from (12.79), (12.85), and (12.86) that

$$\|x_{k-m_0+1} - P_C(x_{k-m_0})\| \le \delta_1/2. \tag{12.87}$$

We show that for all $i = 1, \dots, m_0$,

$$\|x_{i+k-m_0} - P_C^i(x_{k-m_0})\| \le \delta_i/2. \tag{12.88}$$

(Note that in view of (12.87), inequality (12.88) holds for $i = 1$.) Assume that $i \in \{1, \dots, m_0\} \setminus \{m_0\}$ and that (12.88) holds. By (12.74), there exists

$$y_{i+k-m_0} \in B_X(x_{i+k-m_0}, \delta_0) \tag{12.89}$$

such that

$$\|P_C(y_{i+k-m_0}) - x_{i+k-m_0+1}\| \le \delta_0. \tag{12.90}$$

In view of (12.88) and (12.89),

$$\|y_{i+k-m_0} - P_C^i(x_{k-m_0})\| \le \delta_i. \tag{12.91}$$

It follows from (12.80)–(12.82), (12.88), (12.89), and (12.91) that

$$\|P_C(y_{i+k-m_0}) - P_C^{i+1}(x_{k-m_0})\| \le \delta_{i+1}/4. \tag{12.92}$$

It follows from (12.79), (12.90), and (12.92) that

$$\|x_{i+k-m_0+1} - P_C^{i+1}(x_{k-m_0})\| \le \delta_{i+1}/2. \tag{12.93}$$

Thus our assumption holds for $i + 1$ too. Therefore we showed by induction that

$$\|x_{i+k-m_0} - P_C^i(x_{k-m_0})\| \le \delta_i/2$$

for all $i = 1, \dots, m_0$ and in particular (see (12.78))

$$\|x_k - P_C^{m_0}(x_{k-m_0})\| \le \delta_{m_0} = \epsilon/4.$$

Together with (12.83) this implies that

$$d(x_k, C) \le \epsilon/2.$$

Lemma 12.9 is proved. □

12.5 Proof of Theorem 12.1

We may assume without loss of generality that $\epsilon < 1$. In view of Proposition 12.3, there exists a number

$$\bar{\epsilon} \in (0, \min\{\epsilon/8, 1/8\}) \tag{12.94}$$

such that the following property holds:
(iii)

$$\text{if } x \in X,\ d(x, C) \le 2\bar{\epsilon} \text{ and } f(x) \le \inf(f, C) + 2\bar{\epsilon},$$

$$\text{then } d(x, C_{min}) \le \epsilon.$$

In view of (12.1), we may assume without loss of generality that

$$M > 4\bar{K} + 8, \tag{12.95}$$

$$\{x \in X :\ f(x) \le \inf(f, C) + 16\bar{L}\} \subset B_X(0, 2^{-1}M - 1). \tag{12.96}$$

Fix

$$\bar{\epsilon}_1 \in (0, \bar{\epsilon}(64\bar{L})^{-1}). \tag{12.97}$$

Lemmas 12.8 and 12.9 imply that there exist $\delta_1 \in (0, 1)$ and a natural number m_0 such that the following property holds:
(iv) for each integer $n \ge m_0$ and each finite sequence $\{y_i\}_{i=0}^{n} \subset B_X(0, 3M)$ satisfying

$$B_X(y_{i+1}, \delta_1) \cap P_C(B_X(y_i, \delta_1)) \ne \emptyset,\ i = 0, \dots, n-1$$

the inequality

$$d(y_i, C) \le \bar{\epsilon}_1$$

holds for all integers $i \in [m_0, n]$.

Since $\lim_{i\to\infty} \alpha_i = 0$ (see (12.6)) there is an integer $p_0 > 0$ such that for all integers $i \geq p_0$, we have

$$\alpha_i \leq \delta_1/2. \tag{12.98}$$

Fix

$$\bar{x} \in C_{min}. \tag{12.99}$$

Since $\lim_{i\to\infty} \alpha_i = 0$ (see (12.6)) there is an integer

$$p_1 > p_0 \tag{12.100}$$

such that for all integers $i \geq p_1$, we have

$$\alpha_i < (32\bar{L})^{-1}16^{-1}\bar{\epsilon}_1. \tag{12.101}$$

Since $\sum_{i=0}^{\infty} \alpha_i = \infty$ (see (12.6)) there exists a natural number

$$n_0 > p_0 + p_1 + 4 + m_0 \tag{12.102}$$

such that

$$\sum_{i=p_0+p_1+m_0}^{n_0-1} \alpha_i > (3M + \bar{K})^2 128\bar{\epsilon}^{-1}\bar{L} + 1. \tag{12.103}$$

Fix a positive number δ such that

$$\delta(3M + \bar{K} + 3) < 8^{-1}(64\bar{L})^{-1}\bar{\epsilon}_1. \tag{12.104}$$

Assume that an integer $n \geq n_0$ and that

$$\{x_k\}_{k=0}^{n} \subset X, \ \|x_0\| \leq M, \tag{12.105}$$

$$\{\eta_k\}_{k=0}^{n-1}, \ \{\xi_k\}_{k=0}^{n-1} \subset B_X(0, \delta), \tag{12.106}$$

$$v_k \in \partial_\delta f(x_k) \setminus \{0\}, \ k = 0, 1, \ldots, n-1 \tag{12.107}$$

and that for all integers $k = 0, \ldots, n-1$, we have

$$x_{k+1} = P_C(x_k - \alpha_k \|v_k\|^{-1} v_k - \alpha_k \xi_k) - \alpha_k \eta_k. \tag{12.108}$$

In order to prove the theorem it is sufficient to show that

$$d(x_k, C_{min}) \le \epsilon \text{ for all integers } k \text{ satisfying } n_0 \le k \le n.$$

First we show that for all integers $i = 0, \dots, n$,

$$d(x_i, C_{min}) \le 2M. \tag{12.109}$$

In view of (12.105), inequality (12.109) holds for $i = 0$. Assume that

$$i \in \{0, \dots, n\} \setminus \{n\}$$

and that (12.109) is true. There are two cases:

$$f(x_i) \le \inf(f, C) + 16\bar{L}; \tag{12.110}$$

$$f(x_i) > \inf(f, C) + 16\bar{L}. \tag{12.111}$$

Assume that (12.110) holds. In view of (12.96) and (12.110),

$$\|x_i\| \le M/2 - 1. \tag{12.112}$$

By (12.7), (12.8), and (12.99),

$$\|\bar{x}\| \le \bar{K}. \tag{12.113}$$

It follows from (12.112) and (12.113) that

$$\|x_i - \bar{x}\| \le \bar{K} + M/2. \tag{12.114}$$

By (12.3), (12.95), (12.99), (12.104), (12.106), (12.108), and (12.114),

$$\begin{aligned}
&\|x_{i+1} - \bar{x}\| \\
&\le \alpha_i \|\eta_i\| + \|\bar{x} - P_C(x_i - \alpha_i \|v_i\|^{-1} v_i - \alpha_i \xi_i)\| \\
&\le \alpha_i \delta + \|\bar{x} - (x_i - \alpha_i \|v_i\|^{-1} v_i - \alpha_i \xi_i)\| \\
&\le \alpha_i \delta + \|\bar{x} - x_i\| + \alpha_i + \alpha_i \delta \\
&\le \|\bar{x} - x_i\| + 3 \\
&\le \bar{K} + M/2 + 3 < 2M
\end{aligned}$$

and

$$d(x_{i+1}, C_{min}) \le 2M. \tag{12.115}$$

Assume that (12.111) holds. It follows from (12.104), (12.106), (12.107), (12.113), Lemma 12.7 applied with

$$K_0 = 2M,\ \alpha = \alpha_i,\ x = x_i,\ \xi = \xi_i,\ v = v_i,\ y = x_{i+1},\ \eta = \alpha_i \eta_i,$$

(12.95), and (12.109) that

$$d(x_{i+1}, C_{min})^2 \le d(x_i, C_{min})^2 - 2\alpha_i + \alpha_i \delta(4M)$$

$$\le d(x_i, C_{min})^2 \le 4M^2.$$

Thus in both cases

$$d(x_{i+1}, C_{min}) \le 2M.$$

Thus we have shown by induction that for all integers $i = 0, \ldots, n$,

$$d(x_i, C_{min}) \le 2M.$$

Together with (12.7), (12.8), and (12.95) this implies that

$$\|x_i\| \le 3M,\ i = 0, \ldots, n. \tag{12.116}$$

Set

$$y_i = x_{i+p_0},\ i = 0, \ldots, n - p_0. \tag{12.117}$$

By (12.116)–(12.118),

$$\{y_i\}_{i=0}^{n-p_0} \subset B_X(0, 3M). \tag{12.118}$$

In view of (12.102),

$$n - p_0 > m_0. \tag{12.119}$$

It follows from (12.98), (12.104), (12.106), (12.108), and (12.117) that for all $i = 0, \ldots, n - p_0 - 1$,

$$\|y_i - (y_i - \alpha_{i+p_0} \|v_{i+p_0}\|^{-1} v_{i+p_0} - \alpha_{i+p_0} \xi_{i+p_0})\|$$

$$= \|x_{i+p_0} - (x_{i+p_0} - \alpha_{i+p_0} \|v_{i+p_0}\|^{-1} v_{i+p_0} - \alpha_{i+p_0} \xi_{i+p_0})\|$$

$$\le 2\alpha_{i+p_0} \le \delta_1,$$

$$\|y_{i+1} - P_C(y_i - \alpha_{i+p_0}\|v_{i+p_0}\|^{-1}v_{i+p_0} - \alpha_{i+p_0}\xi_{i+p_0})\|$$

$$\leq \alpha_{i+p_0} \leq \delta_1$$

and

$$B_X(y_{i+1}, \delta_1) \cap P_C(B_X(y_i, \delta_1)) \neq \emptyset. \tag{12.120}$$

It follows from (12.118) to (12.120) that

$$d(y_i, C) \leq \bar{\epsilon}_1, \ i = m_0, \ldots, n - p_0.$$

Together with (12.96) and (12.117) this implies that

$$d(x_i, C) \leq \bar{\epsilon}_1 < \bar{\epsilon}, \ i = m_0 + p_0, \ldots, n. \tag{12.121}$$

Assume that an integer

$$k \in [p_0 + p_1 + m_0, n - 1], \tag{12.122}$$

$$f(x_k) > \inf(f, C) + \bar{\epsilon}/8. \tag{12.123}$$

It follows from (12.94), (12.97), (12.99), (12.106), (12.114), (12.116), Lemma 12.5 applied with

$$K_0 = 3M, \ \epsilon = \bar{\epsilon}/8, \ \alpha = \alpha_k, \ x = x_k, \ \xi = \xi_k, \ v = v_k,$$

$$y = x_{k+1}, \ \eta = \alpha_k \eta_k,$$

(12.96), (12.101), (12.104), and (12.122) that

$$\|x_{k+1} - \bar{x}\|^2 \leq \|x_k - \bar{x}\|^2 - \alpha_k(4\bar{L})^{-1}\bar{\epsilon}/8$$

$$+2\alpha_k^2 + \alpha_k^2\|\eta_k\|^2 + 2\|\eta_k\|\alpha_k(3M + \bar{K} + 2)$$

$$\leq \|x_k - \bar{x}\|^2 - \alpha_k(32\bar{L})^{-1}\bar{\epsilon}$$

$$+2\alpha_k^2 + \alpha_k^2\delta^2 + 2\delta\alpha_k(3M + \bar{K} + 2)$$

$$\leq \|x_k - \bar{x}\|^2 - \alpha_k(64\bar{L})^{-1}\bar{\epsilon} + 2\delta\alpha_k(3M + \bar{K} + 3)$$

$$\leq \|x_k - \bar{x}\|^2 - \alpha_k(128\bar{L})^{-1}\bar{\epsilon}.$$

Thus we have shown that the following property holds:

(v) if an integer k satisfies (12.122) and (12.123), then we have

$$\|x_{k+1} - \bar{x}\|^2 \le \|x_k - \bar{x}\|^2 - (128\bar{L})^{-1}\alpha_k \bar{\epsilon}.$$

We claim that there exists an integer $j \in \{p_0 + p_1 + m_0, \dots, n_0\}$ such that

$$f(x_j) \le \inf(f, C) + \bar{\epsilon}/8.$$

Assume the contrary. Then

$$f(x_j) > \inf(f, C) + \bar{\epsilon}/8, \ i = p_0 + p_1 + m_0, \dots, n_0. \tag{12.124}$$

Property (v) and (12.124) imply that

$$\|x_{i+1} - \bar{x}\|^2 \le \|x_i - \bar{x}\|^2 - (128\bar{L})^{-1}\alpha_i \bar{\epsilon},$$

$$i = p_0 + p_1 + m_0, \dots, n_0 - 1. \tag{12.125}$$

Relations (12.7), (12.8), (12.99), (12.116), and (12.125) imply that

$$(3M + \bar{K})^2 \ge \|x_{p_0+p_1+m_0} - \bar{x}\|^2$$

$$\ge \|x_{p_0+p_1+m_0} - \bar{x}\|^2 - \|x_{n_0} - \bar{x}\|^2$$

$$= \sum_{i=p_0+p_1+m_0}^{n_0-1} [\|x_i - \bar{x}\|^2 - \|x_{i+1} - \bar{x}\|^2]$$

$$\ge (128\bar{L})^{-1}\bar{\epsilon} \sum_{i=p_0+p_1+m_0}^{n_0-1} \alpha_i. \tag{12.126}$$

In view of (12.126),

$$\sum_{i=p_0+p_1+m_0}^{n_0-1} \alpha_i \le 128(3M + \bar{K})^2 \bar{L}\bar{\epsilon}^{-1}.$$

This contradicts (12.103). The contradiction we have reached proves that there exists an integer

$$j \in \{p_0 + p_1 + m_0, \dots, n_0\} \tag{12.127}$$

such that

$$f(x_j) \le \inf(f, C) + \bar{\epsilon}/8. \tag{12.128}$$

By (12.121), (12.127), and (12.128), we have

$$d(x_j, C_{min}) \le \epsilon. \tag{12.129}$$

We claim that for all integers i satisfying $j \le i \le n$,

$$d(x_i, C_{min}) \le \epsilon.$$

Assume the contrary. Then there exists an integer $k \in [j, n]$ for which

$$d(x_k, C_{min}) > \epsilon. \tag{12.130}$$

By (12.127), (12.129), and (12.130), we have

$$k > j \ge p_0 + p_1 + m_0. \tag{12.131}$$

We may assume without loss of generality that

$$d(x_i, C_{min}) \le \epsilon \text{ for all integers } i \text{ satisfying } j \le i < k.$$

Thus

$$d(x_{k-1}, C_{min}) \le \epsilon. \tag{12.132}$$

There are two cases:

$$f(x_{k-1}) \le \inf(f, C) + \bar{\epsilon}/8; \tag{12.133}$$

$$f(x_{k-1}) > \inf(f, C) + \bar{\epsilon}/8. \tag{12.134}$$

Assume that (12.133) is valid. It follows from (12.121) and (12.131) that

$$d(x_{k-1}, C) \le \bar{\epsilon}_1. \tag{12.135}$$

By (12.135), there exists a point

$$z \in C \tag{12.136}$$

such that

$$\|x_{k-1} - z\| < 2\bar{\epsilon}_1. \tag{12.137}$$

By (12.3), (12.101), (12.104), (12.106), (12.108), (12.131), (12.136), and (12.137),

$$\|x_k - z\| \le \alpha_{k-1}\delta$$

$$+\|z - P_C(x_{k-1} - \alpha_{k-1}\|v_{k-1}\|^{-1}v_{k-1} - \alpha_{k-1}\xi_{k-1})\|$$

$$\le \delta + \|z - x_{k-1}\| + \alpha_{k-1} + \delta$$

$$\le 2\bar{\epsilon}_1 + 2\delta + \alpha_{k-1} < 3\bar{\epsilon}_1. \tag{12.138}$$

In view of (12.137) and (12.138),

$$\|x_k - x_{k-1}\| \le \|x_k - z\| + \|z - x_{k-1}\| < 5\bar{\epsilon}_1. \tag{12.139}$$

It follows from (12.7), (12.8), (12.94), (12.132), and (12.139) that

$$\|x_{k-1}\| \le \bar{K} + 2\epsilon,$$

$$\|x_k\| \le \|x_{k-1}\| + 5\bar{\epsilon}_1 \le \bar{K} + 3\epsilon$$

and

$$\|x_{k-1}\|,\ \|x_k\| \le \bar{K} + 4.$$

Combined with (12.9) and (12.139) the relation above implies that

$$|f(x_{k-1}) - f(x_k)| \le \bar{L}\|x_{k-1} - x_k\| \le 5\bar{L}\bar{\epsilon}_1.$$

Together with (12.97) and (12.133) this implies that

$$f(x_k) \le f(x_{k-1}) + 5\bar{L}\bar{\epsilon}_1$$

$$\le \inf(f, C) + \bar{\epsilon}/8 + 5\bar{L}\bar{\epsilon}_1 \le \inf(f, C) + \bar{\epsilon}/4. \tag{12.140}$$

Property (iii), (12.96), (12.136), (12.138), and (12.140) imply that

$$d(x_k, C_{min}) \le \epsilon.$$

This inequality contradicts (12.130). The contradiction we have reached proves (12.134). It follows from (12.96), (12.104), (12.106), (12.116), and (12.134) that Lemma 12.6 holds with

$$K_0 = 3M,\ \epsilon = \bar{\epsilon}/8,\ x = x_{k-1},\ y = x_k,$$

$$\xi = \xi_{k-1},\ v = v_{k-1},\ \alpha = \alpha_{k-1},\ \eta = \alpha_{k-1}\eta_{k-1}.$$

Combined with (12.101), (12.104), (12.106), (12.131), and (12.132) this implies that

$$d(x_k, C_{min})^2 \le d(x_{k-1}, C_{min})^2 - \alpha_{k-1}(4\bar{L})^{-1}\bar{\epsilon}/8$$

$$+2\alpha_{k-1}^2 + \delta^2\alpha_{k-1}^2 + 2\alpha_{k-1}\delta(3M + \bar{K} + 2)$$

$$\le d(x_{k-1}, C_{min})^2 - (32\bar{L})^{-1}\alpha_{k-1}\bar{\epsilon} + 4\alpha_{k-1}^2 + 2\alpha_{k-1}\delta(3M + \bar{K} + 2)$$

$$\le d(x_{k-1}, C_{min})^2 - (64\bar{L})^{-1}\alpha_{k-1}\bar{\epsilon} + 2\delta\alpha_{k-1}(3M + \bar{K} + 2)$$

$$\le d(x_{k-1}, C_{min})^2 - (128\bar{L})^{-1}\alpha_{k-1}\bar{\epsilon}$$

and

$$d(x_k, C_{min}) \le d(x_{k-1}, C_{min}) \le \epsilon.$$

This contradicts (12.130). The contradiction we have reached proves that $d(x_i, C_{min}) \le \epsilon$ for all integers i satisfying $j \le i \le n$. This completes the proof of Theorem 12.1. □

12.6 Proof of Theorem 12.2

In view of (12.1), we may assume without loss of generality that

$$\epsilon < 1, \ M > 8\bar{K} + 8 \tag{12.141}$$

and that

$$\{x \in X : f(x) \le \inf(f, C) + 16\bar{L}\} \subset B_X(0, 2^{-1}M - 1). \tag{12.142}$$

Proposition 12.3 implies that there exists a number

$$\bar{\epsilon} \in (0, \epsilon/8) \tag{12.143}$$

such that the following property holds:

(vi)

$$\text{if } x \in X, \ d(x, C) \le 2\bar{\epsilon} \text{ and } f(x) \le \inf(f, C) + 2\bar{\epsilon},$$

$$\text{then } d(x, C_{min}) \le \epsilon/4.$$

Fix

$$\bar{x} \in C_{min} \tag{12.144}$$

and

$$\bar{\epsilon}_1 \in (0, \bar{\epsilon}(64\bar{L})^{-1}). \tag{12.145}$$

Lemmas 2.8 and 2.9 imply that there exist $\delta_1 \in (0, 1)$ and a natural number m_0 such that the following property holds:

(vii) for each integer $n \geq m_0$ and each finite sequence $\{y_i\}_{i=0}^n \subset B_X(0, 3M)$ satisfying

$$B_X(y_{i+1}, \delta_1) \cap P_C(B_X(y_i, \delta_1)) \neq \emptyset, \ i = 0, \dots, n-1$$

the inequality

$$d(y_i, C) \leq \bar{\epsilon}_1$$

holds for all integers $i \in [m_0, n]$.

Choose a positive number β_0 such that

$$\beta_0 \leq \delta_1/2, \ 2\beta_0 < (34\bar{L})^{-1}\bar{\epsilon}_1. \tag{12.146}$$

Let

$$\beta_1 \in (0, \beta_0). \tag{12.147}$$

Fix a natural number n_0 such that

$$n_0 > m_0 + 32^2 M^2 \bar{L}\bar{\epsilon}^{-1}\beta_1^{-1}. \tag{12.148}$$

Fix positive number δ such that

$$\delta(3M + \bar{K} + 3) \leq (128)^{-1}\bar{\epsilon}_1\beta_1. \tag{12.149}$$

Assume that an integer $n \geq n_0$, $\{x_k\}_{k=0}^n \subset X$,

$$\|x_0\| \leq M, \tag{12.150}$$

$$v_k \in \partial_\delta f(x_k) \setminus \{0\}, \ k = 0, 1, \dots, n-1 \tag{12.151}$$

$$\{\alpha_k\}_{k=0}^{n-1} \subset [\beta_1, \beta_0], \tag{12.152}$$

$$\{\eta_k\}_{k=0}^{n-1}, \ \{\xi_k\}_{k=0}^{n-1} \subset B_X(0,\delta) \tag{12.153}$$

and that for all integers $k = 0, \dots, n-1$,

$$x_{k+1} = P_C(x_k - \alpha_k \|v_k\|^{-1} v_k - \alpha_k \xi_k) - \eta_k. \tag{12.154}$$

In order to complete the proof it is sufficient to show that

$$d(x_k, C_{min}) \le \epsilon$$

for all integers k satisfying $n_0 \le k \le n$. First we show that for all integers $i = 0, \dots, n$,

$$d(x_i, C_{min}) \le 2M. \tag{12.155}$$

In view of (12.7), (12.8), (12.141), and (12.150), inequality (12.155) holds for $i = 0$. Assume that

$$i \in \{0, \dots, n\} \setminus \{n\}$$

and that (12.155) is true. There are two cases:

$$f(x_i) \le \inf(f, C) + 16\bar{L}; \tag{12.156}$$

$$f(x_i) > \inf(f, C) + 16\bar{L}. \tag{12.157}$$

Assume that (12.156) holds. In view of (12.142) and (12.156),

$$\|x_i\| \le M/2 - 1. \tag{12.158}$$

By (12.7), (12.8), and (12.144),

$$\|\bar{x}\| \le \bar{K}. \tag{12.159}$$

It follows from (12.158) and (12.159) that

$$\|x_i - \bar{x}\| \le \bar{K} + M/2. \tag{12.160}$$

By (12.3), (12.141), (12.144), (12.146), (12.149), (12.152)–(12.154), and (12.160),

$$d(x_{i+1}, C_{min}) \le \|x_{i+1} - \bar{x}\|$$

$$\le \|\eta_i\| + \|\bar{x} - P_C(x_i - \alpha_i \|v_i\|^{-1} v_i - \alpha_i \xi_i)\|$$

$$\leq \delta + \|\bar{x} - x_i\| + \alpha_i + \alpha_i \delta$$

$$\leq 2\beta_0 + \delta + \bar{K} + M/2$$

$$\leq \bar{K} + M/2 + 3 \leq M. \tag{12.161}$$

Assume that (12.157) holds. In view of (12.141), (12.144), and (12.155),

$$\|x_i\| \leq 3M.$$

It follows from (12.141), (12.146), (12.147), (12.149), (12.151), (12.152), (12.157), and Lemma 12.7 applied with

$$K_0 = 3M,\ \alpha = \alpha_i,\ x = x_i,\ \xi = \xi_i,\ v = v_i,\ y = x_{i+1},\ \eta = \eta_i,$$

that

$$d(x_{i+1}, C_{min})^2 \leq d(x_i, C_{min})^2 - 2\alpha_i + 2\|\eta_i\|(3M + \bar{K} + 3)$$

$$\leq d(x_i, C_{min})^2 - 2\beta_1 + 8\delta M \leq d(x_i, C_{min})^2 - \beta_1$$

and in view of (12.155),

$$d(x_{i+1}, C_{min}) \leq d(x_i, C_{min}) \leq 2M.$$

Thus in both cases

$$d(x_{i+1}, C_{min}) \leq 2M.$$

Thus we have shown by induction that for all integers $i = 0, \dots, n$,

$$d(x_i, C_{min}) \leq 2M. \tag{12.162}$$

By (12.7), (12.8), (12.141), and (12.162),

$$\|x_i\| \leq 3M,\ i = 0, \dots, n. \tag{12.163}$$

It follows from (12.149), (12.152), and (12.153), for all $i = 0, \dots, n-1$,

$$\|x_i - (x_i - \alpha_i \|v_i\|^{-1} v_i - \alpha_i \xi_i)\|$$

$$\leq \alpha_i + \alpha_i \delta \leq 2\alpha_i \leq 2\beta_0 \leq \delta_1,$$

$$\|x_{i+1} - P_C(x_i - \alpha_i \|v_i\|^{-1} v_i - \alpha_i \xi_i)\|$$

$$\leq \|\eta_i\| \leq \delta \leq \beta_0 < \delta_1$$

and

$$B_X(x_{i+1}, \delta_1) \cap P_C(B_X(x_i, \delta_1)) \neq \emptyset. \tag{12.164}$$

Property (vii), (12.148), (12.155), (12.163), and (12.164) imply that

$$d(x_i, C) \leq \bar{\epsilon}_1 < \bar{\epsilon},\ i = m_0, \dots, n. \tag{12.165}$$

Assume that an integer

$$k \in [m_0, n-1],$$

$$f(x_k) > \inf(f, C) + \bar{\epsilon}/8. \tag{12.166}$$

It follows from (12.43)–(12.146), (12.149), (12.151), (12.154), (12.163), and Lemma 12.5 applied with

$$K_0 = 3M,\ \epsilon = \bar{\epsilon}/4,\ \alpha = \alpha_k,\ x = x_k,\ \xi = \xi_k,\ v = v_k,$$

$$y = x_{k+1},\ \eta = \eta_k$$

that

$$\|x_{k+1} - \bar{x}\|^2$$

$$\leq \|x_k - \bar{x}\|^2 - \alpha_k(4\bar{L})^{-1}\bar{\epsilon}/4 + 2\alpha_k^2$$

$$+\|\eta_k\|^2 + 2\|\eta_k\|(3M + \bar{K} + 2)$$

$$\leq \|x_k - \bar{x}\|^2 - \alpha_k(16\bar{L})^{-1}\bar{\epsilon}$$

$$+2\alpha_k^2 + \delta^2 + 2\delta(3M + \bar{K} + 2)$$

$$\leq \|x_k - \bar{x}\|^2 - \alpha_k(32\bar{L})^{-1}\bar{\epsilon} + 2\delta(3M + \bar{K} + 3)$$

$$\leq \|x_k - \bar{x}\|^2 - \beta_1(32\bar{L})^{-1}\bar{\epsilon} + 2\delta(3M + \bar{K} + 3)$$

$$\leq \|x_k - \bar{x}\|^2 - \beta_1(64\bar{L})^{-1}\bar{\epsilon}.$$

Thus we have shown that the following property holds:
(viii) if an integer $k \in [m_0, n-1]$ satisfies

$$f(x_k) > \inf(f, C) + \bar{\epsilon}/4,$$

then

$$\|x_{k+1} - \bar{x}\|^2 \le \|x_k - \bar{x}\|^2 - \beta_1(64\bar{L})^{-1}\bar{\epsilon}.$$

We claim that there exists an integer $j \in \{m_0, \dots, n_0\}$ such that

$$f(x_j) \le \inf(f, C) + \bar{\epsilon}/4.$$

Assume the contrary. Then

$$f(x_j) > \inf(f, C) + \bar{\epsilon}/4, \ \ j = m_0, \dots, n_0. \tag{12.167}$$

Property (viii) and (12.167) imply that for all $k \in \{m_0, \dots, n_0 - 1\}$,

$$\|x_{k+1} - \bar{x}\|^2 \le \|x_k - \bar{x}\|^2 - \beta_1(64\bar{L})^{-1}\bar{\epsilon}. \tag{12.168}$$

Relations (12.7), (12.8), (12.141), (12.144), (12.163), and (12.168) imply that

$$(4M)^2 \ge \|x_{m_0} - \bar{x}\|^2$$

$$\ge \|x_{m_0} - \bar{x}\|^2 - \|x_{n_0} - \bar{x}\|^2$$

$$= \sum_{i=m_0}^{n_0-1} [\|x_i - \bar{x}\|^2 - \|x_{i+1} - \bar{x}\|^2]$$

$$\ge (n_0 - m_0)\beta_1(64\bar{L})^{-1}\bar{\epsilon}$$

and

$$n_0 - m_0 \le 32^2 M^2 \bar{L}\bar{\epsilon}^{-1}\beta_1^{-1}.$$

This contradicts (12.148). The contradiction we have reached proves that there exists an integer

$$j \in \{m_0, \dots, n_0\} \tag{12.169}$$

such that

$$f(x_j) \le \inf(f, C) + \bar{\epsilon}/4. \tag{12.170}$$

Property (vi), (12.145), and (12.170) imply that

$$d(x_j, C_{min}) \le \epsilon/4. \tag{12.171}$$

We claim that for all integers i satisfying $j \le i \le n$,

$$d(x_i, C_{min}) \le \epsilon.$$

Assume the contrary. Then there exists an integer $k \in [j, n]$ for which

$$d(x_k, C_{min}) > \epsilon. \tag{12.172}$$

By (12.171) and (12.172), we have

$$k > j.$$

We may assume without loss of generality that

$$d(x_i, C_{min}) \le \epsilon \text{ for all integers } i \text{ satisfying } j \le i < k.$$

Thus

$$d(x_{k-1}, C_{min}) \le \epsilon. \tag{12.173}$$

There are two cases:

$$f(x_{k-1}) \le \inf(f, C) + \bar{\epsilon}/8; \tag{12.174}$$

$$f(x_{k-1}) > \inf(f, C) + \bar{\epsilon}/8. \tag{12.175}$$

Assume that (12.174) is valid. It follows from (12.165) and (12.169) that

$$d(x_{k-1}, C) \le \bar{\epsilon}_1. \tag{12.176}$$

By (12.176), there exists a point

$$z \in C \tag{12.177}$$

such that

$$\|x_{k-1} - z\| < 2\bar{\epsilon}_1. \tag{12.178}$$

By (12.3), (12.146), (12.149), and (12.152)–(12.154),

$$\|x_k - z\| \le \delta$$

$$+\|z - P_C(x_{k-1} - \alpha_{k-1}\|v_{k-1}\|^{-1}v_{k-1} - \alpha_{k-1}\xi_{k-1})\|$$

$$\le \delta + \|z - x_{k-1}\| + 2\alpha_{k-1}$$

$$\leq 2\bar{\epsilon}_1 + \delta + 2\beta_0 \leq 3\bar{\epsilon}_1. \tag{12.179}$$

In view of (12.179),

$$\|x_k - x_{k-1}\| \leq \|x_k - z\| + \|z - x_{k-1}\| < 5\bar{\epsilon}_1. \tag{12.180}$$

It follows from (12.7), (12.8), and (12.173) that

$$\|x_{k-1}\| \leq \bar{K} + 1. \tag{12.181}$$

By (12.144), (12.180), and (12.181),

$$\|x_k\| \leq \|x_{k-1}\| + 5\bar{\epsilon}_1 \leq \bar{K} + 4. \tag{12.182}$$

Relations (12.9) and (12.80)–(12.82) imply that

$$|f(x_{k-1}) - f(x_k)| \leq \bar{L}\|x_{k-1} - x_k\| \leq 5\bar{L}\bar{\epsilon}_1.$$

Together with (12.145) and (12.174) this implies that

$$f(x_k) \leq f(x_{k-1}) + 5\bar{L}\bar{\epsilon}_1$$

$$\leq \inf(f, C) + \bar{\epsilon}/8 + 5\bar{L}\bar{\epsilon}_1 \leq \inf(f, C) + \bar{\epsilon}/4. \tag{12.183}$$

By (12.145), (12.176), and (12.180),

$$d(x_k, C) \leq 6\bar{\epsilon}_1 < \bar{\epsilon}. \tag{12.184}$$

Property (vi), (12.183), and (12.184) imply that

$$d(x_k, C_{min}) \leq \epsilon.$$

This inequality contradicts (12.172). The contradiction we have reached proves (12.175).

It follows from (12.145), (12.149), (12.151)–(12.154), (12.163), (12.165), (12.169), and (12.175) that Lemma 12.6 holds with

$$K_0 = 3M,\ \epsilon = \bar{\epsilon}/8,\ x = x_{k-1},\ y = x_k,$$

$$\xi = \xi_{k-1},\ v = v_{k-1},\ \alpha = \alpha_{k-1},\ \eta = \eta_{k-1}.$$

Combined with (12.145), (12.146), and (12.152) this implies that

$$d(x_k, C_{min})^2 \leq d(x_{k-1}, C_{min})^2 - \alpha_{k-1}(4\bar{L})^{-1}\bar{\epsilon}/8$$

$$+2\alpha_{k-1}^2 + \|\eta\|^2 + 2\|\eta_{k-1}\|(3M + \bar{K} + 2)$$

$$\le d(x_{k-1}, C_{min})^2 - \alpha_{k-1}((32\bar{L})^{-1}\bar{\epsilon} - 2\beta_0) + \delta^2 + 2\delta(3M + \bar{K} + 2)$$

$$\le d(x_{k-1}, C_{min})^2 - \alpha_{k-1}(64\bar{L})^{-1}\bar{\epsilon} + 2\delta(3M + \bar{K} + 3)$$

$$\le d(x_{k-1}, C_{min})^2 - \beta_1(64\bar{L})^{-1}\bar{\epsilon} + 2\delta(3M + \bar{K} + 3)$$

$$\le d(x_{k-1}, C_{min})^2 - \beta_1(128\bar{L})^{-1}\bar{\epsilon}.$$

In view of (12.173),

$$d(x_k, C_{min}) \le d(x_{k-1}, C_{min}) \le \epsilon.$$

This contradicts (12.172). The contradiction we have reached proves that $d(x_i, C_{min}) \le \epsilon$ for all integers i satisfying $j \le i \le n$. This completes the proof of Theorem 12.2. □

References

1. Alber YI (1971) On minimization of smooth functional by gradient methods. USSR Comp Math Math Phys 11:752–758
2. Alber YI, Iusem AN, Solodov MV (1997) Minimization of nonsmooth convex functionals in Banach spaces. J Convex Anal 4:235–255
3. Alber YI, Iusem AN, Solodov MV (1998) On the projected subgradient method for nonsmooth convex optimization in a Hilbert space. Math Program 81:23–35
4. Alber YI, Yao JC (2009) Another version of the proximal point algorithm in a Banach space. Nonlinear Anal 70:3159–3171
5. Alvarez F, Lopez J, Ramirez CH (2010) Interior proximal algorithm with variable metric for second-order cone programming: applications to structural optimization and support vector machines. Optim Methods Softw 25:859–881
6. Antipin AS (1994) Minimization of convex functions on convex sets by means of differential equations. Differ Equ 30:1365–1375
7. Aragon Artacho FJ, Geoffroy MH (2007) Uniformity and inexact version of a proximal method for metrically regular mappings. J Math Anal Appl 335:168–183
8. Attouch H, Bolte J (2009) On the convergence of the proximal algorithm for nonsmooth functions involving analytic features. Math Program Ser B 116:5–16
9. Baillon JB (1978) Un Exemple Concernant le Comportement Asymptotique de la Solution du Probleme $0 \in du/dt + \partial\phi(u)$. J Funct Anal 28:369–376
10. Barbu V, Precupanu T (2012) Convexity and optimization in Banach spaces. Springer Heidelberg, London, New York
11. Barty K, Roy J-S, Strugarek C (2007) Hilbert-valued perturbed subgradient algorithms. Math Oper Res 32:551–562
12. Bauschke HH, Borwein JM (1996) On projection algorithms for solving convex feasibility problems. SIAM Rev 38:367–426
13. Bauschke HH, and Combettes PL (2011) Convex analysis and monotone operator theory in Hilbert spaces. Springer, New York
14. Bauschke HH, Goebel R, Lucet Y, Wang X (2008) The proximal average: basic theory. SIAM J Optim 19:766–785
15. Bauschke H, Wang C, Wang X, Xu J (2015) On subgradient projectors. SIAM J Optim 25:1064–1082
16. Beck A, Pauwels E, Sabach S (2018) Primal and dual predicted decrease approximation methods. Math Program 167:37–73
17. Beck A, Teboulle M (2003) Mirror descent and nonlinear projected subgradient methods for convex optimization. Oper Res Lett 31:167–175

A. J. Zaslavski, *Convex Optimization with Computational Errors*, Springer Optimization and Its Applications 155, https://doi.org/10.1007/978-3-030-37822-6

18. Beck A, Teboulle M (2009) A fast iterative shrinkage-thresholding algorithm for linear inverse problems. SIAM J Imaging Sci 2:183–202
19. Benker H, Hamel A, Tammer C (1996) A proximal point algorithm for control approximation problems, I. Theoretical background. Math Methods Oper Res 43:261–280
20. Bolte J (2003) Continuous gradient projection method in Hilbert spaces. J Optim Theory Appl 119:235–259
21. Brezis H (1973) Opérateurs maximaux monotones. North Holland, Amsterdam
22. Burachik RS, Grana Drummond LM, Iusem AN, Svaiter BF (1995) Full convergence of the steepest descent method with inexact line searches. Optimization 32:137–146
23. Burachik RS, Iusem AN (1998) A generalized proximal point algorithm for the variational inequality problem in a Hilbert space. SIAM J Optim 8:197–216
24. Burachik RS, Kaya CY, Sabach S (2012) A generalized univariate Newton method motivated by proximal regularization. J Optim Theory Appl 155:923–940
25. Burachik RS, Lopes JO, Da Silva GJP (2009) An inexact interior point proximal method for the variational inequality problem. Comput Appl Math 28:15–36
26. Butnariu D, Kassay G (2008) A proximal-projection method for finding zeros of set-valued operators. SIAM J Control Optim 47:2096–2136
27. Butnariu D, Resmerita E (2002) Averaged subgradient methods for constrained convex optimization and Nash equilibria computation. Optimization 51:863–888
28. Ceng LC, Mordukhovich BS, Yao JC (2010) Hybrid approximate proximal method with auxiliary variational inequality for vector optimization. J Optim Theory Appl 146:267–303
29. Censor Y, Gibali A, Reich S (2011) The subgradient extragradient method for solving variational inequalities in Hilbert space. J Optim Theory Appl 148:318–335
30. Censor Y, Gibali A, Reich S, Sabach S (2012) Common solutions to variational inequalities. Set-Valued Var Anal 20:229–247
31. Censor Y, Zenios SA (1992) The proximal minimization algorithm with D-functions. J Optim Theory Appl 73:451–464
32. Chadli O, Konnov IV, Yao JC (2004) Descent methods for equilibrium problems in a Banach space. Comput Math Appl 48:609–616
33. Chen Z, Zhao K (2009) A proximal-type method for convex vector optimization problem in Banach spaces. Numer Funct Anal Optim 30:70–81
34. Chuong TD, Mordukhovich BS, Yao JC (2011) Hybrid approximate proximal algorithms for efficient solutions in for vector optimization. J Nonlinear Convex Anal 12:861–864
35. Davis D, Drusvyatskiy D, MacPhee KJ, Paquette C (2018) Subgradient methods for sharp weakly convex functions. J Optim Theory Appl 179:962–982
36. Demyanov VF, Vasilyev LV (1985) Nondifferentiable optimization. Optimization Software, New York
37. Drori Y, Sabach S, Teboulle M (2015) A simple algorithm for a class of nonsmooth convex-concave saddle-point problems. Oper Res Lett 43:209–214
38. Facchinei F, Pang J-S (2003) Finite-dimensional variational inequalities and complementarity problems, volume I and volume II. Springer-Verlag, New York
39. Gibali A, Jadamba B, Khan AA, Raciti F, Winkler B (2016) Gradient and extragradient methods for the elasticity imaging inverse problem using an equation error formulation: a comparative numerical study. Nonlinear Anal Optim Contemp Math 659:65–89
40. Gockenbach MS, Jadamba B, Khan AA, Tammer Chr, Winkler B (2015) Proximal methods for the elastography inverse problem of tumor identification using an equation error approach. Adv Var Hemivariational Inequal 33:173–197
41. Gopfert A, Tammer Chr, Riahi, H (1999) Existence and proximal point algorithms for nonlinear monotone complementarity problems. Optimization 45:57–68
42. Grecksch W, Heyde F, Tammer Chr (2000) Proximal point algorithm for an approximated stochastic optimal control problem. Monte Carlo Methods Appl 6:175–189
43. Griva I (2018) Convergence analysis of augmented Lagrangian-fast projected gradient method for convex quadratic problems. Pure Appl Funct Anal 3:417–428

44. Griva I, Polyak R (2011) Proximal point nonlinear rescaling method for convex optimization. Numer Algebra Control Optim 1:283–299
45. Hager WW, Zhang H (2008) Self-adaptive inexact proximal point methods. Comput Optim Appl 39:161–181
46. Hiriart-Urruty J-B, Lemarechal C (1993) Convex analysis and minimization algorithms. Springer, Berlin
47. Iusem A, Nasri M (2007) Inexact proximal point methods for equilibrium problems in Banach spaces. Numer Funct Anal Optim 28:1279–1308
48. Iusem A, Resmerita E (2010) A proximal point method in nonreflexive Banach spaces. Set-Valued Var Anal 18:109–120
49. Kaplan A, Tichatschke R (1998) Proximal point methods and nonconvex optimization. J Global Optim 13:389–406
50. Kaplan A, Tichatschke R (2007) Bregman-like functions and proximal methods for variational problems with nonlinear constraints. Optimization 56:253–265
51. Kassay G (1985) The proximal points algorithm for reflexive Banach spaces. Studia Univ Babes-Bolyai Math 30:9–17
52. Kiwiel KC (1996) Restricted step and Levenberg–Marquardt techniques in proximal bundle methods for nonconvex nondifferentiable optimization. SIAM J Optim 6:227–249
53. Konnov IV (2003) On convergence properties of a subgradient method. Optim Methods Softw 18:53–62
54. Konnov IV (2009) A descent method with inexact linear search for mixed variational inequalities. Russian Math (Iz VUZ) 53:29–35
55. Konnov IV (2018) Simplified versions of the conditional gradient method. Optimization 67:2275–2290
56. Korpelevich GM (1976) The extragradient method for finding saddle points and other problems. Ekon Matem Metody 12:747–756
57. Lemaire B (1989) The proximal algorithm. In: Penot JP (ed) International series of numerical, vol 87. Birkhauser-Verlag, Basel, pp 73–87
58. Mainge P-E (2008) Strong convergence of projected subgradient methods for nonsmooth and nonstrictly convex minimization. Set-Valued Anal 16:899–912
59. Minty GJ (1962) Monotone (nonlinear) operators in Hilbert space. Duke Math J 29:341–346
60. Minty GJ (1964) On the monotonicity of the gradient of a convex function. Pacific J Math 14:243–247
61. Mordukhovich BS (2006) Variational analysis and generalized differentiation, I: I: Basic theory. Springer, Berlin
62. Mordukhovich BS, Nam NM (2014) An easy path to convex analysis and applications. Morgan & Clayton Publishes, San Rafael, CA
63. Moreau JJ (1965) Proximite et dualite dans un espace Hilbertien. Bull Soc Math France 93:273–299
64. Nadezhkina N, Takahashi Wataru (2004) Modified extragradient method for solving variational inequalities in real Hilbert spaces. Nonlinear analysis and convex analysis, pp 359–366. Yokohama Publ., Yokohama
65. Nedic A, Ozdaglar A (2009) Subgradient methods for saddle-point problems. J Optim Theory Appl 142:205–228
66. Nemirovski A, Yudin D (1983) Problem complexity and method efficiency in optimization. Wiley, New York
67. Nesterov Yu (1983) A method for solving the convex programming problem with convergence rate $O(1/k2)$. Dokl Akad Nauk 269:543–547
68. Nesterov Yu (2004) Introductory lectures on convex optimization. Kluwer, Boston
69. Nguyen TP, Pauwels E, Richard E, Suter BW (2018) Extragradient method in optimization: convergence and complexity. J Optim Theory Appl 176:137–162
70. Pallaschke D, Recht P (1985) On the steepest–descent method for a class of quasidifferentiable optimization problems. Nondifferentiable optimization: motivations and applications (Sopron, 1984), pp 252–263. Lecture Notes in Econom. and Math. Systems, vol. 255. Springer, Berlin

71. Polyak BT (1987) Introduction to optimization. Optimization Software, New York
72. Polyak RA (2015) Projected gradient method for non-negative least squares. Contemp Math 636:167–179
73. Qin X, Cho SY, Kang SM (2011) An extragradient-type method for generalized equilibrium problems involving strictly pseudocontractive mappings. J Global Optim 49:679–693
74. Rockafellar RT (1976) Augmented Lagrangians and applications of the proximal point algorithm in convex programming. Math Oper Res 1:97–116
75. Rockafellar RT (1976) Monotone operators and the proximal point algorithm. SIAM J Control Optim 14:877–898
76. Shor NZ (1985) Minimization methods for non-differentiable functions. Springer, Berlin
77. Solodov MV, Svaiter BF (2000) Error bounds for proximal point subproblems and associated inexact proximal point algorithms. Math Program 88:371–389
78. Solodov MV, Svaiter BF (2001) A unified framework for some inexact proximal point algorithms. Numer Funct Anal Optim 22:1013–1035
79. Solodov MV, Zavriev SK (1998) Error stability properties of generalized gradient-type algorithms. J Optim Theory Appl 98:663–680
80. Su M, Xu H-K (2010) Remarks on the gradient-projection algorithm. J Nonlinear Anal Optim 1:35–43
81. Takahashi W (2009) Introduction to nonlinear and convex analysis. Yokohama Publishers, Yokohama
82. Xu H-K (2006) A regularization method for the proximal point algorithm. J Global Optim 36:115–125
83. Xu H-K (2011) Averaged mappings and the gradient-projection algorithm. J Optim Theory Appl 150:360–378
84. Yamashita N, Kanzow C, Morimoto T, Fukushima M (2001) An infeasible interior proximal method for convex programming problems with linear constraints. J Nonlinear Convex Anal 2:139–156
85. Zaslavski AJ (2010) The projected subgradient method for nonsmooth convex optimization in the presence of computational errors. Numer Funct Anal Optim 31:616–633
86. Zaslavski AJ (2010) Convergence of a proximal method in the presence of computational errors in Hilbert spaces. SIAM J Optim 20:2413–2421
87. Zaslavski AJ (2011) Inexact proximal point methods in metric spaces. Set-Valued Var Anal 19:589–608
88. Zaslavski AJ (2011) Maximal monotone operators and the proximal point algorithm in the presence of computational errors. J Optim Theory Appl 150:20–32
89. Zaslavski AJ (2012) The extragradient method for convex optimization in the presence of computational errors. Numer Funct Anal Optim 33:1399–1412
90. Zaslavski AJ (2012) The extragradient method for solving variational inequalities in the presence of computational errors. J Optim Theory Appl 153:602–618
91. Zaslavski AJ (2013) The extragradient method for finding a common solution of a finite family of variational inequalities and a finite family of fixed point problems in the presence of computational errors. J Math Anal Appl 400:651–663
92. Zaslavski AJ (2016) Numerical optimization with computational errors. Springer, Cham
93. Zaslavski AJ (2016) Approximate solutions of common fixed point problems, Springer optimization and its applications. Springer, Cham
94. Zeng LC, Yao JC (2006) Strong convergence theorem by an extragradient method for fixed point problems and variational inequality problems. Taiwanese J Math 10:1293–1303

Index

A. J. Zaslavski, *Convex Optimization with Computational Errors*, Springer Optimization and Its Applications 155, https://doi.org/10.1007/978-3-030-37822-6

Printed in the United States
By Bookmasters